DUAL-FUEL
DIESEL ENGINES

DUAL-FUEL
DIESEL ENGINES

GHAZI A. KARIM

CRC Press
Taylor & Francis Group
Boca Raton London New York

CRC Press is an imprint of the
Taylor & Francis Group, an **informa** business

CRC Press
Taylor & Francis Group
6000 Broken Sound Parkway NW, Suite 300
Boca Raton, FL 33487-2742

First issued in paperback 2021

ISBN-13: 978-0-367-78358-7 (pbk)
ISBN-13: 978-1-4987-0308-6 (hbk)

Visit the Taylor & Francis Web site at
http://www.taylorandfrancis.com

and the CRC Press Web site at
http://www.crcpress.com

*To my teachers, students, and associates
for all their contributions over the years.*

Contents

Preface

The rate of consumption of the energy resources available continues to increase in spite of the significant progress being made to enhance the efficiency of their usage. Contributions to this increase in consumption include the rise in world population, improvements in the average standard of living, and the increase in average life expectancy. It is expected that the combustion of fossil fuels will continue to be the prime source to rely on for energy for some time to come. This will be constrained, however, by the continued depletion of crude petroleum resources of quality, the need for ever-cleaner exhaust emissions, and the rapid progress being made in the development of renewable resources. On the other hand, it is becoming increasingly evident, especially more recently, that the availability of natural gas, as well as other gaseous fuels, is increasing. This has come about mainly as a result of the improvements in the long-range transportation of gas and the increased availability of unconventional sources, such as those of shale gas, biogases, and coal bed–derived gases. Moreover, the increased exploitation of unconventional resources of liquid hydrocarbon fuels, such as tar sands, is bringing with it the potential production of large amounts of gaseous fuel mixtures, albeit of varied quality and having lower heating values than those of the conventional processed pipeline natural gas. On this basis, there are many opportunities to increasingly exploit such gaseous fuel resources while at the same time economizing on the consumption of other nonrenewable type liquid fossil fuel resources. Such approaches would need to develop further new means for the production of power while continuing the overall reduction of exhaust emissions, including those associated with their contribution to global warming.

Much benefit has been derived from the continuing investment in research and development directed at supporting and improving the performance of the wide variety of power-producing devices of the internal combustion type. These have progressed rapidly over the years to a current level where high-quality performance is achieved with reliability while spanning a wide diversity of designs, sizes, and fields of application. In recent decades in particular, reciprocating internal combustion engines became the dominant versatile and convenient form of power generation systems, including at the individual consumer level. Moreover, the diesel engine, especially through the rapid progress made in recent years, has come to be recognized as the prime class of combustion engines for the economic production of power, for both stationary and mobile applications. The modern diesel engine nowadays is of a superbly high efficiency, combined with very high specific power output and excellent reliability and durability, and favorable output torque and exhaust emission characteristics. The conventional diesel engine is distinguished additionally by its ability to be developed in widely different sizes and ranges of power output. These can vary by orders of magnitude, from those producing merely a few kilowatts to those capable of producing several thousands of kilowatts per engine cylinder. A wide spectrum of fields of application that spans small sizes such as those powering lawn mowers to huge ones that may power a super-tanker or provide the main electrical energy supply to a small town is also covered.

An important positive feature of the compression ignition engine of the diesel type, by virtue of its mode of combustion, is its tolerance to accept a wider variation in the fuel type used, more so than the spark ignition engine. Historically, diesel engine applications have employed a wide range of fuels that include not only those that are liquid, but also those that are gaseous. On this basis, the usage of diesel engines will continue to increase, benefiting from the improvements in performance, economy, emissions, and reliability. Moreover, this type of engine probably will continue to command the major sector of power generation via internal combustion prime movers, aided by the inherent capacity of the diesel engine to relatively easily tolerate changes in the type of fuel employed, including those fuels derived directly or indirectly from biosources or from the processing of natural gas.

It is also evident that there is a need to develop approaches to the production of power that capitalize on the relative superiority of the diesel engine as a power producer, in comparison to other prime movers, while exploiting the increasingly available reserves of natural and other fuel gases. At the same time, this will reduce the consumption of the depleting reserves of liquid petroleum fuels in general, and in particular those associated with the high-quality liquid fuels recently required for diesel engine operation. Accordingly, these requirements point to the gas-fueled diesel engine, commonly known as the dual-fuel engine, as the most favorable device to employ. The future of such applications appears to be very promising, supported by the rapid technical advances made in recent years that lead to the superior optimization of their operation and control.

Engines that operate in the dual-fuel mode are normally conventional diesel engines converted so as to be capable of burning a gaseous fuel while using the conventional liquid fuel injection system of the engine to introduce only a relatively small amount of liquid diesel fuel to provide consistent ignition. These engines that require a minimum of modification from the diesel version can be operated to consume a relatively wide range of gaseous fuels while often retaining the full capacity to operate whenever desired as a conventional diesel engine. They economically produce power at high thermal efficiencies that can exceed those of diesel engines, with favorable levels of exhaust emissions and reliability.

The objective of the present contribution is to present to engineering students, practicing engineers, technologists, and scientists at all levels a comprehensive and well-integrated review of the relevant fundamentals and practices of the operation of gas-fueled diesel engines of the dual-fuel type in their variety of sizes and fields of applications. Both the positive features and limitations are highlighted, together with an outline of measures for optimizing their operation and overcoming some of their apparent potential limitations. The latest developments in this rapidly expanding field, especially for the transportation sector, are reviewed, together with a discussion of possible future development in this economically and environmentally important field.

Acknowledgments

The author is thankful that he could draw from the many jointly published contributions of his following associates who helped to contribute so much over the years to dual-fuel engine science and technology: M. Abraham, A.I. Ali, N. Amoozigar, A.Z. Attar, D. Azzouz, O.M. Bade Shrestha, O. Badr, J. Beck, K.S. Burn, K. Chen, M.V. D'Souza, D. Gee, C. Gunea, R. Gustaphson, K. Ito, L. Jensen, W.G. Jones, E. Khalil, M.O. Khan, S.L. Khanna, M.G. Kibrya, S.R. Klat, R. Lapucha, H. Li, B. Liu, C. Liu, Z. Liu, M. Metwally, E. Milkens, E. Mirosh, N.P.W. Moore, R.R. Raine, M. Razhavi, A. Rogers, P. Samuel, R. Satterford, A. Sohrabi, B. Soriano, S. Thiessen, H.C. Watson, I. Wierzba, P. Wierzba, F. Xiao, and Y. Zhaoda.

The financial support for dual-fuel engine research by the Canadian Natural Sciences and Engineering Research Council over the years is also acknowledged.

About the Author

Ghazi A. Karim earned a DSc, PhD, and DIC from London University (Imperial College) and a BSc (Hons) from Durham University, United Kingdom. He has been a university professor for more than four decades both at the University of Calgary and earlier at the Imperial College of Science Medicine and Technology, London University. He is a professional engineer, fellow of SAE, and fellow of Engineers Canada. Dr. Karim has taught numerous engineering courses at the graduate, undergraduate, and continuing education levels; conducted research; and published extensively, notably in topics relating to energy conversion, fuel combustion, and the environment. His contributions to the dual-fuel engine field are especially significant. He has acted on numerous occasions as a consultant in these areas to a variety of public and private institutions and bodies.

Nomenclature

ATDC	After top dead center
BMEP	Brake mean effective pressure
BP	Brake power/net power
BSEC	Brake specific energy consumption
BSFC	Brake specific fuel consumption
BSU	Bosch smoke unit
BTDC	Before top dead center
CFD	Computational fuel dynamics
CFR	Cooperative fuel research
CI	Compression ignition
CNG	Compressed natural gas
COV	Coefficient of variation
CR	Compression ratio
DF	Dual fuel
DI	Direct injection
EGR	Exhaust gas recirculation
ER	Equivalence ratio
FTD	Fischer–Tropsch diesel
HCCI	Homogeneous charge compression ignition
HHV	Higher heating value
HSU	Hartridge smoke unit
HV	Heating value/calorific value
ICE	Internal combustion engine
IDI	Indirect injection
IMEP	Indicated mean effective pressure
IP	Indicated power/gross power
ISEC	Indicated specific energy consumption
KL power	Knock-limited power output
LFL	Lower/lean flammability limit
LHV	Lower heating value
LNG	Liquefied natural gas
LPG	Liquefied petroleum gas
MN	Methane number
ON	Octane number
PLC	Programmable logic control
PM	Particulate matter
SCR	Selective catalytic reduction
SI	Spark ignition
SOF	Soluble organic fraction
T	Absolute temperature

TDC	Top dead center
WN	Wobbe number
3D-CFD	Three-dimensional fluid dynamics
β	Mole fraction of heptane in mixtures with methane
Φ	Equivalence ratio on fuel-to-air basis
λ	Equivalence ratio on air-to-fuel basis

1 Introduction

1.1 PROLOGUE

The conventional internal combustion engine or clones of its mode of working will be with us for quite some time to come. At the same time, the demand for higher-output efficiencies, greater specific power output, increased reliability, and ever-reduced emissions will continue to rise in intensity. There is an increased need also, whether for environmental, economic, or resource conservation reasons, to operate on a multitude of natural and processed gaseous fuels. Of course, there is much continuing research and development being expended worldwide to provide further improvements, often through making incremental progress toward achieving the desired goals. It is evident that a main effort needed is through improvements to better understanding and controlling the relatively complex physical and chemical processes of combustion in the engine environment. Many of the remedial technologies that have been successfully developed over the years were, until relatively recently, largely the product of empiricism, with the rate of progress remaining rather insufficient to fulfill the demanding long-term objectives. These, in addition to achieving high efficiency, minimum pollutant emissions, and high specific power output, need to include the capacity to respond automatically to variations in the fuel quality without losses in output, efficiency, or reliability.

It is to be shown that the gas-fueled diesel engine, when operated in the dual-fuel mode, can be associated with exhaust gas emissions that are less detrimental to the environment than those of the other forms of internal combustion engines, including the diesel engine. For example, carbon monoxide concentrations are relatively low, with NOx emissions no higher than those emitted by a comparable diesel engine, and often they are even much lower. The emissions of oxides of sulfur are negligible since the main fuels employed are usually processed to become virtually free of any sulfur compounds, except perhaps for the exceedingly small amounts arising from the combustion of the odorant additives, such as mercaptans, commonly added to gaseous fuels for safety reasons. The unburned hydrocarbons emitted are essentially nontoxic and hardly reactive in the atmosphere to produce photochemical smog or acid rain. However, since methane is the main component of the fuel gases commonly available, such emissions, when not suitably treated, may be potential contributors to greenhouse gas emissions, and they need stricter controls. Particulates, which represent a serious concern in the exhaust emissions of common diesel engines, are hardly a concern in dual-fuel engines. This is largely because of dual-fuel engines overall employing lean fuel–air mixtures throughout, the gaseous nature of their dominant fuel, and that the very small amounts of diesel fuel that are employed are usually for ignition purposes only.

1.2   RELEVANT TERMINOLOGY

There has been some confusion over the years surrounding the term *dual-fuel engines*, whether in the technical literature or in the popular press. The term, when used, describes compression ignition engines that burn *simultaneously* two entirely different fuels in varying proportions. These two fuels are usually made up of a gaseous fuel, which supplies much of the energy released through combustion, and a second fuel, which is a liquid employed mainly to provide the energy needed for ignition and the remaining fraction of the energy release by the engine.

Engines that utilize two different fuels *alternately* while having the ignition provided by an external source of energy, such as an electric spark, are better described as *bi-fuel engines.* A common example of these is engines that are diesel engines in origin that have their high compression ratios reduced, consuming fuel gas with ignition carried out via an electric spark. Accordingly, they are strictly spark ignition engines made to operate alternately on either a liquid fuel such as gasoline or a gaseous fuel such as propane or natural gas. Since vehicular travel may be conducted over long distances, there has been a common tendency in transport applications to retain the bi-fuel capacity of gas-fueled vehicles. This trend has continued, and perhaps will until such time when the infrastructure needed for refueling with gaseous fuel supplies becomes sufficiently widely available.

Another term sometimes used in gas-fueled engine applications is the *gas-diesel engine.* These have tended to relate to dual-fuel engines where the gaseous fuel is injected directly into the cylinder either during the early stages of compression or sometimes after the injection of the liquid fuel toward the end of compression. The gaseous fuel does not autoignite on its own via compression ignition, but usually burns with the assistance of the injected liquid-fueled ignition processes. Accordingly, these engines are dual-fuel engines that employ different forms of introduction of the gaseous fuel that does not undergo autoignition on its own.

The term *multifuel engines* tends to be used to relate to liquid-fueled diesel engines of the compression ignition type that have been suitably modified so that they can operate alternately on different liquid fuels, including those that may not necessarily be of the conventional diesel type. An example of such engines is those having the capacity to operate in the conventional diesel operational mode while using suitably modified liquid fuels, such as gasoline or jet fuel. Such approaches are especially attractive for ensuring increased flexibility of operation on a multitude of fuels, such as in military applications or in remote locations with limited accessibility to liquid fuel supplies.

The term *alternative fuels* has come to indicate any fuel that may be used in engine applications other than those conventional liquid fuels, such as gasoline or diesel fuel. They may include the following fuel groups:

- Pipeline and wellhead gases, methane, liquefied natural gas (LNG), compressed natural gas (CNG), biogases, landfill gases, and coal bed methane
- Liquefied petroleum gases (LPGs), propane, and butane
- Industrially processed fuel-gas mixtures such as those of coal gas, coke oven gas, blast furnace gas, synthesis gas, and producer gas

- Hydrogen, as both a gas and a liquid
- A range of liquid fuels that includes alcohols and some forms of bio-derived fuels

The operation of dual-fuel engines depends in a large way on how the gaseous fuel is introduced. The common approach is to introduce the gaseous fuel into the incoming air well ahead of the intake valve, as commonly accomplished in gasoline-fueled spark ignition engines. The gas may also be introduced at the beginning of the compression process so that much of the gaseous fuel becomes thoroughly mixed with the air before pilot fuel injection. These engines can be described as premixed dual-fuel engines, and this mode of gas introduction as fumigation. On the other hand, there are other approaches, such as those where the gaseous fuel is injected under very high pressure directly into the cylinder, either just prior to or after the injection and ignition of the pilot liquid fuel. Such engines are described as high-pressure direct injection gaseous-fueled dual-fuel engines.

The relatively small quantity of the liquid diesel fuel injected to provide controlled ignition in dual-fuel engines is commonly described as the *pilot*. The term *size of pilot* can be defined on the basis of the combustion energy released by the combustion of the pilot fuel relative to the total thermal energy to be released by combustion at full load. The size of the pilot may be alternatively defined relative to the energy to be released by a stoichiometric mixture when operating in the diesel mode. There are also other approaches for quoting the size of the pilot, such as that of the mass of the liquid fuel injected in grams per cubic meter of engine cylinder swept volume. This latter definition can be useful in comparing the size of the pilot directly with pure diesel operation for different engine sizes and speeds.

1.3   THE EQUIVALENCE RATIO

A stoichiometric mixture contains exactly the theoretical amount of oxygen needed for complete combustion. In practice, however, there is no assurance that combustion will be completed simply by providing the correct fuel-to-air ratio. The terms *lean* and *rich mixtures* usually relate to mixtures leaner or richer in fuel than the corresponding stoichiometric amounts. For most common fuels, although their heating value on volume basis can vary widely, the energy release on combustion per unit of stoichiometric fuel–air mixture tends not to be so widely variable. The term *equivalence ratio*, which indicates the ratio of the actual fuel-to-air relative to the corresponding stoichiometric value by mass, moles, or volume, becomes a very useful indicator of the relative energy release on combustion of a mixture of fuel and air. Accordingly, the equivalence ratio of the gaseous fuel–air charge admitted becomes a key operating parameter in dual-fuel engine applications. Another common term in the case of dual-engine applications is the *total equivalence ratio*, which is based on the combined contributions of both the pilot and the gaseous fuels used. It is in general equal to the amount of stoichiometric oxygen needed for the complete combustion of the two fuels divided by the total available oxygen, i.e.,

(Equivalence ratio)$_{pilot}$ =
   (Mass of pilot fuel/Mass of available air)/(Mass of pilot fuel/Mass of air)$_{stoich}$

(Equivalence ratio)$_{fuel\ gas}$ =
   (Mass of fuel gas/Mass of available air)/(Mass of gaseous fuel/Mass of air)$_{stoich}$

(Equivalence ratio)$_{total\ fuel}$ =
   (Masses of gaseous fuel and pilot/Mass of available air)/(Masses of gaseous
   fuel and pilot/Mass of air)$_{stoich}$ = (Masses of stoichiometric air for the
                   gaseous fuel and pilot)/(Mass of available air)

For mixtures containing some oxygen in addition to that of the air:

(Equivalence ratio)$_{total\ fuel}$ =
   (Masses of stoichiometric oxygen for the combustion of the gaseous fuel
                   and pilot)/(Mass of oxygen available)

Occasionally, in lean mixture engine applications the inverse of the equivalence
ratio is used and given the symbol β, while the symbol Φ is used for its inverse,
i.e., relating to the corresponding fuel-to-air equivalence ratio.

The extent of emissions of any component in the engine exhaust gas may be
shown relative to the brake power output as g/kW.h or g/kJ (or g/hp.h). However,
often, to examine the trends for the observed variation in the concentration of a certain species emitted, it may be reported merely on molar (volumetric) concentration
basis in the dry exhaust. In general, it is important to establish whether any concentrations quoted are on volume, mass, dry, or wet bases relative to power output under
steady, rated, or transient operation. It is also necessary to watch whether any plots
of concentration values are on a logarithmic or linear scale basis.

The term *excess air* usually relates to the amount of air supplied beyond that ideally needed to fully oxidize the fuel. Similarly, *excess fuel* is the extra fuel supplied
beyond the amount that can be oxidized completely by the available oxygen.

Example

A diesel engine is supplemented with fumigated propane vapor. Calculate the
equivalence ratio and the rate of excess air employed when 0.255 kg/h diesel
fuel, 2.12 kg/h propane, and 49.53 kg/h air are used. Assume the diesel fuel to be
represented by cetane ($C_{16}H_{34}$).

ANSWER

Write the overall combustion reaction in moles:

Molecular weight of cetane ($C_{16}H_{34}$) = $16 \times 12 + 34 = 226$ kg/kmol

Molecular weight of propane (C_3H_8) = $3 \times 12 + 8 = 44$ kg/kmol

Effective molecular weight of air = 0.21 × 32 + 0.79 × 28 = 28.9 kg/kmol

The overall ideal reaction equation for a stoichiometric mixture in air (in moles) is

$$\frac{0.255}{226}C_{16}H_{34} + 2.\frac{12}{44}C_3H_8 + \lambda(0.21O_2 + 0.79N_2) \rightarrow aCO_2 + bH_2O + dN_2$$

Employ elemental mass balance equations to find the unknowns, λ, a, b, and d.

Carbon balance $= \dfrac{0.255}{226} \times 16 + \dfrac{2.12}{44} \times 3 = 0.1626$

Hydrogen balance $= \dfrac{0.255}{226} \times 17 + \dfrac{2.12}{44} \times 4 = 0.2119$

Oxygen balance = $0.21\lambda = a + b/2 = 0.1626 + 0.10595 = 0.2686$

$$\lambda = 1.279$$

Mass of stoichiometric air = 1.279 (0.21 × 32 + 0.79 × 28) = 36.87 kg/h

But the air supplied is 49.53 kg/h.

Excess air = 49.53 − 36.87 = 12.66 kg/h

The equivalence ratio is

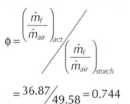

$$\phi = \frac{\left(\dfrac{\dot{m}_f}{\dot{m}_{air}}\right)_{act}}{\left(\dfrac{\dot{m}_f}{\dot{m}_{air}}\right)_{stoich}}$$

$$= {36.87}/{49.58} = 0.744$$

Since in dual-fuel operation the combustion of more than a single fuel is involved, the use of the overall or total equivalence ratio values sometimes tends to be considered insufficient. Knowledge of the equivalence ratio value on the basis of the gaseous fuel only may also be employed.

1.4 EFFICIENCY CONSIDERATIONS

The chemical energy of the fuel when converted by combustion directly into heat becomes degraded and less amenable to universal conversion into useful work energy, a more organized and universally convertible energy form. In principle, the chemical energy of the fuel may be converted fully into work when it is not converted into heat first. The corresponding work production efficiency then, ideally under the right conditions, can be very high, and indeed may even in principle approach 100%.

Depending on the prevailing temperature, only a fraction of the heat energy can be converted directly into work, while rejecting the remainder as unconverted heat to the surroundings. The maximum ideal extent of conversion of heat energy into work depends on the temperatures of the heat supplied and the rejection environment, and it is established via the well-known Carnot cycle efficiency consideration. Various other forms of nonheat energy, e.g., electrical, chemical, high pressure, etc., are also more amenable to full conversion into useful work. Ideally, if no conversion of such energies takes place into heat and without losses, then they are mutually convertible. Of course, work can readily be fully converted wastefully into heat. On this basis, the internal combustion engine in principle has the advantage of a superior level of conversion of the chemical energy of the fuel into useful work. Figures 1.1 and 1.2 show schematic representations of two different approaches to produce work from fuel energy. The first converts the energy initially into heat, followed by its fractional conversion into work, while the second, such as in an internal combustion engine, can ideally avoid going through the heat energy mode, directly converting the fuel chemical energy into work through chemical reactions within the flow process.

The term *efficiency* is universally employed for the evaluation of the performance of energy devices and processes. In engine applications it is commonly used to indicate the net output relative to the ideal or maximum output. The definition employed will depend on the specific field of application and the objective of the intended evaluation.

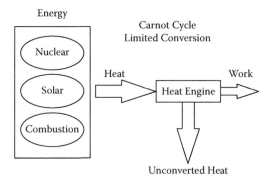

FIGURE 1.1   Only a fraction of the input from a heat energy source is converted into work.

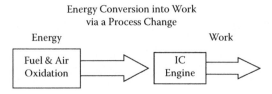

FIGURE 1.2   An internal combustion engine has the potential to convert the fuel energy input more fully into work.

For internal combustion engines, the following term is used:

$$\text{Work production efficiency} = \frac{\text{Work output}}{\text{Max work output}}$$

or

$$= \frac{\text{Work output}}{\text{Exergy change}}$$

or

$$= \frac{\text{Work output}}{\text{Gibbs function change}}$$

For the combustion process of most common fuels, this efficiency term approximates to

$$\approx \frac{\text{Work output}}{\text{Fuel mass consumed} \times \text{Heating value}}$$

In principle, the value of this type of efficiency will be high and may even conceivably ideally reach 100%.

Instead of *efficiency*, the term *rate of specific fuel consumption* may be commonly used. It relates the mass of fuel consumed to the corresponding useful power produced in kg_{fuel}/kJ or $kg_{fuel}/kW.h$. However, in dual-fuel engine applications where two fuels are consumed with different heating values, the term *rate of specific energy consumption* is more suitable when evaluating the performance of dual-fuel combustion devices. It is evident that the value of the work production efficiency will be directly related to the inverse of the corresponding specific energy fuel consumption.

It is also convenient and more informative, when comparing the relative power outputs of engines of different sizes and speeds, to employ the term *mean effective pressure*, which is the rate of specific power output per unit of engine cylinder size:

Mean effective pressure = Volumetric specific power output
                        = Power output/(Engine swept volume × Working cycles per time)

A high value of the mean effective pressure indicates a high power production density with respect to the engine cylinder size. Both indicated and brake effective pressure values are widely used based correspondingly on the indicated or brake power output, respectively. It can be also shown that the value of the mean effective pressure is directly proportional to the torque produced by the engine.

The heating value of a fuel gas is quoted in terms of either its lower or higher values. Normally, the lower value, which is associated with the case where the water vapor in the combustion products remains in the vapor state, is employed in engine applications. This is mainly because it is acknowledged that on combustion in engines the products are not cooled down to the temperature of the surroundings. This would result in the relatively small concentrations of the water vapor not condensing. Also, the heating value per unit volume or mole of the gas is usually quoted for gaseous fuels at specified temperature and pressure, when the gas is dry. Occasionally, the heating value of the stoichiometric mixture of the fuel and the required air is quoted. Such a value tends to be useful in the analysis of the relative thermal contribution of different fuels.

1.5 COGENERATION

All internal combustion engines can manage to convert only a relatively moderate fraction, typically, on average, just over a third of the chemical energy of the fuel released through combustion to net available mechanical or electrical work. The bulk of the energy is then dissipated in the form of environmental thermal pollution. This is via both the high enthalpy exhaust gases discharged and the heat transferred to the external environment of the engine, such as through the circulating cooling water, air, and lubricating oil and through dissipation via radiation transfer. The proportional distribution of this dissipation of energy between the various possible paths can vary significantly, depending on the engine type, its design, fuel employed, and operating conditions. Much effort is expended normally at all stages of engine design and operation to increase its work production capacity and its associated efficiency. However, there are practical and theoretical barriers that limit such increases. Accordingly, there is increasingly a greater effort to utilize some of this rejected heat energy. This is employed especially in power generation stationary engine installations to provide relatively low temperature heat to produce hot water, low-temperature steam, or hot air. This simultaneous production of power and utilization of waste exhaust gas and circulating water heat is known commonly as cogeneration. It is increasingly employed to cut down on fuel consumption, associated costs, and reduce the overall discharge of thermal as well as exhaust emissions, particularly carbon dioxide, the greenhouse gas. A common example of some cogeneration action is in automobiles, where the heating of the interior of the vehicle is carried out by using some of the thermal energy rejected by the engine, usually to the circulating water. Another common example is the improvement to the engine power output and efficiency through fitting exhaust gas turbochargers.

There is also an increasingly wide range of cogeneration installations with an assortment of heating requirements, associated designs, and complexity, depending on the type of fuel used, engine power, heating demands, and heat recovery conditions required. These combined add much to the capital and operational costs of the installation and increase the complexity of its control. Typically, as shown in Figure 1.3, a waste heat recovery system has a number of interconnected water circuits. Heat is transferred from the engine cooling water to the secondary water circuit by means of a heat exchanger. This heated water is then passed through a form of waste heat boiler, where further heat is recovered, this time from the engine exhaust gases before they are discharged into the atmosphere.

The dual-fuel engine is well suited for cogeneration applications. For example, it has been widely used in biomass and sewage processing works to produce electric

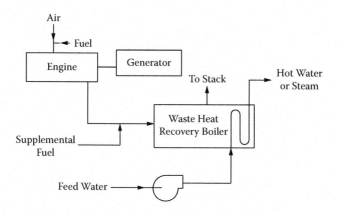

FIGURE 1.3 A schematic arrangement of a cogeneration system producing power and hot water or steam supply simultaneously.

power using the sewage gas produced, and can provide hot water to improve the effectiveness of the anaerobic sewage digestion processes. This way, the gas yield and the overall thermal efficiency can be improved substantially. Also, gas-fueled engines with cogeneration are widely used in some countries not only to maintain the warmth of greenhouses, but also to independently provide electrical lighting over long periods of up to 24 hours per day while suitably modifying the concentration of carbon dioxide in the local environment. Such approaches can promote more vigorous growth of plants while reducing greenhouse emissions. There are also applications where engines can be made to provide electrical power and cooling for air conditioning through the use of chillers and engine waste heat.

REFERENCES AND RECOMMENDED READING

Bechtold, R.L., *Alternative Fuels Guidebook*, SAE, Warrendale, PA, 1997.

Brekken, M., and Durbin, E., *An Analysis of the True Efficiency of Alternative Vehicle Power Plants and Alternative Fuels*, SAE Paper 981399, 1998.

Challen, B., and Barnescu, R., *Diesel Engine Handbook*, 2nd ed., SAE, Warrendale, PA, 1999.

Hosseinzadeh, A., and Saray, R.K., An Availability Analysis of Dual-Fuel Engines at Part Loads: The Effects of Pilot Fuel Quantity on Availability Terms, *Journal of Power and Energy*, 223(8), 903–912, 2009.

Hsu, D., *Practical Diesel Engine Combustion Analysis*, SAE, Warrendale, PA, 2002.

Karim, G.A., *Dual Fuel Engines of the Compression Ignition Type—Prospects, Problems and Solutions—A Review*, SAE Paper 830173, 1983.

Karim, G.A., Combustion in Gas Fueled Compression Ignition Engines of the Dual Fuel Type, *ASME Journal of Engineering for Gas Turbines and Power*, 125, 827–836, 2003.

Karim, G.A., *Fuels, Energy and the Environment*, CRC Press, Boca Raton, FL, 2012.

Maxwell, T., and Jones, J., *Alternative Fuels: Emissions, Economics and Performance*, SAE, Warrendale, PA, 1995.

Stone, R., *Introduction to Internal Combustion Engines*, 2nd ed., SAE, Warrendale, PA, 1995.

Turner, S.H., and Weaver, C.S., *Dual-Fuel Natural/Diesel Engines: Technology, Performance and Emissions*, No. GRI-94/0094, Topical Report Gas Research Institute, November 1994.

2 The Internal Combustion Engine

2.1 THE RECIPROCATING INTERNAL COMBUSTION ENGINE

The internal combustion engine (ICE) may be considered one of the most significant inventions that changed human life in recent times. It provides prompt and simple control of power generation while consuming a variety of commonly available fuels. Enormous research effort and resources have increasingly been expended to improve its performance while reducing its undesirable impact on the environment. Over the years billions of units have been produced, and the internal combustion engine has come to be the prime device at present for the production of work through fuel combustion. Such devices, which are mostly of the reciprocating type, employ intermittent combustion, or they may be of rotary action with continuous combustion. Their distinctly different modes of operation and associated combustion processes often dictate a restrictive choice of design features and the types of fuels that can be used by them. A broad classification of the very wide range of devices developed over the years of this type may be according to the listing shown in Table 2.1. The interrelated engine type devices that manage to a varying degree of success to convert the fuel chemical energy into useful work through controlled combustion are represented in Figure 2.1.

The common internal combustion engine is expected to simultaneously satisfy an increasingly large number of key requirements. These would depend in complex and interrelated ways on numerous engine design features, operating conditions, and fuel requirements. This would require engine designers, developers, and operators to spend much time, effort, and resources to ensure satisfactory and dependable operation for any specific engine and fuel combination.

It is expected that the internal combustion engine in one form or another will remain with us for quite some time to come. However, it must increasingly satisfy more challenging demands for higher-output efficiencies and power per unit of cylinder volume, superior reliability, reduced costs, and all forms of exhaust emissions.

TABLE 2.1
Unsteady Combustion Equipment

Fuel	Form of Introduction to Equipment	Main Equipment
Gasolines	Sprayed or vaporized	Spark ignition engines
Diesel fuels	Sprayed	Compression ignition engines
Gaseous fuels	Fumigated or injected	Spark and compression ignition engines

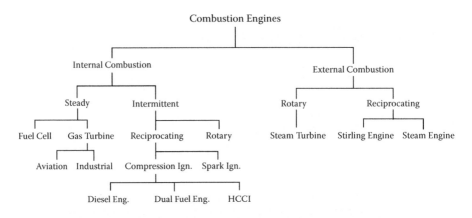

FIGURE 2.1 Schematic representation of the various devices that produce power through the combustion of fuel.

There is also an increasing need, whether for environmental, economic, or resource availability reasons, to operate satisfactorily on a multitude of natural and processed fuels. Accordingly, there is much continued intense research and development proceeding globally, at least to make incremental progress toward meeting these demanding objectives.

The factors exerting the greatest influence on the economics, life, and performance of any power producing installation include unit size, plant load factors, and total hours operated per year. Of critical importance are their cost and possible future changes required. These include the costs of capital investment, fuels, labor, taxes, operation, and maintenance. In the electric power generation field, the availability and electric utility costs are decisive influencing factors. Often, the economics of engine plant operation can be made more favorable by having some of the heat rejected by the engine recovered and utilized to save in overall costs and achieve high overall energy efficiency. It is especially attractive to be able to consume some of the abundant and cheap gaseous fuels for power production, cogeneration, and transport applications.

Prime examples of the reciprocating type engines, shown schematically in Figures 2.2 and 2.3, are the spark ignition and compression ignition diesel engines, respectively.

There are some major variations in these types of engines, such as those of the premixed dual-fuel engine, shown schematically in Figure 2.4, which aims at utilizing gaseous fuels as the prime source of its energy while using only a relatively small amount of liquid diesel fuel injection, mainly to provide ignition.

More recently, interest has been shown in the development of an engine application that operates on the principle of having homogeneously fed fuel–air mixtures that undergo controlled compression ignition. This type of device, which has been coined homogeneous charge compression ignition (HCCI), when it works successfully, has the potential for further improvements in efficiency and reduction of emissions. Figure 2.5 shows a schematic representation of the working of an HCCI engine.

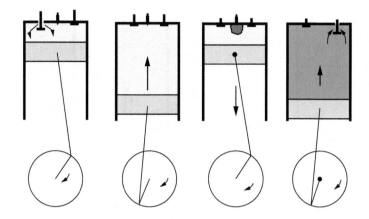

FIGURE 2.2 Schematic of a spark ignition engine operating on premixed fuel–air.

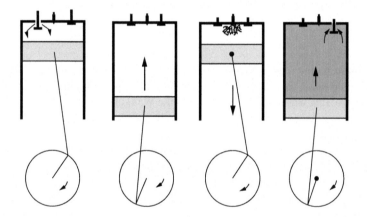

FIGURE 2.3 Schematic of compression ignition diesel engine using liquid fuel injection.

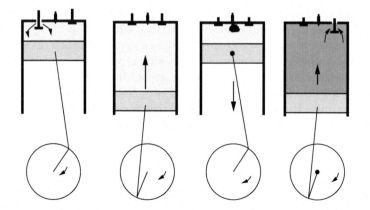

FIGURE 2.4 Compression ignition engine of the dual-fuel type with pilot liquid fuel ignition and premixed gaseous fuel charge.

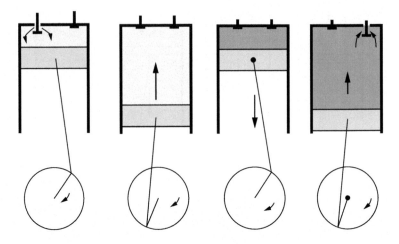

FIGURE 2.5   Homogeneous charge compression ignition (HCCI) engine.

The main rotary type internal combustion engine is the gas turbine, which provides jet propulsion for aviation and power for land applications. The rotary reciprocating engine that was commonly known as the Wankel engine has not shown, on the whole, the necessary promise needed for it to be a serious alternative to the common reciprocating type.

Modern engine installations of recent design tend to be of complex construction and working. This has been mainly driven by the continuing need to increase their specific power output, efficiency, flexibility of operation, and control while continuing to reduce to a minimum their undesirable exhaust emissions. The need to be able to effectively burn a wider range of fuel types is also an important consideration. Most modern internal combustion engines, especially those of large-size diesel type, have been made to usefully exploit a significant fraction of the thermal energy of the exhaust gases in a turbocharger, which provides compression work for the incoming engine intake air and increases the intake total mass and specific power output. Moreover, in recent years some controlled exhaust gas recirculation (EGR) has been increasingly employed. This approach mainly aims at providing an additional control of the composition of the exhaust emissions, especially those of oxides of nitrogen. Figure 2.6 shows a schematic arrangement of an internal combustion engine installation where a turbocharger with EGR is fitted. The recirculated gases, as shown, may be cooled before their introduction or kept partially warm.

It is to be shown that the economics of the dual-fuel engine can be made quite attractive, and its usage of two fuels is a positive feature since it permits whenever desired the changeover back to conventional diesel operation without interruption of power production. It allows the efficient exploitation of interruptible fuel supplies, such as in transport applications, where the limitations imposed to the range that can be traveled by a vehicle between fuel refillings by the limited portability of the gaseous fuel can be relaxed.

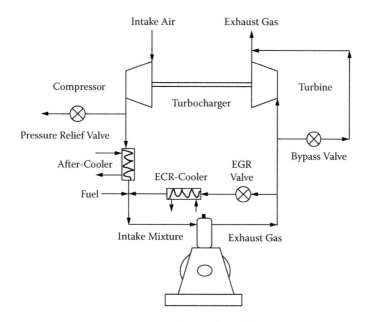

FIGURE 2.6 Schematic representation of a turbocharged engine installation employing exhaust gas recirculation (EGR).

2.2 HYBRID ENGINE APPLICATIONS

The internal combustion engine in general, especially for transport applications, is called upon to satisfy numerous key requirements during its operation under a variety of operating conditions, both steady and transient. Often, these would have the engine not confined to operate over a more limited set of operating conditions that would permit optimal performance, especially in relation to the desired torque ~ speed characteristics, enhanced efficiency, and reduced exhaust emissions. Accordingly, the concept of the hybrid operation of an engine, shown schematically in Figure 2.7, relates typically to an engine that is coupled to an electric motor/generator/storage system, which can controllably provide power through these components. This hybrid approach has been developed in a variety of forms and applications in general, including in a number of examples involving dual-fuel engine applications. A well-established example of a hybrid operation is the diesel-electric locomotive, where the engine is coupled to an electric generator that transfers its electrical output to electric motors that ultimately drive the wheels. In recent years the use of hybrid vehicles in automotive applications has increased significantly, driven mainly by the associated reduction in running fuel costs and emissions, but with increased capital cost and complexity of controls. The dual-fuel engine, just like the common diesel engine, is eminently suited for hybrid engine applications where the optimization of engine operation can be improved, especially in transport applications. However, there are a number of limitations to such potential benefits. These include increased

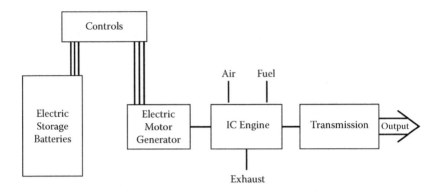

FIGURE 2.7 Arrangement of a hybrid type vehicle employing an IC engine. (From Karim, G.A., *Fuels, Energy, and the Environment*, CRC Press, Boca Raton, FL, 2012.)

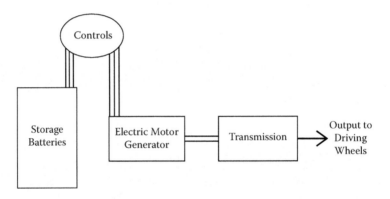

FIGURE 2.8 Schematic arrangement of an electric vehicle system. (From Karim, G.A., *Fuels, Energy, and the Environment*, CRC Press, Boca Raton, FL, 2012.)

capital and operational costs, difficulties in control, increased vehicle mass, and a potential increase in electrical fire and safety hazards in general. Electric-powered vehicles (Figure 2.8) with a sufficient suitable energy storage uncouple the short-term power requirement of the vehicle from the load seen by the engine to permit it to operate at settings to produce high efficiency and low emissions. The size of the engine may be reduced and equipment life improved, but with increased mass, mainly due to the electric storage system.

2.3 SELECTION OF ENGINE POWER SYSTEMS

In the decision to use an engine type and choose its mode of power production, a variety of factors need to be considered. Some of these are of critical importance, while others may be less so, but nevertheless requiring satisfactory consideration. To choose an engine-based power system to perform a specific task such as generating electrical power in stationary units requires the consideration of numerous factors beyond those of merely cost, thermodynamic, or combustion origins. Often, there are other overriding factors that must be catered to. Examples of some of these factors that need to be evaluated and their potential impact on output, cost, and emissions follow:

- Fuels availability and their required quality; current supply costs and possible future trends; availability and potential changes in market demand for the power and any by-products produced, taxes, subsidies, and any incentives from different sources that may become available
- Total costs, such as those relating to capital equipment, design, related civil engineering work, land acquisition, insurance, legal, operation, maintenance, storage, and infrastructure
- Current requirements and future potential changes in social acceptance, workers' availability and safety, exhaust emissions and environmental impact, noise and vibration, water supply, quality, and disposal requirements
- Time needed from conception, design, construction, commissioning and production, ease of starting the equipment, and shutdown; manpower quality requirements and whether there is specialized technical support readily available when needed
- Durability, expected working life, maintenance frequency and costs, materials compatibility, the tendency to depreciate, such as through corrosion and increased wear, ease of replacement of parts and equipment, retrofitting possibilities, recycling and secondhand value of equipment and facilities
- Complexity of instillation, operation, control, and whether remote control and compatibility with other similar installations are possible
- Availability and costs of any auxiliary power and water of the right quality needed, including insufficient quantities in emergency

It can be suggested that when a well-designed and well-operated high-compression-ratio diesel engine is operated on gaseous fuels, and especially natural gas, few, if any, contemporary power sources can successfully compete with it on the basis of economic returns per capital invested. The economics of the combination of cheap fuels and a well-developed prime mover often prove to be a very attractive investment opportunity while offering long-term environmental and resource advantages.

REFERENCES AND RECOMMENDED READING

Baur, H., ed., *Diesel Engine Management*, 2nd ed., SAE, Warrendale, PA, 1999.
Challen, B., and Barnescu, R., *Diesel Engine Handbook*, 2nd ed., SAE, Warrendale, PA, 1999.
Karim, G.A., *Dual Fuel Engines of the Compression Ignition Type—Prospects, Problems and Solutions—A Review*, SAE Paper 830173, 1983.
Karim, G.A., *Fuels, Energy and the Environment*, CRC Press, Boca Raton, FL, 2012.
Li, H., and Karim, G.A., Modeling the Performance of a Turbo-Charged Spark Ignition Natural Gas Engine with Cooled Exhaust Gas Recirculation, *ASME Journal of Engineering for Gas Turbines and Power*, 130, 328041, 2008.
Obert, E.E., *Internal Combustion Engines and Air Pollution*, Harper & Row, New York, 1973.
Packer, J.P., Advanced Packaged Co-generation, Gas Engines and Co-generation, presented at Proceedings of the Institute of Mechanical Engineers, 1980, pp. 25–32.

Stone, R., *Introduction to Internal Combustion Engines*, 2nd ed., SAE, Warrendale, PA, 1995.

Yonetani, H., Hara, K., and Fukatani, I., *Hybrid Combustion Engine with Premixed Gasoline Homogeneous Charge and Ignition by Injected Diesel Fuel—Exhaust Emission Characteristics*, SAE Paper 940268, 1994.

3 The Compression Ignition Engine

3.1 THE DIESEL ENGINE

Conventional diesel engines rely on compression ignition of an atomized liquid fuel jet injected into the high-temperature and high-pressure cylinder air charge toward the end of the compression stroke of a high-compression-ratio unthrottled recipro-cating piston engine. A brief review of the main specific features of these engines and how they differ from the common spark ignition type follows:

- Diesel engines need to have sufficiently high compression ratios to ensure reliable, prompt, and well-controlled autoignition of the injected fuel. They involve nonhomogenous fuel–air mixtures leading mainly to heterogeneous diffusion type combustion. The rapid energy releases produced require robust engine construction to withstand the resulting high mechanical and thermal loading rates.
- The engine requires suitable liquid fuels that, in comparison to those of the spark ignition, are more prone to autoignition and have high cetane num-bers, but at the same time very low octane numbers. Excess air operation is employed throughout with intense turbulence and swirling fluid action provided so as to aid in the rapid atomization, vaporization, and subsequent mixing of the fuel and air.
- The engine normally operates unthrottled at lower speeds than the spark igni-tion types and is relatively harder to start unaided under cold weather and intake air conditions. But, it tends to have superior work production efficiency and torque characteristics. The engine is well suited for high levels of turbo-charging and can be made of extremely large size and power capacity.
- The exhaust emissions in diesel engines tend to show normally low levels of carbon monoxide and unburned hydrocarbons, but relatively high levels of NOx and particulates. At present, to render their exhaust gas emissions environmentally acceptable, most engine types require special treatment.

The many improvements made in recent years in diesel engine design, manu-facture, and performance, such as in their superior fuel injection systems, are very impressive. They led to significant improvements in key performance features, such as power production, brake torque characteristics, driving dynamics, fuel utilization, increased reliability, and reduced exhaust emissions. The engine has been recog-nized almost from its inception as having the capacity to operate on a multitude of fuels, varying from gaseous to even solid-liquid slurries. The engine was originally conceived as capable of using cheap low-grade liquid fuels, but with the more recent stricter requirements to achieve sufficiently low exhaust emissions combined with high efficiency, ultra-low sulfur-distilled liquid fuels are required.

The four-stroke compression ignition engine of the diesel type, in comparison to its corresponding two-stroke type, tends to be relatively bulkier and more expensive and complex. The two-stroke version, in comparison, yields more average power per engine revolution while often employing simpler piston porting. The engine tends to have poorer exhaust emissions, noisier operation, and somewhat lower efficiency, is more difficult to lubricate, and overheats more readily than the four-stroke type. The two-stroke engine has a greater tendency for a portion of its charge, unless special remedial action is taken, to short-circuit the combustion process by proceeding directly to the exhaust stage.

The operation of diesel engines with swirl chambers tends to be less sensitive to the quality of the fuel in comparison to the direct injection type, with an easier-to-form, relatively less heterogeneous fuel–air mixture. Also, when gaseous fuels are introduced, the mixing of the fuel and air in the combustion chamber can be achieved more readily.

The diesel engine, which spreads over a wide variety of sizes and types, represents one of the major widely used prime movers throughout the world. It is anticipated that its application will continue to have an increasingly commanding lead. However, concerns may be raised about the environmental impact of its emissions and the increasing need to provide economically suitable fuel supplies of the right quality. On the other hand, world resources of natural gas, in comparison to those of petroleum liquids, appear to be enormous, relatively widespread, and have been increasing quite rapidly in recent years. In fact, natural gas is increasingly being touted as capable of potentially ensuring the long-term continued supply of fuel energy. It is to be shown that the gas-fueled diesel engine of the dual-fuel type offers an excellent and flexible means for employing the well-established and highly developed diesel engine while primarily and economically consuming abundantly available gaseous fuels. However, modifications to engine design and operation need to be effected to render diesel engines suitable for gaseous fuel consumption. These changes vary widely in complexity and effectiveness, depending on the fields of their applications, the fuels employed, and the associated operating conditions.

Some common diesel engines, including those of large capacity, were sometimes converted to operate as spark ignition engines so as to consume common gaseous fuels. This was performed through eliminating direct liquid fuel injection and using electric spark ignition instead. In order to avoid the onset of spark knock and any associated uncontrolled autoignition, the high compression ratio normally associated with the diesel engine was substantially reduced. Such conversion of diesel engines was often made largely on an ad hoc basis, potentially wasting fuel energy and contributing to increased undesirable exhaust gas emissions. Moreover, some gas-fueled diesel engines operating in the field were of varying types, ages, and design, and invariably equipped with the minimum of devices for metering and control. Traditionally, this was considered acceptable mainly because of the relatively low cost of traditional gaseous fuels, including natural gas. With the continuing increase in the relative cost of fuels in general and the increasing need to more strictly curb all forms of exhaust gas emissions, including those of the greenhouse gases, remedial effective measures have been increasingly introduced in recent years.

3.2 OPERATIONAL FEATURES OF DIESEL ENGINES

The direct injection (DI) diesel engine (Figure 3.1), which has an open combustion chamber, is widely used, especially in large-bore stationary engine applications. The combustion process in these engines mostly involves the interaction of the injected liquid fuel jet with the bulk air, which is brought about normally by suitably sited and shaped inlet ports directing the incoming air into a swirling and circulating motion. This is subsequently modified by the motion of the piston, residual gases present, changing geometry of the cylinder, and heat transfer effects. Squish in the form of a toroidal air motion is produced by piston motion and modified by piston shape and its mode of travel. The interaction of the swirling fluids cannot be controlled normally well over the wide speed and load ranges of the engine. At any instant of engine operation there are no significant differences in pressure between the different regions of the chamber.

In the indirect injection (IDI) or divided chamber engines (Figure 3.1), the fuel is injected in a separate chamber attached and open to the main cylinder. Fuel injection and some of the liquid jet breakup, atomization, and vaporization take place in this prechamber. During compression, when the main chamber is decreasing in volume, air is forced into the prechamber. In this type of engine, when employed in small-bore engines, the prechamber can represent a significant fraction of the engine clearance volume. Engine designs usually aim at having a smooth communicating passage of the right shape, orientation, and size so as to reduce the frictional, thermal, and aerodynamic losses. After ignition, the rich mixture, products of combustion, partial oxidation products, and unconverted fuel within the prechamber are forced as a burning jet into the much leaner surroundings and air of the main chamber, permitting burning to proceed more vigorously. Since the fuel–air mixing processes, in comparison to direct injection engines, are less dependent on the momentum of the jet, relatively lower fuel injection pressures can be used.

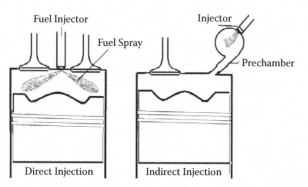

FIGURE 3.1   Schematic representation of direct and indirect injection diesel engines.

Moreover, lower maximum pressures, rates of pressure rise, and noise are associated with indirect injection engines. However, these engines tend to be associated with relatively lower efficiencies, higher thermal stresses, and later burning, and can be harder to start. These trends are due largely to the high heat transfer and the work dissipation by the flow of gases through the restriction between the prechamber and main chamber. Undesirable emissions may be lower than in the direct injection engine counterpart, largely aided by the mode of combustion where rich combustion is followed by lean combustion. A typical diesel engine of recent design is expected to have the following operational and design characteristics:

- Maintain high efficiency with minimized emissions while delivering increased power output per unit of cylinder swept volume aided by turbocharging and higher engine speeds
- Tolerate some variations in the type of fuel burned while retaining good cold starting capacity and ensuring excess air is provided throughout
- Provide sufficient fuel jet penetration while having good atomization and vaporization and avoiding wall impingement
- Have increased life and reliability with reduced requirements for service and maintenance

As a testimony of the superior performance of diesel engines is the diversity of the fields of their successful applications. For example, some of the main fields of their applications are

- Road transport systems, especially heavy-duty trucks and buses, delivery vans, and taxis, as well as increasingly in private passenger automobiles
- Stationary applications, such as for the production of electrical power, either steadily or for standby duties applications, with a wide range of applications involving cogeneration duties
- Military and agricultural applications, such as in tractors and harvesters
- A wide variety of industrial duties, such as driving pumps, compressors, and earth-moving and drilling rigs in the energy industry
- Rail traction and marine systems, especially those of very large power capacity

For example, the relatively slow speed diesel engine driving electric generation sets has become one of the most efficient practical convertors of fuel energy into electricity, with overall efficiency increasingly approaching the 50% mark. Also, its reliable operation in transport applications is virtually unmatched by other power systems. It is becoming increasingly apparent that a properly developed dual-fuel engine can equally replace the conventional diesel engine in almost all its fields of application, especially whenever the limitations associated with the portability of the fuel gas can be resolved.

Diesel engines tend to be more expensive than their equivalent spark ignition gasoline type. The relatively lower fuel prices in some countries in the past made the buying of engines mainly on fuel economy advantages alone less likely. However, this situation is gradually changing in recent years with the rapid increases in fuel costs.

Diesel engine performance is closely controlled by numerous key design and operational variables that must be fully considered in any application, such as the following:

- Fuel type, quality, and properties, both physical and chemical
- Various features of engine design and operation, including turbocharging, EGR, cogeneration, throttling, and variable valve timings employed
- Combustion chamber fluid flow and turbulence characteristics
- Fuel injection characteristics and timing
- Engine size, compression ratio, number of cylinders, and speed range
- Specific operating conditions, such as intake temperature and pressure, overall equivalence ratio, percentage of total load, and water jacket temperature

Much research and development in recent years have been directed successfully at reducing diesel engine emissions. These included, for example, improving the quality of the fuels used, combustion chamber geometry, and the fuel injection characteristics. Controlling parameters included variable ultra-high fuel injection pressures and injection timing with changes to nozzle geometry. Measures such as these allowed the diesel engine to make significant inroads in the relatively light duty and transport applications. Of particular additional significance is the capacity of the diesel engine to effectively utilize gaseous fuel resources for the production of power without contributing to increased emissions or loss of efficiency.

The matching of mixture motion with the fuel injection characteristics in diesel engines is of prime importance for ensuring good emissions and engine load ~ speed characteristics, particularly in direct injection type engines. The increasing employment in recent years of very high pressure liquid fuel injection, together with the intense turbulent swirling airflows, has resulted in improved mixing of the fuel and air that is so essential for superior combustion and thus performance.

3.3 THE IGNITION DELAY

There is a readily observable delay in diesel engines between the commencement of fuel injection and the first detection of ignition. The length of this ignition delay is of paramount importance and controls the progress of the combustion process and affects most aspects of engine performance. Excessively long ignition delays bring about reductions in torque and efficiency with increased emissions. The delay period has been viewed for convenience to be made up of two overlapping parts due to the times needed to satisfy the physical and chemical processes that bring about successful ignition. Generally, the chemical part of the delay tends to be longer than the physical part, but the chemical reaction events are sped up exponentially at high temperatures, while those processes that are physical in nature tend to increase with temperature relatively to a much lesser extent, almost linearly. Figure 3.2 shows schematically pressure development records for three widely different ignition delay values. Figure 3.3 shows typical variation of the ignition delay with changes in the amount of fuel injected, while Figure 3.4 shows the reduction in the delay period length with increases in the partial pressure of oxygen in the intake charge.

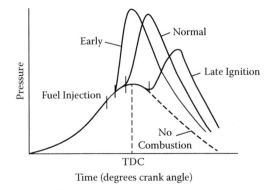

FIGURE 3.2 Schematic variation of the cylinder pressure with time for three cases having different lengths of delay. (From Karim, G.A., *Fuels, Energy, and the Environment*, CRC Press, Boca Raton, FL, 2012.)

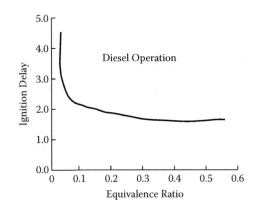

FIGURE 3.3 Typical variation of the ignition delay of a normally aspirated DI diesel engine with equivalence ratio of CR = 14.2:1. (Adapted from Badr, O. et al., *Journal of Energy Resources Technology*, 118, 2, 159–163, 1996.)

The ignition delay is usually defined as the time between the commencement of fuel injection and the first formation of a detectable ignition center somewhere within the chamber. This is normally detected through the first discernable pressure rise from the normal trend associated with the compression process. The end of the delay is easy to determine from the rapid variations in the rate of pressure rise. Another approach is to define the beginning of ignition with the instant at which the rate of change of the energy release rate just becomes positive. In most diesel engine applications, the ignition delay is usually shorter than the fuel injection period. However, in the case of the dual-fuel engines, where only a small quantity of diesel fuel is employed, this may often not be the case.

The development of the liquid fuel spray in diesel engines is mainly controlled by a number of important physical factors, such as atomization, penetration, entrainment, vaporization, and wall impingement. The entrainment rate of fuel jets in the

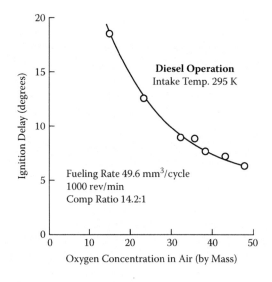

FIGURE 3.4 Variation of the ignition delay of a DI diesel engine with the oxygen concentration in the air at ambient intake operating conditions. (Adapted from Gee, D., and Karim, G. A., *The Engineer*, 222, 473–479, 1966.)

presence of the intense cross-flow within the chamber is, of course, a complex function of the combined effects of the multicomponents of the relative velocity of the jet material and the surrounding fluid. The effects of some chemical reaction activity tend at least, in the very early stages of fuel injection, in comparison, to be less significant. Moreover, cyclic variation in the spray structure and penetration characteristics, especially of modern diesel engines under steady-state operating conditions, tends to be less significant than it used to be.

Most diesel engines of recent design are turbocharged, which tends to reduce the length of the delay. The pressure and temperature of the intake air are increased, with substantial increases in the mass flow of engine air, which permit the burning of much more fuel, leading to substantial increases in the total energy release, power output, and torque. However, unless remedial action is taken, turbocharging results in increased engine thermal load, bringing about the onset of knock. Accordingly, the compression ratio of the engine may be reduced slightly, which can affect the engine capacity to deliver its maximum power potential. Some partial remedies can be provided by employing measures such as intercooling after compression, exhaust gas recirculation (EGR), and variable geometry turbochargers. The proper matching of the turbocharger characteristics with those of the engine in the dual-fuel mode is a requirement that may take much effort by engine manufacturers to accomplish properly.

3.4 DIESEL ENGINE FUELS

The suitability of a fuel for diesel engine applications is evaluated through a number of properties in accordance with standard specifications. They include the cetane number, density, distillation characteristics, viscosity, heating value, flash, cloud and pour points, sulfur and nitrogen contents, carbon and ash residues, water and sediment contents, and overall composition. Most of these, which are defined and discussed in fuel standards such as those of the American Society for Testing and Materials (ASTM), have been briefly defined and listed within the glossary at the end of this book.

One of the prime properties for the suitability of a liquid fuel for acceptable ignition in a diesel engine is the empirically devised cetane number scale. This rating is mainly based on the test fuel having the same ignition delay value as that of a specific binary liquid fuel blend made up of cetane (n-hexa-decane) and iso-hexadecane (heptamethylnonane) when tested in a standard engine under the same standard operating conditions. In this rating scale, the n-hexa-decane is given a rating of a 100 cetane number, while isohexadecane is given a rating of 15, i.e., cetane number = n-cetane (%) + 0.15 heptamethylnonane (%).

A low cetane number indicates that a fuel is hard to ignite within the available time in the engine settings, and thus it is less suitable as a diesel engine fuel. A fuel with a high octane number, which reflects resistance to autoignition, should have a correspondingly low cetane number value. Most common commercial diesel fuels typically have a cetane number within the range of 45 to 60. The long-chain normal hydrocarbons that are easily oxidized tend to have high values of cetane number, while aromatic fuels, which are comparatively less prone to autoignition reactions, such as aromatics, have low cetane numbers. Suitable additives are often employed to improve the properties of fuels destined for diesel engine applications. Following the refining of a typical crude petroleum, the yield of the diesel fuel can represent only a fraction of the products, as shown in Figure 3.5.

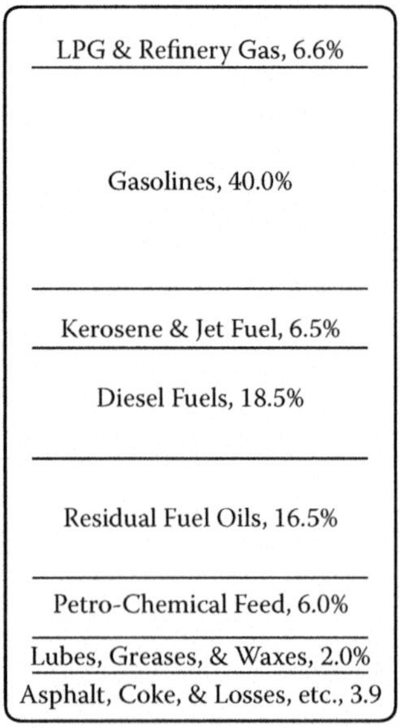

LPG & Refinery Gas, 6.6%

Gasolines, 40.0%

Kerosene & Jet Fuel, 6.5%

Diesel Fuels, 18.5%

Residual Fuel Oils, 16.5%

Petro-Chemical Feed, 6.0%

Lubes, Greases, & Waxes, 2.0%

Asphalt, Coke, & Losses, etc., 3.9

FIGURE 3.5 Typical distribution of the products of refining crude petroleum. (From Karim, G.A., *Fuels, Energy, and the Environment*, CRC Press, Boca Raton, FL, 2012.)

The employment of biodiesel fuels, which are derived from renewable resources, is not new, and some of the earliest diesel engines were developed to run on vegetable oils. These fuels may produce lower emissions of particulates, improved cetane number values, and high fuel lubricity, but they have tended to be more expensive than petroleum-derived fuels. Some of their negative features include high viscosity, poor cold weather performance, higher NOx emissions, slight power loss, and poor materials compatibility, which may include increased corrosion of parts of the fuel system and increased wear. A wide variety of additives have been employed to reduce the impact of some of these negative features. Common biodiesels are usually employed in the form of blends of conventional diesel fuel with a fatty acid methyl ester made from a variety of sources of vegetable oils or animal fats. Their use as a fuel supplement is growing steadily, particularly with the increasing cost of conventional diesel fuel and the need for ultra-low sulfur fuels, but typically much less than 20% by mass. The vegetable component, for example, may be derived from rape seed, palm, or coconut oil.

The production of biodiesel from biological sources is usually done by reacting them with an alcohol in the presence of a catalyst. A biodiesel blend with petroleum-derived diesel fuel, designated B10 as an example, indicates that the blend is made up of 10% biodiesel, and the rest is a conventional diesel fuel, on a liquid volume basis.

Liquid biofuels have sometimes been championed as reducing energy dependence, boosting farm revenues, and contributing toward reducing global warming. However, such claims need to be moderated and qualified suitably. In fact, biofuel usage can be viewed to negatively impact the environment and affect food sources availability and pricing worldwide, especially when factors such as acidification, fertilizer use, biodiversity loss, and toxicity of agricultural pesticides are taken into account.

A synthetic liquid fuel suitable for diesel engine applications is the Fischer–Tropsch diesel (FTD) fuel. It is produced from synthesis gas, which is a mixture of primarily hydrogen and carbon monoxide. The gas is commonly produced by the catalytic reforming of coal or natural gas with steam or by partial oxidation. The product of the FTD synthesis is a mixture of hydrocarbons of different size molecules, which undergo further cracking to produce diesel fuel. The FTD is considered a cleaner-burning fuel and of better ignition quality than conventional diesel fuels. This is mainly because it has a high fraction of straight-chain high molecular weight components that are readily reactive and require lower ignition temperatures. It also has virtually no sulfur. However, its production remains somewhat costly, with the fuel produced from coal tending to be relatively more energy wasteful and more complex than if it were produced from natural gas. The FTD fuel can be used in engines without the need to modify existing engines. For example, the diesel fuel produced from coal tends to be relatively energy wasteful and more expensive and complex than if it were produced from natural gas.

3.5 HOMOGENEOUS CHARGE COMPRESSION IGNITION (HCCI) ENGINES

A practice followed in the past mainly for research and developmental purposes was to feed fuel–air mixtures to engines in the absence of an external ignition source or auxiliary pilot fuel autoignition. This was done mainly to examine various aspects of the thermal, physical, and chemical behavior of fuel–air mixtures under the high-temperature and high-pressure conditions in a firing engine cylinder. Often such a mode of operation would not necessarily lead to autoignition or even the release of sufficient energy on reaction to overcome the frictional and other operational losses. Under such conditions, the engines were partially motor driven using an external power source until the operating conditions used were conducive for independent engine running and the generation of sufficient net positive mechanical work to dispense with the external drive. Such operation, which used to be described as motored engine operation, was employed, for example, to observe the pre- and post-autoignition processes of the fuel–air mixture. However, relatively recently, mainly with the drive to develop engines with ultra-low emissions, especially those of NOx, and the desirability to operate on lean mixtures, such an approach has been receiving more attention and further development. It became known as homogeneous charge compression ignition (HCCI). This mode of engine operation is akin to that of the motored engine approach, except ensuring that the power output is sufficiently positive in excess of that needed to overcome engine running losses. On this basis, the HCCI operation can also be viewed as a form of dual-fuel engine operation that is employing essentially a zero-pilot quantity.

Through the potential improvements to the operational control of engines, there have been numerous variations on such approaches in recent years. These include the deliberate lack of homogeneity of the charge where the introduction and mixing of the fuel with the air are carried out in a variety of approaches and stages during the intake and compression strokes. These have been pursued since they are potentially capable of overall lean mixture operation, which has ultra-low NOx emissions while having high efficiency at low loads, albeit with lower specific power output than in comparable diesel or spark ignition engines. To effectively control the progress of the combustion process and the associated energy releases in HCCI engines in the absence of a deliberate source of external ignition such as a spark or pilot fuel injection remains with a number of challenges. For example, the autoignition reaction rates tend to be self-accelerating, resulting in the heat release rates over a number of consecutive cycles becoming rapidly too high for acceptable operation. This takes place in association with other troublesome factors, such as increased heat loss, cyclic variations, noise, and vibrations. Thus, the elimination of deliberate externally controlled sources of ignition can lead to a loss in the effective and reliable control of the autoignition timing, with exceedingly high associated combustion rates.

REFERENCES AND RECOMMENDED READING

Anon., SAE Handbook: *Engine, Fuel, Lubricants, Emissions, and Noise*, vol. 3, SAE, Warrendale, PA, 1993.

Badr, O., Elsayed, N., and Karim, G.A., An Investigation of the Lean Operational Limits of Gas-Fueld Spark Ignition Engines, Transactions of the ASME, *Journal of Energy Resources Technology*, 118, 2, 159–163, 1996.

Badr, O., Karim, G.A., and Liu, B., An Examination of the Flame Spread Limits in a Dual Fuel Engine, *Applied Thermal Engineering*, 19, 1071–1080, 1999.

Baur, H., ed., *Diesel Engine Management*, 2nd ed., SAE, Warrendale, PA, 1999.

Borman, G.I., and Ragland, K., *Combustion Engineering*, int. ed., McGraw Hill, New York, 1998.

Challen, B., and Barnescu, R., *Diesel Engine Handbook*, 2nd ed., SAE, Warrendale, PA, 1999.

Cummins, L., *Internal Fire*, SAE, Warrendale, PA, 1989.

Gee, D., and Karim, G.A., Heat Release in a Compression Ignition Engine, *The Engineer*, 222, 473–479, 1966.

Heywood, J., *Internal Combustion Engines*, McGraw Hill, New York, 1988.

Hsu, D., *Practical Diesel Engine Combustion Analysis*, SAE, Warrendale, PA, 2002.

Karim, G.A., *Fuels, Energy and the Environment*, CRC Press, Boca Raton, FL, 2012.

Karim, G.A., and Ward, S., The Examination of the Combustion Processes in a Compression-Ignition Engine by Changing the Partial Pressure of Oxygen in the Intake Charge, *Transactions of SAE*, 77, 3008–3016, 1968.

Obert, E.E., *Internal Combustion Engines and Air Pollution*, Harper & Row, New York, 1973.

Patterson, D.J., and Henein, N.A., *Emissions from Combustion Engines and Their Control*, Ann Arbor Science Publishers, Ann Arbor, MI, 1972.

Pounder, C.C., ed., *Diesel Engine Principles and Practice*, George Newens Ltd., London, 1955.

Ricardo, H., *The High Speed Internal Combustion Engine*, 4th ed., Blackie and Son, London, 1953.

Stone, R., *Introduction to Internal Combustion Engines*, 2nd ed., SAE, Warrendale, PA, 1995.

Taylor, C.F., and Taylor, E.S., *The Internal Combustion Engine*, International Textbook Co., Scranton, PA, 1962.

Wood, C.D., *Alternative Fuels in Diesel Engines: A Review*, SAE Paper 810248, 1981.

4 Gas-Fueled Engines

4.1 MERITS OF OPERATING ON GASEOUS FUELS

Fuels that are gaseous under normal ambient conditions are classified as gaseous. Notable examples of these include natural gas, propane, and hydrogen. The gaseous nature of such fuels provides distinct advantages in their utilization in combustion devices in comparison to the combustion of common liquid or solid fuels.

The price of natural gas relative to that of diesel or gasoline can vary widely from time to time and from one location to another. Generally, on an energy basis, natural gas and liquefied petroleum gas (LPG) sell significantly cheaper than diesel fuel and gasoline. Also, to store an equivalent amount of energy on board a vehicle, the volume of a liquefied natural gas (LNG) tank may be approximately 1.7 times that for diesel fuel, and the corresponding volume of a typical compressed natural gas (CNG) tank is approximately 4 times. Of course, in engine applications, when evaluating the merits of a gaseous fuel, it is the volumetric heating value of the combustible fuel–air mixture used that is rather more important than merely the consideration of its conventional heating value. Much care is also needed when considering prospects, benefits, and limitations of the use of alternative fuels in general to ensure that available information is not colored by possible interests of lobbyists on behalf of either producers or consumers. Reducing costs, retaining jobs, and exploiting available indigenous fuel resources remain compelling factors. Table 4.1 shows the relative values of the specific energy storage for a number of common fuels. The superiority of the fuels as potential energy carriers, as compared to batteries.

TABLE 4.1
Some Energy Storage Mediums

Energy Storage Medium	Specific Energy (MJ/kg)
Gasoline	42.4
Diesel fuel	42.5
Methanol	19.7
Ethanol	26.8
Hydrogen (gas)	119.9
Methane	50.0
Lead acid battery	0.19

Source: Bols, R.E., and Tuve, G.L., eds., *Handbook of Tables for Applied Engineering Science*, CRC Press, Cleveland, OH, 1970.

An apparent limiting feature of the use of gaseous fuels in engine applications is the fact that they enter the engine generally as a gas, which displaces some of the air that would otherwise be inducted by the engine. This would reflect adversely then on the extent of power that can be produced by the engine. This is in contrast to liquid fuel applications, such as gasoline, which enters the engine initially mostly as a liquid spray, and its vaporizing helps lower the overall intake temperature and increase the density of the intake mixture. However, the slight lowering of the effective volumetric efficiency of the gas-fueled engine may not be a serious constraint since the lowering of the capacity to produce power can be offset by the associated positive combustion features and turbocharging. Moreover, operation on gaseous fuels has an important positive influence on engine maintenance and its lubrication system, while any small shortfall in power output from the corresponding operation on liquid fuel can be made up through employing a slightly larger or faster version.

Any undesirable substances that may be introduced into an engine with the gas supply vary widely with the type of fuel gas used, the processing it received, and local operating conditions. For example, "wet" natural gas, which normally contains small concentrations of higher hydrocarbons, will contribute to an increased tendency for the incidence of knocking, valve sticking, and increased deposits. However, "dry" processed natural gas, although it may be considered to be clean, can carry with it some very fine particles, such as those of rust, along the supply lines. Also, producer and coke oven gases tend to contain some undesirable particles, such as those of carbon black and particles of sulfur compounds, while fuel gases originating from the steel industry can additionally contain some hydrogen sulfide.

Large fleet operators such as those of taxis, road trucks, delivery vans, buses, and locomotives tend to have large expenditures for fuel and maintenance. Their use of natural gas and other gaseous fuels often will save a considerable amount of money. Additionally, there are numerous positive features that may be considered to be associated with the employment of gaseous fuels in combustion systems in general and internal combustion piston engines in particular, in comparison to operation on liquid fuels. For example, the composition of gaseous fuels generally tends to be much simpler in structure in comparison to common liquid fuels, which are made of complex mixtures of a wide range of large molecular weight hydrocarbons. Thus, the properties of a liquid fuel are varying averages that can change with the source of the fuel and the processes that have gone through its refining, with optimum engine performance being less certain than in the corresponding employment of gaseous fuels that require no atomization or vaporization for their combustion. This renders gaseous fuels introduction into the engine both simple and more efficient. It helps to ensure good mixing in the desired proportions with air and good mixture quality distribution between the various engine cylinders while requiring relatively simpler manifold design. Engine idling can be also improved with easier cold starting. In addition, an important positive feature contributing to the widespread utilization of gaseous fuels, and notably in the form of natural gas, is the ease of distribution and availability of the fuel over considerable distances from points of production and processing to consumers. Usually, in the case of stationary engine applications, there is no need for fuel storage and associated infrastructure since a clean, convenient, and constant supply of fuel gas is assured from the mains to the combustion device and flows easily with automatic and remote regulation provided.

Some other main positive features that can be cited for the employment of gaseous fuels in combustion systems in general and in piston engines in particular in comparison to liquid fuels are the following:

- Gaseous fuels are generally associated with high combustion efficiency over wide firing ranges with improved flame stability limits. They need less time for proper mixing with the necessary air for combustion, because of their gaseous nature.
- The burning of gaseous fuels is associated with the production of relatively cleaner product gases with hardly any solid pollutants emitted, such as ash, particulates, or soot. They normally have a lower tendency to initiate corrosion.
- Simpler and cheaper combustion equipment can be developed with better low-temperature operation. The fuel supply is easy to control, with finer control of excess combustion air and larger turn-down ratios.
- Design, operation, and the control of gaseous fuel combustion systems that depend on the properties of the fuel, and the mixing needed between the fuel and air following their introduction, tend to be relatively simpler and easier for initiating combustion.

A serious limitation associated with gaseous fuels usage is the tendency for increased fire, explosion, and toxic hazards, especially following a leak. However, over the years there have been much experience and strict regulatory controls established for the safe handling and usage of gaseous fuels.

4.2 MODES OF OPERATION OF GAS-FUELED ENGINES

The premixed dual-fuel engine is basically a conventional compression ignition engine of the diesel type where the injection of some liquid fuel, often in quite small dosages, is used to provide the source for ignition. The cylinder charge is made up mainly of lean mixtures of a gaseous fuel and air (Figure 4.1). There are a number of variations of this mode of operation, such as having the gaseous fuel injected at very high supply pressures directly into the engine cylinder so that the fuel burns into the wake of the earlier injected and already ignited liquid fuel jet (Figure 4.2).

Normally in dual-fuel engine applications, mainly for economic reasons, much of the energy release comes from the combustion of the usually cheaper gaseous fuel, while only a small amount of diesel liquid fuel is injected to provide ignition through timed cylinder injection in the usual way as takes place in conventional diesel engines. Such an operation, with optimum conversion methods, has been shown to have the potential to provide operational characteristics that are often comparable or even superior to those of conventional liquid-fueled diesel or gas-fueled spark ignition engines. This may be achieved while displaying improved emission characteristics and quiet, smooth, and improved low-ambient-temperature operation with reduced thermal loading. Such superior performance may be achieved only when

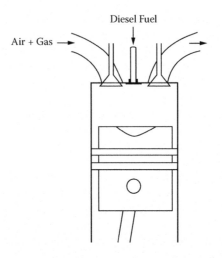

FIGURE 4.1 A schematic representation of a premixed dual-fuel engine with diesel injection to serve as the pilot for ignition.

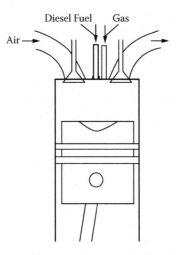

FIGURE 4.2 Schematic representation of a dual-fuel engine where the fuel gas is injected directly into the chamber and ignition is obtained with pilot fuel injection.

sufficiently effective measures are ensured, such as, for example, the avoidance of knock at high loads and the excessively incomplete gaseous fuel utilization at relatively light loads. Usually, a main aim while retaining alternatively acceptable diesel operation is to maximize the replacement of the diesel fuel by a usually cheaper and more abundant gaseous fuel while maintaining acceptable levels of exhaust emissions and engine performance.

Dual-fuel engines are expected to display sufficiently high power density and efficiency, and lower heat rejection, with multifuel consumption capability. Their operation may be viewed to be falling into broad categories, largely depending on the amount of the gaseous fuel employed relative to that of the diesel liquid fuel. The more common and simple strategy is to have only a relatively small pilot quantity of liquid diesel fuel injected at a fixed rate over the whole load and speed ranges. This is mainly to provide timed ignition of the cylinder contents made up usually of lean premixed gaseous fuel–air mixtures, as shown schematically in Figure 4.3. The bulk of the energy release then comes increasingly from the combustion of the gaseous fuel component.

Another category may be viewed as merely supplementing the variable supplies of liquid fuel of a fully operational diesel engine, at a stage well beyond the light load region, with an increasing addition of a gaseous fuel to the engine air, as shown schematically in Figure 4.4.

However, it would be more beneficial to optimally vary the liquid fuel quantity relative to that of the gaseous fuel and its supply rates with operating conditions and the type of fuels employed. Usually, there is no need then to maintain relatively large pilots with the increase in engine load (Figure 4.5). This way it becomes possible to achieve superior performance over the whole load and speed ranges. This form of approach may not be incorporated in some dual-fuel engine installations, mainly because of the relatively demanding complexity of its control.

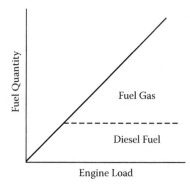

FIGURE 4.3   Schematic representation of the changes in gaseous fuel quantity relative to a constant liquid pilot fuel quantity.

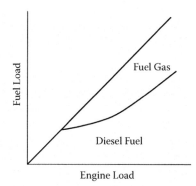

FIGURE 4.4   Schematic representation of the supplementing of diesel liquid fuel by increasing the amount of fuel gas.

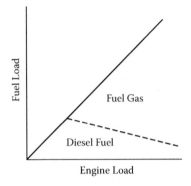

FIGURE 4.5 Variation in the liquid pilot quantity with load.

The modern high-compression-ratio diesel engine is imminently suited to dual-fuel operation, particularly when using a fuel such as methane, the main component of a wide range of natural and biogases. Largely by virtue of the provision of a high compression ratio with excess air and an adequate sized, consistently timed pilot liquid fuel injection, almost any gaseous fuel or vapor can be utilized to a varying degree of success. A wide range of pure gaseous fuels, such as methane, ethane, propane, butane, hydrogen, and ethylene, have been employed in dual-fuel engine applications. Moreover, various gaseous fuel mixtures, which include, besides natural gases, liquefied petroleum gases and those containing excessive concentrations of diluents such as biogases, landfill gases, coal bed methane, and blast furnace gases, are being successfully used. Information is also available in the open literature about the performance of dual-fuel engines when a range of liquid fuels, including liquefied natural gas, gasoline, and alcohols, are introduced into the incoming engine air or injected subsequently into the cylinder during the early part of the compression stroke with diesel fuel pilot ignition.

The economics of dual-fuel engine operation is quite attractive. Its usage of two separate fuel systems may be viewed as a positive feature since it would permit the changeover to conventional diesel operation without interruption of power production whenever required. With this feature it would allow the efficient exploitation of interruptible fuel supplies such as in transport applications, where the limitations imposed to the range that can be traveled by a vehicle between refuelings due to the limited portability of the gaseous fuel can be waved. It can be suggested that when the well-designed and well-operated high-compression-ratio diesel engine is operated on gaseous fuels, and especially pipeline natural gas, few, if any, contemporary power sources can compete successfully on the basis of economic returns per capital invested. The combination of cheap fuels and a well-developed prime mover proves a very attractive investment opportunity while offering environmental and resource clear advantages.

Most diesel engines of recent design that are turbocharged can perform equally well when converted to dual-fuel operation. Diesel engines of the two-stroke type normally require excess scavenging air to clear the cylinder adequately of residual gases. In their corresponding dual-fuel applications, when the gaseous fuel is fumigated into the intake air before entry into the cylinder, special procedures need to be taken to ensure the elimination of the discharge of some unburned fuel into the exhaust. Otherwise, there will be both much fuel wastage and increased contribution to exhaust emissions.

4.3   THE APPLICATIONS OF GAS-FUELED ENGINES

The applications of gas-fueled diesel engines can be found in numerous fields. Stationary engine applications have included electric power generations, driving compressors, pumps, and cogeneration sets. The engines have also been extensively employed in mobile applications, such as for driving of fleet vehicles, heavy-duty trucks, buses, rail locomotives, and marine vessels, and in various applications within the construction and agricultural industries. Often, it is desirable to have the diesel engine converted to fuel gas operation capable of accommodating some variation in the composition of the gaseous fuel.

Common diesel engines tend to endure successfully long operational lives. They vary widely in type, size, applications, and operational ranges of load and speed. These make their wide conversion to gas-fueled operation quite demanding, especially when optimal performance is aimed for. There is, for example, a consistent need to develop operational approaches in conjunction with the gaseous fuel combustion characteristics, such as by a careful choice of pilot fuel size and its injection timing so as, for example, to reduce oxides of nitrogen emissions. Moreover, the operation of these engines, especially when employing some gaseous fuel mixtures, can be seriously limited by the onset of knock. Its avoidance, while retaining the high-compression ratios of the efficient diesel engine, with some gaseous fuels, often remains a challenge and requires proper expertise.

The employment of dual-fuel engines for the production of power, despite its numerous operational, environmental, and economic positive features, has not seen as much wide application in the past as would be expected. A number of contributory factors can be suggested that tend to confine their application to power generation in relatively large, mostly stationary engines. Relatively fewer applications were related to the transport sector. Some of the contributory factors to these trends could have included the following:

- The combustion process in the gas-fueled diesel engine of the dual-fuel type is quite complex. It displays some of the features and associated problems of the compression ignition diesel engine, as well as those of the premixed spark ignition type.
- The exhaust emissions of dual-fuel operation, until relatively recently, were not easy to deal with satisfactorily when in combination with the widely varying quality of the fuel systems possible, over the whole ranges of load and speed. However, many recent advances have been made, making it easier to deal with these difficulties.
- To obtain the potential benefits associated with gas-fueled operation, many influencing operational and design factors require careful optimization and control. Suitable computer controls of the many engine variables were not widely available in the past, while relying mainly on mechanical type

control devices. This tended to make transport applications harder to implement satisfactorily and lacked the versatility and reliability required for the
frequent transients in performance. The need to provide two complex sets
of controls and adequate portable storage for two sets of fuel systems of
widely different properties has also contributed to the challenges.

- There has been a continuing need to retain the capacity of gas-fueled
engines to revert to the diesel operational mode promptly and smoothly
when needed. This facility is required to be retained without undermining
the performance of the engine as a diesel. This sometimes represented a
challenge since diesel engines are required to satisfy increasingly stricter
and more challenging requirements of performance and controls.

- Until recently, relatively small diesel engines were less operationally flexible and tended to be more costly in comparison to their spark ignition
counterparts. Accordingly, the exclusive conversion to fuel gas spark ignition operation was easier to accomplish and more satisfactorily performed,
especially while retaining the bi-fuel operational capacity.

- Large output stationary engines, such as those employed exclusively for the
generation of electric power, consume very large quantities of fuel. The need
to lower fuel costs and maintain high efficiencies is an important consideration. These tended to favor employing the efficient and reliable liquid-fueled
diesel engines, which in recent years have reached a very advanced stage of
sophistication and reliability. They are readily turbocharged, and their performance is well optimized. However, only when relatively cheap and abundant
fuel gas supplies were available were dual-fuel engines employed.

- A consequence of operating diesel engines in the premixed dual-fuel mode
is that some of the intake air is displaced by the gaseous fuel induced. The
amount of air displaced increases with the decrease in average molecular
weight of the fuel. For example, for stoichiometric mixtures, the decrease by
volume is 3.13% for butane, 4.03% for propane, 5.66% for ethane, 9.51% for
methane, and 29.6% for hydrogen. This air displacement can have a significant influence on the volumetric efficiency, emissions, and associated power
output of the engine, requiring remedial measures to recoup the potential
shortfall in engine power output and performance. Otherwise, relatively bigger engines are needed with increased capital and operational costs.

- The low efficiency and sensitivity to changes in the fuel gas composition
of the low-compression-ratio spark ignition gas engine made it fall increasingly out of favor in comparison to the gas-fueled diesel engine. However,
diesel engines, with their continued improvements in the control of their
emissions, efficiency, and reliability, tended to relegate the dual-fuel engine
to a secondary role, confining it increasingly to special applications where
economic advantages can be assured through the exploitation of much
cheaper gaseous fuels. The increased attention in recent years to reduce
the emission of greenhouse gases is aiding in the widespread application of
dual-fuel engines in transport applications.

REFERENCES AND RECOMMENDED READING

Barbour, T.R., Crouse, M.E., and Lestz, S.S., *Gaseous Fuel Utilization in a Light Duty Diesel Engine*, SAE Paper 860070, 1986, p. 13.

Bechtold, R.L., *Alternative Fuels Guidebook*, SAE, Warrendale, PA, 1997.

Bols, R.E., and Tuve, G.L., eds. *Handbook of Tables for Applied Engineering Science*, CRC Press, Cleveland, OH, 1970.

Challen, B., and Barnescu, R., *Diesel Engine Handbook*, 2nd ed., SAE, Warrendale, PA, 1999.

Daisho, Y., Takahashi, Y.I., Iwashiro, Y., Nakayama, S., and Saito, T., *Controlling Combustion and Exhaust Emissions in a Direct-Injection Diesel Engine Dual Fuelled with Natural Gas*, SAE Paper 952436, 1995.

Danyluk, P.R., Development of a High Output Dual Fuel Engine, *ASME Journal of Engineering for Gas Turbines and Power*, 115, 728–733, 1993.

D'Souza, M.V., and Karim, G.A., The Combustion of Methane with Reference to Its Utilization in Power Systems, *Journal of the Institute of Fuel*, 335–339, 1972.

Ebert, K., Beck, N.J., Barkhimer, R.L., and Wong, H., *Strategies to Improve Combustion and Emission Characteristics of Dual Fuel Pilot Ignited Natural Gas Engines*, SAE Paper 971712, 1997.

Eke, P., and Walker, J.H., Gas as an Engine Fuel, in *Proceedings of the Institute of Gas Engineers*, London, UK, 1970, pp. 121–138.

Erickson, R., Campbell, K., and Morgan, D.L., Application of Dual Fuel Engine Technology for on Highway Vehicles, presented at Proceedings of ASME Internal Combustion Engine Division, Salzburg, Austria, 2003, Paper ICES2003-586.

Heenan, J., and Gettel, L., *Dual Fueling Diesel/NGV Technology*, SAE Paper 881655, 1988.

Karim, G., Combustion in Gas-Fueled Compression Ignition Engines of the Dual Fuel Type, in *Handbook of Combustion*, ed. M. Lackner, F. Winter, and A. Agarwal, vol. 3, Wiley Publishers, Weinheim, Germany, 2010, pp. 213–235.

Karim, G.A., A Review of Combustion Processes in the Dual Fuel Engine—The Gas Diesel Engine, *Progress in Energy and Combustion Science*, 6, 277–285, 1980.

Karim, G.A., *Dual Fuel Engines of the Compression Ignition Type—Prospects, Problems and Solutions—A Review*, SAE Paper 830173, 1983.

Liss, W.E., and Thrasher, W.H., *Natural Gas as a Stationary Engine and Vehicular Fuel*, SAE Paper 912364, 1991.

Maxwell, T., and Jones, J., *Alternative Fuels: Emissions, Economics and Performance*, SAE, Warrendale, PA, 1995.

Turner, S.H., and Weaver, C.S., *Dual-Fuel Natural/Diesel Engines: Technology, Performance and Emissions*, No. GRI-94/0094, Topical Report Gas Research Institute, November 1994.

Wood, C.D., *Alternative Fuels in Diesel Engines: A Review*, SAE Paper 810248, 1981.

5 The Premixed Dual-Fuel Engine

5.1 HISTORICAL DEVELOPMENT

Much of the early development of the internal combustion engine was based on employing gaseous fuels. The widespread use at the time of coal gas, produced by the processing of coal for lighting, domestic, and industrial applications, also encouraged its use in engines, despite the gas being highly toxic and explosive. There were also some engine applications employing industrially derived fuel gases such as those in the steelmaking industry or in the processing of municipal sewage and biomass. After the rapid success of the four-stroke liquid-fueled internal combustion engine, gas-fueled engines tended to be confined primarily to special applications and the stationary engine fields.

It was in 1901 that Rudolf Diesel obtained a U.S. patent covering the concept of the dual-fuel engine. A mixture of fuel and air was to be compressed to a temperature below that needed for its autoignition. Ignition was then produced through the injection of a second, more reactive fuel that has a lower ignition temperature than the fuel gas. Others followed later with patents describing the operation of high-compression-ratio internal combustion engines that operate on both oil and a gaseous fuel. Regular ignition was also provided through the injection of a suitable liquid fuel.

Some applications of gas-fueled engines could be seen before the Second World War in countries such as Italy, Russia, Germany, Holland, the United States, and the U.K. During the war there was much activity to utilize gaseous fuels for engine applications as supplies of liquid fuels of the right quality became increasingly scarce, especially in countries such as Germany. There were also attempts to use some low-quality gasolines instead of gas in engines that operated on the dual-fuel mode due to the shortage of diesel fuels. It was mainly before the Second World War that some dual-fuel engines were supercharged and others operated on high-pressure natural gas that was injected directly into the cylinder at near TDC position and ignited by diesel fuel injection. These approaches, however, did not see wide applications in the energy industry until much later on.

Natural gas-fueled transport applications declined considerably after the war, mainly due to the increasing availability at the time of relatively cheap oil supplies. However, dual-fuel engines were widely employed for special stationary applications, mainly for the generation of power, such as by the oil industry in the field or in municipal sewage works aiding, through cogeneration, the production of sewage gas, which was also used as the fuel for the engine. It has only been within the last few decades, driven by the ever-stricter air pollution controls, the rise in the cost of fuels in general, and the relatively reduced availability of liquid fuels of the right quality, that gas-fueled engine applications began increasing, mainly in the stationary electric power generation and commercial vehicle sectors. Also, the increasingly wide availability of bulk supplies of natural gas through long-distance pipeline transport and supplies of liquefied natural gas (LNG) transported by sea from remote parts of the world made the running of engines on natural gas increasingly possible and economically attractive. This is particularly true in countries that are relatively well endowed with cheap natural gas supplies, and for bus transport in some highly polluted cities. The increased availability of LNG supplies also encouraged the development of engines that can operate either directly on LNG or its boil-off, including in the large engines of marine tankers transporting the liquefied fuel to markets. Currently, there are many commercially available dual-fuel systems offered by various manufacturers in many parts of the world. However, although these systems work adequately, they do not necessarily have their operation sufficiently well optimized, particularly for the specific application. Moreover, the localized electric power generation by dual-fuel engines has been made somewhat less common by the increased availability in many countries of electric power supplies through the grid systems. With the increased recent uncertainties surrounding the dependability of the electric power supplies, together with the increased availability of cheap and relatively abundant supplies of natural gas, increased price of diesel fuel, and potential limitations to its supplies, and with the ever stricter controls on emissions, the dual-fuel engine is expected to undergo significant gains in popularity. Meanwhile, the normal diesel engine continues to gain in efficiency, power density, and reliability.

In the early days of the diesel engine air-blast injection was commonly applied, mainly in stationary and marine applications, to assist in the rapid atomization of the injected liquid fuel. Compressed air around 65–70 bars was supplied from storage air cylinders charged by a separate compressor, which was often driven by the engine. At a certain point toward the end of the compression stroke a high-pressure blast of air drove the fuel at a great velocity into the cylinder to mix with the combustion air and later ignite. The blast of air usually constituted around 2–3% of the total air and contributed to lengthening the fuel injection and ignition delay periods. The whole system was cumbersome, costly, and troublesome, absorbing a considerable amount of power, which contributed to reducing the overall efficiency of the engine and its available power output. The high-pressure air storage cylinders were especially too cumbersome for transport applications. When more effective fuel injection systems were successfully developed later, such as through employing the jerk pump, the air-blast method of fuel injection was replaced and rendered obsolete. In recent years, some remnants of this technology have again found their way in the development of high-pressure fuel gas injection directly into the combustion chamber of dual-fuel engines.

Some limited attempts were made in using slurries of liquid and powdered solid fuels as supplementary fuels in engines injected through suitably modified injectors. Some other attempts of direct injection of dry powdered solid fuel into the engine cylinder were made, but on the whole they were operationally or economically not viable.

5.2 DUAL-FUEL ENGINE OPERATION

To render common diesel engines capable of operation on a range of gaseous fuels, whether for stationary or mobile applications, has tended in the past to be made often without achieving some of the potentially attractive features of the engine. Moreover, the operational and design features of converted engines were on occasion reported in the open literature, while based on experience gained mainly from prototype installations or under laboratory conditions. Such features may need to be modified significantly later on through subsequent design changes made or through additional operational experience. Accordingly, it is not uncommon that some of the published reports about dual-fuel performance, especially in the popular or trade press, appear to be in conflict with others.

Currently, there are many commercially available dual-fuel engine systems offered by various manufacturers in many parts of the world, usually in parallel with their diesel engine lines. There is also an increased successful entry of dual-fuel engines in the transportation sector targeting primarily the public and industrial fields. However, although these systems have been operating satisfactorily and economically, there is still room for reducing costs and enhancing their performance further, whether in terms of efficiency, power production, maximizing diesel fuel replacement, displaying a wider tolerance to changes in gaseous fuel composition, or ensuring additional improvements in exhaust gas emissions. Indeed, dual-fuel engines, when dedicated properly primarily to fuel gas operation, can have some aspects of their performance equal or superior to those of the common diesel engine while economically exploiting a wider variety of gaseous fuel resources.

The satisfactory operation of dual-fuel engines depends on numerous operating and design variables that tend to be greater in number than those controlling the performance of conventional spark ignition or diesel engines. Accordingly, the level of power production potential and the associated levels of exhaust emissions of engines operating on various gaseous fuels in the dual-fuel mode will depend on how well the effects of the following design and operational variables have been dealt with.

- The type of fuel used, its composition and heating value, the physical and chemical properties of the fuel, and their variations with temperature and pressure; values of the effective flammability limits and burning rates, especially at the high temperatures and pressures encountered in engines at the end of compression and initiation of the combustion process
- Intake and exhaust temperatures and pressures, the equivalence ratio values used, the presence of any diluents in the fuel or air supplied, the autoignition characteristics, ignition energy and limits, and their corresponding variations with equivalence ratio and temperature, and operational knock limits
- The associated operational volumetric efficiency with throttling when employed, turbocharging, and the extent of exhaust gas recirculation (EGR) used; the nature and extent of any fuel and temperature stratification, the type of pilot fuel employed, its relative size, injection characteristics, and timing
- Values of the engine compression ratio, bore, and stroke, combustion chamber shape, engine speed range, and any surface activity

In the case of turbocharged dual-fuel engines, their output can be controlled through employing waste-gates and variable geometry turbines. The waste-gate is a valve fitted in the exhaust gas path that can be actuated to divert some of the exhaust gas to bypass the turbine altogether, thereby reducing the available work for compression. In engines fitted with a variable geometry charger, the turbine nozzles are adjusted to vary the pressure drop across the turbine to better match the pressure rise needed for the compressor part. This would also allow, for example, the torque to rise beyond that of the corresponding diesel engine version, which is traditionally limited by the emission of unacceptably high density smoke at peak torque speeds. When sufficiently high pressure fuel gas supply is provided, turbocharged dual-fuel engines often have their fuel gas introduced beyond the outlet of the compressor. However, for some applications the fuel gas carburetion may be made ahead of the turbocharger.

Of course, throughout, more than just adequate measures must be taken to ensure the safety of operation of dual-fuel engines irrespective of the type of fuel gas, engine, and load employed. Transient operating engine conditions are usually more likely to provide a greater risk, especially when highly turbocharged engines operating with exhaust gas recirculation are employed.

Diesel engines with their high compression ratios are associated with relatively high work production efficiencies. However, high NOx and particulate emissions are produced mainly due to the diffusion types of combustion of the liquid fuel associated with multiple ignition sites and the many localized regions of high combustion temperature.

The performance of the gas-fueled dual-fuel engines in comparison to that of the conventional premixed spark ignition gas engine displays a number of positive features. For example, a wider selection of gaseous fuels can be employed, with a much lesser tendency to encounter the incidence of knock. This is aided by the overall lean mixtures normally employed while having superior control and safety characteristics. The dual-fuel engine displays less cyclic variation even at light load, arising mainly from the deliberate and reliable source of the controlled energetic pilot ignition employed. The engine that is normally of diesel engine origin is of more robust construction and can better withstand shock and knock loadings at high loads with low heat loss and good cogeneration capacities. The control of power output in dual-fuel engines is normally obtained by quality control, with changes made in the fuel concentration rather than through throttling, as is the common practice in spark ignition engines. The engine can then operate over a wider overall range of mixture equivalence ratios and turbocharging intensity. The high compression ratios and torque of the diesel engine can be retained in dual fueling while requiring less variation in pilot fuel injection timing in comparison to the usually large changes in the timings of electric spark ignition, with changes in speed and load needed by the spark ignition engine. In recent years, gas engines of the spark ignition type have tended to be stationary and with large size and multicylinders. They are rather slow and less efficient with their performance, severely restricted by their tendency to encounter knocking, especially with some gaseous fuels, such as those of raw unprocessed natural gases. On this basis, the gas-fueled spark ignition engine has fallen increasingly out of favor to the dual-fuel engine. However, conventional diesel engines of recent design have tended to overtake dual-fuel engines in some applications with the continuing improvement in the effective control of their exhaust emissions and their high efficiency and reliability. The dual-fuel engine remains favored for applications where its economic advantages are considered to be very important and clearly assured.

Diesel engines converted to unthrottled premixed dual-fuel operation always employ at light loads very lean gaseous fuel–air mixtures, which can lead to the production of significant amounts of unconverted gaseous fuel and carbon monoxide emissions. The catalytic treatment of the exhaust gas also tends to be less straightforward than in spark ignition engines or when burning fuels other than methane. Care is also needed to minimize the size of the pilot liquid fuel component relative to that of the fuel gas. Depending on operating conditions, it may lead at high loads to increased exhaust emissions, encourage the incidence of knock, and intensify cyclic variations. However, the emission of particulates tends throughout to be of much less significance than for comparable diesel engine operation.

There is a complex interaction between the liquid fuel spray and the bulk premixed gaseous fuel–air charge. This interaction not only is thermal, but also has a significant chemical kinetic component, which can bring about an extension to the ignition delay and increased emissions. The direct injection of the gaseous fuel into the cylinder, especially as implemented in some recent engine systems, represents an added complexity. A sufficiently high pressure gas supply needs to be generated on board and the fuel gas introduced at the appropriate instant directly into the engine cylinder. The relative phasing of the introduction and mixing processes of the injected fuel gas and those of the liquid fuel pilot require very careful sequencing, control, and implementation. All these efforts are being supported by the increasingly successful modeling of dual-fuel combustion while using comprehensive three-dimensional computational fluid dynamics (3D CFD) approaches with increasingly more realistic detailed chemical kinetic simulation and aided by advanced diagnostic experimental methods. However, as will be shown in the following chapters, the ignition and combustion processes of dual-fuel engines in comparison to those of diesel or spark ignition engines remain relatively complex. This is because they include diesel injection and autoignition processes together with premixed and diffusional modes of combustion of the fuel gas.

5.3 THE CONVERSION OF DIESEL ENGINES TO DUAL-FUEL OPERATION

In general, to convert diesel engines to dual-fuel operation, the control system must be suitably matched to the operational and design characteristics of the diesel engine to be converted, which imposes limitations to the application of universal engine conversion kits. This can become quite restrictive, especially since the corresponding operation of diesel engines of recent design is the subject of demanding performance requirements, especially in relation to emissions, fuel quality, and efficiency controls.

Some of the recent advances in the operational dual-fuel technology include the improved control of exhaust gas emissions through transient catalytic treatment of the exhaust gases and EGR, high-pressure direct in-cylinder gas injection, enhanced capacity to operate on a multitude of gaseous fuels and some of their mixtures, employing when needed multiple timed fuel injections, turbocharging, ultra-high boost intake pressures, employment of "micro-size" liquid fuel pilots, cogeneration, and extensive safety and operational computer controls. All these features have been supported by the recent advances made for the effective predictive modeling of the complex combustion processes, exhaust emissions, and performance.

It appears so far that there are not many examples of engines designed and developed wholly dedicated to dual-fuel operation. Accordingly, the conversion of common diesel engines to dual-fuel operation does not require significant modifications to the engine, substantially reducing capital, operational, and maintenance costs. The diesel engine converted to fuel gas operation can also continue to operate without undermining its operational life. It may even have it extended beyond that expected of conventional diesel engines. The overall specific energy/fuel consumption may also be lower, or at least not much in excess of that associated with the correspondingly loaded normal diesel operation. Similarly, the engine power output and speed ranges are kept around those values associated with diesel operation. Turbocharged dual-fuel engine operation can also continue to provide effective power supplement over the whole operational range while maintaining a high level of operational safety. The extent and mode of associated exhaust gas emissions may be on the whole superior to those encountered with diesel operation, especially with respect to the hard-to-deal-with emission of particulates and NOx. Additionally, dual-fuel engine operation can be made to allow some variations in the type and composition of the gaseous fuel used, while permitting controlled variations in the relative size of the pilot and its injection characteristics so as to maintain a superior performance. The fitted gaseous fuel intake delivery system also does not produce unacceptably high pressure drop or excessive cooling within the intake manifold, so as not to undermine the volumetric efficiency of the converted engine to fuel gas operation. Of course, safety of operation on gaseous fuels remains of paramount importance, with the idle and overspeed governing remaining available throughout, while ensuring fail-safe operation in case of an emergency, malfunction, or sudden interruption to either of the two individual fuel supplies.

In principle, some of the main potential positive features to be anticipated out of the proper conversion of diesel engines to gas fuel operation in the dual-fuel mode may be the following:

- The conversion need not be too costly and aims to produce an engine that has an overall specific energy consumption not exceeding that associated with its operation as a diesel. The corresponding power output and the extent and mode of exhaust gas emissions need not be inferior to those associated with normal diesel operation. The converted engine would be able to operate without reductions in its operational life.
- The volumetric efficiency of the engine should not be reduced by the added gaseous fuel delivery system, with safe turbocharged operation maintained over the whole operational range.
- It is preferable to have the size and injection timing of the pilot varied independently so as to contribute toward achieving and maintaining optimum performance in the dual-fuel mode. Also, the developed controls should maintain acceptable engine performance when some relatively minor

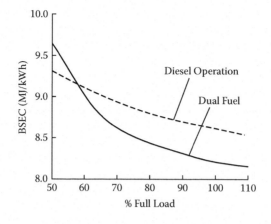

FIGURE 5.1 Variation of brake specific energy consumption (BSEC) with load for a dual-fuel engine operating on natural gas, and the corresponding values when operating as a diesel. (From Danyluk, P.R., *ASME Journal of Engineering for Gas Turbines and Power*, 115, 728–733, 1993.)

variations in the composition of the gaseous fuel supplied are encountered. It has been a common practice to have the engine start and stop while operating on liquid diesel fuel only. Figure 5.1 shows an example of the superior performance of a highly rated dual-fuel engine in comparison to the corresponding diesel operation.

5.4 THE EMPLOYMENT OF DUAL-FUEL ENGINES

The dual-fuel engine is particularly attractive where the fuel gas is considered superfluous, such as in coal mines or the waste gases from refining processes. The conversion then requires relatively little investment with low fuel costs, good durability, and low maintenance. These features have often tended in the past to provide insufficient monitoring and control systems and dispensing with the optimization of the different systems.

The economics of conversion of diesel engines to dual-fuel operation benefits from the combination of a cheap fuel and an efficient well-developed and readily available prime mover in the form of diesel engines. It often proves to be an attractive investment opportunity while offering clear environmental and resource advantages. However, the dual-fuel engine has not seen, until relatively recently, the wide application it should have commanded. The majority of its fields of application tended to be electric power generation, with fewer applications within the smaller engines of the transport sector. In this sector, fleets of trucks and vans that perform routine tasks such as delivery, pickup, and other repetitive daily tasks do not usually require the generation of much power, but instead favor having high reliability, low specific fuel consumption, and low operational costs. These vehicles also tend to have experienced drivers with access to regular maintenance, and technical attention should it be needed. However, there are a number of associated potential limitations. These include the limited range of driving distances with increased vehicle capital costs, higher weight, and some of the storage space occupied by the cylinders containing the fuel gas. They can be associated with increased costs, some deterioration in vehicle handling, decreased net payload, and may lead even to some increases in fuel consumption. The limitations imposed by the distance traveled after a fuel refill represent a significant limitation to the flexibility of vehicle operation on gaseous fuels. Additionally, there is often a limited availability of refueling facilities. In general, these can contribute to increased costs of engine conversion and operating costs. In the economic evaluation of the viability of the application of gaseous fuels for a transport system such as diesel-powered buses, numerous factors need to be considered and evaluated. These would include the capital cost of the converted engines and buses, fuel, operational and maintenance costs, fleet size, infrastructure for fuel supplies, and its capital; with routes' characteristics, taxes, and subsidies, if available. There is also the need to provide the gaseous fuel at high pressure within sufficiently reliable fuel containers with safety issues attended to effectively. The cost of the gas cylinders remains quite substantial and can add considerably to the weight of the vehicle and may occupy a substantial volume. Another consideration is the fact that the composition of the fuel gas, when originating from different supply points and times, may vary unless special measures are taken in the engine operational controls to tolerate such changes in composition or to suitably process the fuel gas before its entry into the engine.

The compression work expended in compressing the fuel gas is not recovered or much of it exploited usefully. The expansion of the gas to the much lower engine intake pressure produces cooling to engine parts and controls, and may require the provision of suitable warming. Otherwise, it may adversely affect engine performance. Of course, whenever the supply of the specific fuel gas being used is not constant or is inadequate, then the continuity of power generation may have to be ensured by using an increased diesel-to-fuel gas ratio or switching over to wholly diesel operation.

Another prominent common application for stationary dual-fuel engines has been for many decades the exploitation of sewage biogases, where a gas of around two-thirds of the heating value of methane is produced following the processing of sewage. Such installations not only can produce electric power and heat, but also can, through the fitting of exhaust gas boilers, warm the sludge material and enhance the rate of gas production.

Dual-fuel engines usually can develop their diesel output when operating on gaseous fuels that are high in methane content. For operation on gaseous fuels that are high in hydrogen and low in methane content, such as producer or coal gases, the engine controls need to be suitably adjusted, and its power output may need to be suitably derated. For engines of the two-stroke type, the fuel gas needs to be injected directly into the engine cylinder to avoid discharging some of the fuel gas directly unburned into the exhaust manifold.

Among many power generation applications of dual-fuel engines are those arranged as packaged skid mounted to serve as mini-power production installations, which can be easily transported and may operate on the fuel supplies and electric power available locally. Such arrangements tend to be sufficiently versatile and economic both operationally and capital cost-wise for providing direct power supply where it is needed, including in remote locations, without incurring otherwise high costs and significant transmission losses. Such units can be also employed to provide cogeneration in a variety of forms and applications. They may also be employed as standby generators in case of power cuts or when excess power is needed on occasion at a specific site. They use combinations of equipment that include the engines, electric generators and motors, heat exchangers, controls and fuels, and water supply facilities.

Dual-fuel engines in general are well suited to multicylinder applications having high-output capacity, which makes them attractive both operationally and economically, while maintaining the flexibility to revert to wholly diesel operation. With proper adjustments they can be made to also operate satisfactorily on a range of gaseous fuels with an equally wide range of fuel characteristics. Examples of the fields of applications of stationary dual-fuel engines include power generation with cogeneration, pumping duties for irrigation, driving compressors, drilling rigs, and forklift trucks. Additionally, there are applications for marine transport via ships, tankers, submarines, fishing trawlers, and motorboats. On the other hand, transport applications, while employing relatively smaller engines, are largely for relatively short haul trucks, vans, buses, taxis, and school transport applications, and until recently, seldom for private individual vehicle transport.

An important consideration for converted engines is whether they need to maintain the capacity for delivering the diesel full load output when needed, or they are merely destined exclusively for dedicated dual-fuel application, and diesel operation is merely a short-time measure in case of interruption to dual-fuel operation. Hence, kits for the proper conversion of engines usually are not made of universal and generic design-type. They target specific diesel engine and fuel gas types with a knowledge of the expected operating conditions. Such requirements have tended to impose limitations to the number of conversions made. Low-technology conversion of engines to dual-fuel operation has tended to be unreliable. It results in excessive emissions, poor fuel economy, tolerating less variation in fuel gas composition variations, and tendency to encounter knock, with reduced power and reliability, in comparison to the corresponding diesel version.

In general, the conversion of a diesel engine to dual-fuel operation should maintain the following desirable features:

- Major modifications to the engine need to be avoided while retaining the facility to operate as a diesel engine. Some partial derating of the engine output may need to be tolerated.
- Peak cylinder pressure values are to be kept within the acceptable limits for the engine as a diesel. It is preferable to have the size of the liquid fuel pilot relative to that of the gaseous fuel-made variable, depending on the load, speed, and other prevailing operating conditions.
- Safety of operation with gaseous fuels must remain of paramount importance with the diesel idle, and fail-safe and overspeed governing are to remain available in the dual-fuel mode.

5.5   SOME ECONOMIC CONSIDERATIONS

The cost of conversion of diesel engines to dual-fuel operation depends on many factors that include the type and number of engines to be converted and whether they are new or used, turbocharged or normally aspirated, two-stroke or four-stroke, of multi- or single cylinder, and directly or indirectly liquid fuel injected; the field and type of application, such as whether for transport or power production; and what type of exhaust gas treatment is needed, and whether cogeneration and EGR are to be applied. There are other key factors affecting such costs that include the type and composition of the gaseous fuel used and whether the engine is intended for multiple fuel applications, pilot fuel size minimizing, or to be implemented merely for topping up with some gaseous fuel the power output of diesel operation. The cost advantage with fuel gas operation relative to that with conventional diesel fuel must be maintained so as to make the additional costs associated, for example, with fuel storage, its compression, and engine conversion economically worthwhile.

The capital cost of diesel engines tends to be higher than that of their counterpart spark ignition engines, which in the past often discouraged further investment in fuel-saving hardware, especially when the engine is not meant to be operated near full load for long periods. Usually large engines are favored since they operate at low speeds, permitting the operation of slow-burning fuels without the production of excessive deposits, fouling, or rapid deterioration of lubricants. Close control of the temperatures of injection nozzles, piston crowns, and cylinder liners is easier to implement than in small high-speed engines that have limited cooling capacity. Accordingly, the cost of conversion of diesel engines to operate as dual-fuel engines is widely varying and would depend on the many influencing factors.

Many conversions in the past have been made largely on an ad hoc basis, indicating there is a continuing need for further progress to bring about a more cost-effective and trouble-free conversion of the much improved diesel engines of recent years. Of course, it pays to remember that the conventional internal combustion engine in general and the diesel engine in particular had benefited from many decades of research, development, and experience to reach the present state of reliability, sophistication, and superior performance.

Nowadays many cities around the globe are increasingly embarking on converting their diesel buses so as to operate on the flexible dual-fuel system while employing mostly compressed natural gas. This is mainly on the grounds of the associated low operating cost and the reduced emissions, especially since buses operate in thickly populated areas of large cities. Fuel tankage and its protection against the possibility of an impact leakage or fire, as shown typically in Figure 5.2, have been dealt with quite satisfactorily. Nevertheless, having to carry the fuel gas in the form of compressed gas within heavy metallic cylinders does represent a limitation by taking considerable space, added mass, and increased costs.

In general, there is a great uncertainty in calculating the economy of natural gas-fueled vehicles, with values varying considerably, depending on the application. The investment for conversion can often be recovered typically around 8 years or less, depending on the engine type, mode of conversion, field of application, fuel gas employed, number of engines to be converted, and average load factor. The

FIGURE 5.2 An example of a city bus powered with a dual-fuel engine by having the cylinders of compressed natural gas installed on its roof.

periods between oil changes and engine overhaul tend to be significantly increased. However, to produce an optimum control of engine units tends to be more costly at present given the relatively low volume of dual-fuel engines produced in comparison to those mass-produced conventional liquid-fueled engines. This is in addition to the increasing constraints being imposed by emissions regulations and the variation in the quality of the gaseous fuels that may be available for consumption.

The application of dual-fuel control strategies to transport applications has tended to be more challenging than their application to stationary engines, which often run at one speed and may be even at one load. However, engines for transport applications usually operate over wide speed and load ranges, requiring more complex controls, such as to avoid knock, overfueling, and unacceptably high levels of emissions. In contrast, the conversion of diesel engines to spark ignition, mainly to consume cheap gaseous fuels, tends to still be rather expensive and less flexible. It would require extensive and permanent changes to the engine installation while losing the option to maintain the facility to diesel operation. The diesel engine liquid fuel injection system and associated controls and fuel storage need to be removed and replaced with metering and control systems and storage for the gas fuel. The cylinder head needs to be modified to remove injectors and install spark ignition plugs, while the valves, pistons, and ring systems are suitably modified or replaced. The high compression ratio of the engine is reduced, and the turbocharging system, if it is to be retained suitably, changed or modified. An ignition system and its controls need to be installed, and a throttle valve with a suitable catalytic exhaust cleaning system added. Suitable changes must also be made for the modified engine to comply with the corresponding engine's emissions standards.

Figure 5.3 shows a dual-fuel engine where an independently controlled supply of the fuel gas is made just outside the intake valve, while Figure 5.4 shows an engine where the fuel gas supply is made independently via a prechamber directly into the cylinder.

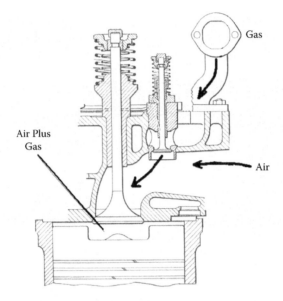

FIGURE 5.3 An engine where the fuel gas supply is made independently just outside the intake valve. (From Karim, G., Combustion in Gas-Fueled Compression Ignition Engines of the Dual-Fuel Type, in *Handbook of Combustion*, ed. M. Lackner, F. Winter, and A. Agarwal, vol. 3, Wiley Publishers, Weinheim, Germany, 2010.)

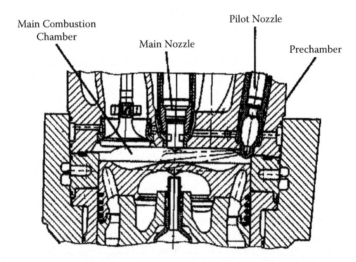

FIGURE 5.4 An engine with independent injection of the pilot directly into a prechamber attached to the engine cylinder. (From Schiffgens, H.J., Brandt, D., Rieck, K., and Heider, G., Low NOx-Gas Engines from MAN B&W, in *CIMAC Congress*, Copenhagen, Denmark, 1998, pp. 1399–1414.)

REFERENCES AND RECOMMENDED READING

Aisho, Y., Yaeo, T., Koseki, T., Saito, T., and Kihara, R., *Combustion and Exhaust Emissions in a Direct Injection Diesel Engine Dual-Fueled with Natural Gas*, SAE Paper 950465, 1995.

Boyer, R.L., *Status of Dual Fuel Engine Development*, SAE Paper 490018, 1949.

Callahan, T., Survey of Gas Engine Performance and Future Trends, presented at Proceedings of ASME Conference, ICE Division, Salzburg, Austria, 2003, Paper ICES2003-628.

Challen, B., and Barnescu, R., *Diesel Engine Handbook*, 2nd ed., SAE, Warrendale, PA, 1999.

Cummins, L., *Internal Fire*, SAE, Warrendale, PA, 1989.

Danyluk, P.R., Development of a High Output Dual Fuel Engine, *ASME Journal of Engineering for Gas Turbines and Power*, 115, 728–733, 1993.

Diesel, R., Method of Igniting and Regulating Combustion for Internal Combustion Engines, U.S. Patent 673,160, April 1901.

Eke, P., and Walker, J.H., Gas as an Engine Fuel, in *Proceedings of the Institute of Gas Engineers*, London, UK, 1970, pp. 121–138.

Ishyama, T., Shioji, M., Mitani, S., Shibata, H., and Ikegami, M., *Improvement of Performance and Exhaust Emissions in a Converted Dual Fuel Natural Gas Engine*, SAE Paper 2000-01-1866, 2000.

Karim, G.A., Combustion in Gas Fueled Compression Ignition Engines of the Dual Fuel Type, *ASME Journal of Engineering for Gas Turbines and Power*, 125, 827–836, 2003.

Karim, G., Combustion in Gas-Fueled Compression Ignition Engines of the Dual Fuel Type, in *Handbook of Combustion*, ed. M. Lackner, F. Winter, and A. Agarwal, vol. 3, Wiley Publishers, Weinheim, Germany, 2010, pp. 213–235.

Karim, G.A., *Fuels, Energy and the Environment*, CRC Press, Boca Raton, FL, 2012.

Kusaka, J., Daisho, Y., Shimonagata, T., Kihara, R., and Saito, T., Combustion and Exhaust Characteristics of a Diesel Engine Dual-Fuelled with Natural Gas, in *Proceedings of the 7th International Conference and Exhibition on Natural Gas Vehicles*, Yokohama, Japan, October 17–19, 2000, pp. 23–31.

Maxwell, T., and Jones, J., *Alternative Fuels: Emissions, Economics and Performance*, SAE, Warrendale, PA, 1995.

Ricardo, H., *The High Speed Internal Combustion Engine*, 4th ed., Blackie and Sons, London, 1953.

Schiffgens, H.J., Brandt, D., Rieck, K., and Heider, G., Low NOx-Gas Engines from MAN B&W, in *CIMAC Congress*, Copenhagen, 1998, pp. 1399–1414.

Steiger, A., Large Bore Sulzer Dual Fuel Engines: Their Development, Construction and Fields of Application, *Sulzer Technical Review*, 3, 1–8, 1970.

Turner, S.H., and Weaver, C.S., *Dual-Fuel Natural/Diesel Engines: Technology, Performance and Emissions*, No. GRI-94/0094, Topical Report Gas Research Institute, Chicago, November 1994.

Wong, W.Y., Midkiff, K.C., and Bell, S.R., *Performance and Emissions of a Natural Gas Fuelled Indirect Injected Diesel Engine*, SAE Paper 911766, 1991.

6 Methane and Natural Gas as Engine Fuels

6.1 NATURAL GAS AS A FUEL

Natural gas is usually the volatile portion of crude petroleum. It normally occupies under high pressure the porous rocks of oil reservoirs above the liquid fuel zone. The gas is also found in dry form or nonassociated with oil gas fields. Initially, when the prime objective was the production of oil, the gas was traditionally considered a nuisance and was often wasted and flared off. Unfortunately, some significant amounts of gas are still being flared when the gas cannot be effectively utilized locally, pumped back into wells to enhance oil recovery, or transported to potential markets via pipelines over long distances.

Natural gas has been known since ancient times, mainly through its fires following its ignition when it escaped through fractures and fissures in the earth, such as in Persia, Iraq, and China, producing the so-called eternal fires that were revered by some ancient civilizations. Its industrial exploitation began mainly in the 19th century. It was used initially for street lighting and domestic heating. In around the 1920s and later, high-pressure gas pipelines were laid over long distances, making the gas increasingly available to millions of domestic and industrial users. After the Second World War, long pipeline networks with large diameters, well in excess of 1 m, and internal pressures in excess of 70 bars were laid in many parts of the world, such as Russia, Canada, and the United States.

Rapid progress has been made worldwide in recent years in the discovery of new natural gas deposits and its transportation over the globe, both as a gas and in its cryogenic liquid state, liquefied natural gas (LNG). Its increased availability, the need to meet increasingly lower emission controls, and its relatively low cost have tended to increase its usage as a fuel in a wide variety of applications. The gas has been increasingly viewed as a premium fuel that is in much demand, and may well be for quite some time in the future a prime source of usable fuel energy.

The transmission of natural gas via pipelines represents a very efficient and economic highly developed distribution system that can be controlled remotely over the whole integrated network. More recently, through the expenditure of much capital and advances in technology, the gas resources have been increasingly exploited via liquefaction, so as to permit the transport of gas via suitable ships as LNG.

Natural gas is also a prime feedstock in the petrochemical industry, such as for the production of a wide range of economically important materials. Above all, it is increasingly being used to produce the hydrogen needed for the upgrading of common fossil fuels, especially those that are of low quality, to make them more

TABLE 6.1
Total World Energy Supply (Mtoe)

Fuel Type	1973	2011
Natural gas	16.0%	21.3%
Oil	46.0%	31.5%
Coal/peat	24.6%	28.8%
Biofuels and waste	10.6%	10.0%
Hydro	1.8%	2.3%
Nuclear	0.9%	5.1%
Other	0.1%	1.0%
Total	6109	13,113

Source: Adapted from U.S. Energy Information Administration, U.S. Department of Energy, Washington, DC, 2014, http://www.eia.doe.gov/emeu/international/reserves.html.

acceptable operationally and ecologically. Table 6.1 shows the rapid increase in energy consumption over four recent decades and displays the corresponding big increase in the relative consumption of natural gas in comparison to other fuels.

The composition of the natural gas delivered through pipelines to consumers could vary somewhat with time and location. Such variations in gas composition can affect gas-fueled engine performance and its emissions, such as through changes in fuel metering characteristics and the resistance to autoignition and knock. This is especially the case since high-output, low-emissions engines are operated increasingly close to the knock-limited power for economic reasons, and to take advantage of the high resistance to knock of methane, the main component of natural gas.

Much attention has recently been given to natural gases that are sometimes found in low-permeability tight, deep underground rock formations. The exploitation of these gas resources is associated with high capital, operational, and environmental costs since it usually requires extensive hydraulic fracturing of the reservoir structure to improve the mobility of the gas.

The composition of raw natural gas varies significantly, depending on its source and whether it has been processed for pipeline distribution and consumption or not. Typically, natural gas as delivered to consumers is suitably processed and composed of about 90% methane, and the remainder is of various concentrations of ethane, propane, butane, and non-fuel-diluent gases, such as nitrogen and carbon dioxide. Methane, a much lighter than air gas, is odorless, nontoxic, and colorless. Odorants, such as mercaptans (e.g., ethyl mercaptans) are added to pipeline natural gas, usually at its distribution points, to ensure easy detection down to very low concentrations in case of a leak. The processing of natural gas removes almost all the hydrogen sulfide that may be present, and reduces the water contents to an acceptable level; also, it reduces much of the inert gases and the economically valuable higher hydrocarbon vapors present in much smaller concentrations.

The processed gas is normally sold mainly on the basis of its heating value, often with a lesser consideration of its specific chemical composition. For thermal applications such as in furnaces, the main properties that are considered of much significance are its specific gravity and heating value through its Wobbe number, which is an indication of the rate of energy release by combustion of a fuel when discharged through an orifice such as that of a burner under the action of a specified constant pressure head. However, the detailed nature of the constituents of a natural gas has an important significance for its use as a chemical feedstock and as an engine fuel, including for dual-fuel applications. For example, the detailed nature of the gas composition can affect very markedly the resistance of the gas engine to knocking and the nature of its emissions. Engine operating conditions, design, and controls need to be increasingly optimized so as to cope with the consequences of any variation in the composition of the fuel gas.

The potential for methane participating in the formation of photochemical smog is small. Its presence in the open atmosphere, on the other hand, represents a source of greenhouse gases with a serious contribution to global warming that is requiring increasingly effective control. Much of the emissions of methane arise from escapes and releases of the gas described as fugitive gas emissions, whether naturally, such as through biodegrading processes, or industrial releases, especially those associated with the functioning of the oil and gas industry.

The sulfur content of processed natural gas is reduced to extremely low values, leading to virtually no sulfur remaining in the fuel. Any small presence of sulfur in processed natural gas is usually that originating from the very small concentrations of the mercaptans added for safety to the gas to serve as odorants. Also, much attention is given to controlling water in the feed fuel gas to reduce the potential for corrosion and the formation of solid hydrates, such as in pipes and storage vessels. In peak shaving applications employed to adjust for some of the variation in the heating value and other properties of the gas, its composition may be varied seasonally by including very small amounts of a higher-density fuel, such as propane, or of air. Any remaining presence of some higher hydrocarbon in the gas, such as propane, needs appropriate accounting and control since it can affect engine performance significantly, especially in spark ignition gas engines and, to a lesser extent, in dual-fuel engines. Table 6.2 shows a listing of some major properties of the main hydrocarbon fuel gases.

The expansion of the compressed high-pressure gas, whether during fueling or prior to its mixing with the air for combustion, will result in substantial cooling of the expanded gas. Not only will the associated temperature drop lead to cooling of parts of the combustion device, but also the resulting change in density, unless accounted for, can produce variations in the quality of the mixture reaching the engine cylinder. Suitable remedial measures are sometimes incorporated, especially in large-capacity engines and in transport applications. For example, some heating is provided to gas pressure regulators or to locate the first stage of regulation near the fuel tank, while the second is located closer to the intake manifold so that heat can be picked up by the fuel flow along the piping system.

There have been various schemes to suitably process natural gas into other industrially important fuel mixtures, such as synthesis gas (H_2 and CO) that is employed to produce liquid fuels such as alcohols, gasoline, and diesel fuel for transport applications. However, recent attempts to use them as a fuel tend in general to be less economically attractive.

Methane, by virtue of its simple chemical structure, is not rapidly reactive in comparison to other common fuels. This feature helps to render methane a very attractive engine

TABLE 6.2

Properties of the Main Hydrocarbon Fuel Gases

	Methane	Ethane	Propane	Propene	Butane	Iso-Butane	Butene
Energy content (LHV) (MJ/kg)	50.01	47.48	46.35	45.78	45.74	45.59	45.32
Liquid density (kg/L)	0.466	0.572	0.501	0.519	0.601	0.549	0.607
Liquid energy density (MJ/L)	23.30	27.16	23.22	23.76	27.49	25.03	27.51
Gas energy density (MJ/m³)	32.6	58.4	84.4	79.4	111.4	110.4	113.0
Gas specific gravity (@ 25C)	0.55	1.05	1.55	1.47	2.07	2.06	1.93
Boiling point (°C)	−164	−89	−42	−47	−0.5	−12	−6.3 to 3.7

Source: Adapted from Rose, J.W., and Cooper, J.R., eds., *Technical Data on Fuels*, 7th ed., British National Committee of World Energy Conference, London, UK, 1977.

fuel with exceptionally high knock-resistant properties. It also renders the fuel, when present in the exhaust gases of engines, rather hard to oxidize through the use of common catalysts. To complete its oxidation would require special metallic catalysts and higher exhaust gas temperatures to activate the catalyst than with other common hydrocarbon fuels.

A lowering of the gas temperature can result in condensation of the excess water, which may contribute to increased tendency for corrosion of retaining vessels, equipment, and pipes. Increasing the gas pressure will lower the relative retention of the water vapor by the gas. The use of activated carbon has been tried for storing natural gas and reducing the negative effects of the portability problems of the fuel. However, this approach has so far been shown to not be practically suitable, with a tendency for any higher hydrocarbons in the gas to be adsorbed preferentially by the solid carbon, which reduces the desorption capacity of the methane gas. Thus, with each refueling these strongly adsorbed compounds build up on the surface, gradually reducing the capacity of the activated carbon to adsorb fuel. This tendency has sometimes been described as the poisoning of the solid with repeated usage.

The methane number, which is an empirical criterion, has been employed by some to denote the relative tendency of the gaseous fuel to encounter knock in engines. Such an approach has its limitations, especially when applied indiscriminately to all possible gaseous fuels, including those of nonnatural gas types or those containing many diluents.

6.2 COMPRESSED NATURAL GAS (CNG)

The fuel gas system on a vehicle faces a number of demanding requirements. The fuel must be safely stored at high pressures and then safely and accurately delivered to the engine. Variations in the fuel pressure and temperature can present operational and control challenges. These problems would be compounded by the fact that there may be some variations in the fuel composition from one filling to another unless rigorously regulated.

The inability to carry a sufficient amount of natural gas in transport applications represents a major limitation to employing the widely available fuel. Much progress has been made in recent years in the construction of high-pressure compact cylindrical containers of light weight for carrying compressed natural gas, making it the preferred mode of carrying the gas on board vehicles. Many of the CNG storage tanks used are constructed of aluminum or steel liners with fiberglass or carbon fiber overwraps to increase the strength while minimizing the weight. Gas pressures of 200 to 240 bars are increasingly common, which typically correspond on an energy basis to an equivalent amount of gasoline of merely around 30% by volume. The weight, bulk, and cost of the fuel storage tanks remain a serious limitation to the wide application of CNG, particularly for transport. They can typically constitute a significant fraction of the cost of the conversion to fuel gas operation.

Composite cylinders can have masses of only a fraction of the mass of the corresponding equivalent all-steel cylinder. Also, CNG fuel tanks cannot be made of irregular shapes, as with gasoline applications, to fit into available spaces in vehicles (Figures 6.1 and 6.2). This would contribute to their relative bulkiness and

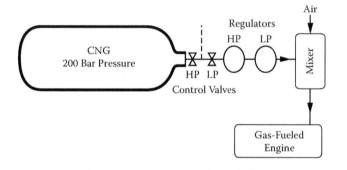

FIGURE 6.1   A typical arrangement of the supply of CNG to an engine.

FIGURE 6.2   Examples of CNG containers used in transport applications. (From Gas Research Institute, *Technology Today*, Chicago, IL, 1991.)

wastage of vehicle-carrying volume capacity. Moreover, as long as CNG is used, then the associated cost should include the high cost of compression work, which is usually wasted by simply expanding the high-pressure gas to around that of the atmosphere as it enters the intake section of the engine. For buses, the CNG fuel tanks are often located on the roof while connected in series. The fuel cylinders are mounted with brackets designed to withstand better crash loading, and they are not removed during refueling.

Local distribution and capital costs of high-pressure CNG vary greatly with the type of station served. Fuel dispensers are of either the fast or slow fills. Fast-fill installations may have sufficiently fast filling time, comparable to that of filling with liquid fuels, but they require storage vessels that are of very substantial capacity. Compression heating of cylinders during a quick fill is usually small and quickly dissipates (Figure 6.3).

There can be some operational and materials compatibility problems with the use of CNG, arising mainly from the possible presence of lubricating oil droplets and vapors out of the compression system. Also, the presence of any water vapor in the gas can result in freezing the water, especially following the rapid expansion of the high-pressure gas for use. This will lead to corrosion problems, particularly with the presence in the gas of very small concentrations of hydrogen sulfide or mercaptans, and it can lead to the formation of hydrates. Effective natural gas dryers and filters are employed normally to ensure sufficient removal of any water vapor in the gas prior to and during its compression and storage. Also, the odorant mercaptans are normally added to CNG for safety and leak detection.

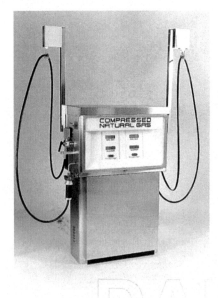

FIGURE 6.3 A CNG dispensing unit for transport applications. (From Pollock, E., ed., *NGV Resource Guide*, RP Publishing, Denver, CO, 1985.)

In typical transport vehicle applications the tank is equipped with relief valves designed to vent the gas in the event of a fire near the tanks. Rupture disc type pressure relief valves designed to vent at excessive buildup of pressure are also fitted. A shutoff valve is installed in each tank and is located on the outside, which can isolate each tank from the rest of the fuel system. Stainless steel tubing is usually used throughout the high-pressure side of the fuel system. The entire fuel storage is isolated from the engine fuel delivery system by a shutoff valve. The high-pressure gas is reduced from the service high pressure down to a pressure of around several atmospheres through regulators that can be water warmed when needed and located near the fuel tanks to eliminate freezing of the regulators as the gas expands down to its working pressure. The lower-pressure gas is fed to secondary regulators located near the engine, where it is reduced further. On this basis, the largest retrofit item in engine conversion to CNG operation tends to be the fuel tanks.

Most compressed natural gas tanks in use are restricted to a cylindrical shape with two hemispherical ends since they must be symmetrical in order to contain the high-pressure gas effectively and safely. CNG fuel tanks are typically located in the beds of trucks, in the trunks of cars, and on the tops of buses. The number of tanks carried depends on the range to be traveled and how much extra weight can be tolerated by the vehicle. Compressor costs depend on factors that include compressor electrical demand, electrical unit cost, and compressor pressure and its capacity.

CNG for transport applications has had relatively limited usage in the past, especially in North America. Some of the main factors contributing to that are the following:

- Until relatively recently, price differentials between liquid and gaseous fuels in the form of CNG tended to make its usage in the transport sector insufficiently attractive economically. The increased availability and rapid relative reductions in recent years in the price of natural gas have helped to increase CNG usage.
- The average distances of common travel in North America are relatively large, and the size of engines that are typically used, in comparison to other world locations, are also large. This made it less attractive to operate with the insufficient capacity of the CNG fuel tanks, especially with the lack of support of the needed fuel supplies' infrastructure.
- Efficient electrical power grid distribution has been widespread, which often made the production of power using gas-fueled combustion engines in different locations unnecessary.
- The combined costs of the engine, vehicle, and high-pressure gas compression, storage, and carrying often tended to be too high.

On this basis, these factors resulted in the need for a much higher fuel price differential relative to the corresponding liquid fuels. However, in recent years, this has been increasingly satisfied, helped by the relatively low price of the increasingly abundant natural gas supplies, while the cost of refined liquid fuels of the right quality needed to meet current and future environmental requirements continues to escalate.

The development of economic and safe small home compression facilities for refilling gas cylinders under sufficiently high pressure, although attractive in principle, has not been proven sufficiently well so far as a safe, reliable, and economic alternative to large-sized compression in suitable supply stations.

6.3   LIQUEFIED NATURAL GAS (LNG)

Natural gas is liquefied in very large installations close to the points of gas source, which permits its long-distance transportation, including over oceans and between continents. The liquefied natural gas (LNG) is then transported mainly via large specialized tankers to points of usage and distribution, often after its vaporization into the gas delivery pipes network. At present, it is still somewhat less likely, for both economic and technical reasons, that the widespread liquefaction of the gas can be made conveniently in relatively small units at points of gas consumption. Hence, its relative limited availability remains a problem associated with its wider cost-effective employment as a fuel in combustion devices in general and engine transportation applications in particular. It is in locations where LNG is stored or distributed, such as at marine terminals, and where it is unloaded that some of it may be diverted for use as a fuel nearby.

Often for simplicity, liquid methane is taken to represent LNG. However, the heating value and composition of LNG can vary slightly from one batch to another, depending on the concentrations of some of its minor nonmethane components that can remain in solution. Table 6.3 shows the main properties of liquefied methane representing LNG.

The specially designed and built marine tankers that are suitable for safely transporting the cryogenic liquid long distances may use dual-fuel engines for their propulsion. These engines invariably consume some of the boil-off fraction of the LNG and may be supplemented when needed by conventional liquid fuels.

LNG production is capital-intensive, yet its production from natural gas, which contains in the liquid phase some higher hydrocarbon components, notably ethane and propane, is still economically attractive in many situations. Also, the lead time between the discovery of natural gas at a certain location and its export in the form of LNG is usually quite long. In spite of these limitations, the demand for LNG, whether for domestic heating applications or industrial use, including its use for

---

TABLE 6.3

Properties of LNG When Considered as Pure Liquid Methane

Specific gravity	0.415
Density	426.1 kg/m^3
Boiling point	111.8 K
Critical temperature	191.2 K
Critical pressure	46.3 bar
Heating value	55.83 MJ/kg

Source: Zabetakis, M., *Flammability Characteristics of Combustible Gases and Vapors*, Bureau of Mines Bulletin 627, U.S. Department of the Interior, Washington, DC, 1965.

---

power generation in power stations or transport engines, continues to increase. The high capital cost to liquefy the gas and the associated cost of the energy requirements contribute to the relatively high costs of production, operation, and transport of LNG. Moreover, LNG projects usually require very large gas reserves to make them economically viable and return the high initial capital and operating costs.

Tanks storing LNG are normally double-walled with an inner vessel made up of stainless steel and another one made up of carbon steel. The LNG is usually stored at pressures only slightly higher than atmospheric. A typical tank costs substantially more than those for CNG, and of course many times more than for gasoline or diesel fuel. Heat transfer from the environment into the well-insulated LNG tank will boil off increasingly more fuel, raising the pressure within the tank. Modern tank designs will not release the remains of the gas through insulated relief valves until there is a significant buildup of pressure to a prescribed value over a protracted period of time, typically such as 2 weeks elapsing. Figure 6.4 shows a schematic representation of a typical container of LNG for transport applications.

On arrival in marine terminals, LNG is stored in specialized tanks and then vaporized when needed for transport by pipelines as a gas to augment conventional pipeline natural gas supplies. The conversion of natural gas to the liquid state somewhat changes its composition and properties, including its heating value. All the sour components, such as H_2S and CO_2, and any water vapor are removed. Some hydrocarbon molecules that are larger than methane can still remain in small concentrations.

The transportation, fueling, and storage of LNG represent a substantial safety hazard due to its extremely low temperature (111 K at atmospheric pressure), with the potential for fire and explosion following a leakage or spillage and its subsequent vapor spreading. At present, there are only a relatively limited number of locations where LNG receiving terminals are sanctioned.

The use of LNG in transport engine applications is attractive because much more energy is stored per unit volume than as high-pressure, compressed natural gas (CNG). The advantages of using LNG compared to CNG in transport applications include lower fuel system weight, shorter refueling times, lower refueling facility cost, and a more uniform and relatively higher quality fuel. The liquid nature of LNG permits its pressurization to high levels with cryogenic pumps. However, normally

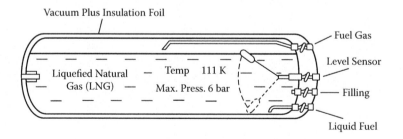

FIGURE 6.4 A typical container of LNG for transport applications. (From Bergman, H., and Busenthur, B., *Facts Concerning the Utilization of Gaseous Fuels in Heavy Duty Vehicles*, in *Proceedings of the Conference on Gaseous Fuels for Transportation*, August 1986, pp. 813–849.)

LNG engines use the fuel in its gaseous state by vaporizing it through heat transfer before entry into the cylinder. The engine coolant is usually used to supply the heat needed to vaporize and warm the fuel sufficiently. The fuel pressure is further regulated before feeding the gas to the engine, where it is metered and mixed with the air.

In applications of natural gas the temperatures of the air and gas have a marked effect on engine performance and the resulting emissions. The introduction of LNG into the intake air of an engine, whether as a liquid or when vaporized, will result, upon mixing with the incoming air, in a marked decrease in the temperature of the resulting mixture. For example, as shown in Figure 6.5, a stoichiometric mixture of LNG cold vapor and ambient temperature air can produce a mixture at a temperature as low as –26°C. The extent of increase in the observed length of the ignition delay tends to be very sensitive to the effective value of the temperature of the intake mixture. The low-temperature mixtures, when ignited with small pilot quantities, tend to produce increased emissions and cyclic variations.

The operation of a dual-fuel engine on LNG requires an adequate capacity cryogenic fuel tank equipped with sufficient insulation. The diesel fuel tank capacity in turn may be reduced to partially offset the increase in bulk associated with the LNG tank. Adequate safety measures are also provided with protection against the likelihood of fuel spillage. There may also be a case for employing a better quality diesel fuel when used as the pilot, such as having lower viscosity and vapor pressure with a higher cetane number. Often with dual-fuel engines operating on LNG, the engine is started and shut down while operating as a diesel. The cold of the LNG may be used to provide, if needed, cooling to the diesel fuel injector or to maintain refrigerated

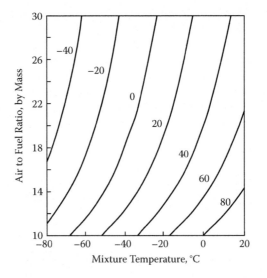

FIGURE 6.5 The resulting mixture temperature with the feed air-to-fuel ratio following the evaporation of LNG into air at different temperatures. (Adapted from Ali, A.I., and Karim, G.A., Combustion, Knock, and Emission Characteristics of a Natural Gas Fuelled Spark Ignition Engine with Particular Reference to Low Intake Temperature Conditions, in *Proceedings of the Institute of Mechanical Engineers*, London, vol. 189, 1975, pp. 139–147.)

cargo in the case of transport trucks. Some work extraction may also be possible, such as to reduce the compression work for the fuel injected. These measures represent added complexities and costs that need to be justified. With the many possible associated variables affecting the performance of LNG-fueled engines, computer controls are being relied on and used.

The provision of a suitable heat source is required to evaporate the LNG. This is done through a heat exchanger such as that using hot circulating cylinder jacket water. Arrangements are usually made to bypass this heat exchanger whenever needed. The gas injection system is also made such that it is capable of coping with the very cold aerosol of methane, especially at the high flow rates of the fuel near engine full load conditions. The hot exhaust gas tends to be a less convenient source for such fuel heating, which has to be accomplished without much reduction in the water jacket temperature. An excessive reduction in this temperature will adversely affect engine performance, such as through increased ignition delay, noise, and exhaust emissions of notably unburnt hydrocarbons and carbon monoxide, with the potential for increases in smoke and particulates. The water jacket temperature may be suitably controlled according to the load. The injection of the gas from the boiling liquid fuel into the intake manifold or directly into the cylinder would tend to enhance the volumetric efficiency of the engine through the resulting increase in the density of the charge, leading to an improvement in power output.

Some engine modifications are needed with dual-fuel LNG operation. These can include the potential to employ higher compression ratios since there is less likelihood for the onset of knock then, and a further reduction in NOx emissions. The injection timing of the pilot may benefit from a certain degree of advance to compensate somewhat for the increased cooling of the intake charge, which leads to slightly lower temperatures at the end of compression. In transport applications, the size of the radiator may be reduced, while greater attention is given to increased lubricating oil potential problems, increased corrosion in the exhaust system, and some possible material compatibility problems. Some icing problems may need to be attended to also in LNG engine operation, particularly in locations where high humidity air is normally encountered.

6.4 FUEL MIXTURES CONTAINING METHANE WITH DILUENT GASES

It is common to consider carbon dioxide, nitrogen, or water vapor, when present with a fuel gas such as biogas with methane, as diluents, although they may not necessarily remain entirely neutral chemically, especially at high temperatures. The processed pipeline natural gas is usually made up primarily of methane and may include in small concentrations the gases nitrogen and carbon dioxide with some water vapor. However, the composition of raw natural gases varies widely, depending on the source, nature, and level of treatment they may have undergone. The presence of such diluent components with methane can bring about significant changes to the combustion characteristics of the fuel mixture. It may undermine its effective utilization as an engine fuel altogether, depending on the composition and type of engine and associated operating conditions.

There are numerous sources of combustible gas mixtures besides natural gas that are made up of mainly methane, with varying concentrations of carbon dioxide and nitrogen, such as in the case of biogases. Such gas mixtures have lower overall heating values in comparison to that of pure methane, which has commonly come to be considered a typical representation of processed pipeline natural gas. The reactivity and key characteristics of such gaseous fuel mixtures will depend on their composition and not necessarily, as sometimes assumed, on their heating value. For example, the presence of a small amount of hydrogen in a gaseous fuel mixture may not necessarily increase the heating value significantly as the corresponding presence of a similar concentration of propane. However, the combustion characteristics and performance as engine fuels can be significantly different for the two sets of gas mixtures under similar conditions. Hence, the composition of the different biogases needs to be considered carefully and may require special processing to render them suitable for utilization in engines and avoid operational problems affecting engine performance and durability. For example, a fuel mixture of methane and nitrogen would be more likely to permit more flexible variations in combustion controlling parameters than a similar mixture of methane and carbon dioxide having the same heating value. The combustion of fuel mixtures containing yet higher concentrations of diluents can be made with the use of oxygen-enriched air.

The presence of diluents with a fuel such as methane proportionally reduces the effective heating value of the resulting fuel mixture. The energy released through combustion of the fuel component will be shared with the diluents present. The relative fraction of this energy release taken by these diluents will increase as the temperature is increased since the thermodynamic properties, especially the enthalpy and internal energy, as shown in Figure 6.6, increase rapidly with temperature. The rates of increase for carbon dioxide are greater than those for nitrogen. Carbon dioxide, water vapor, and, to a much lesser extent, nitrogen will also tend to increasingly undergo endothermic dissociation reactions that increase rapidly in intensity as high

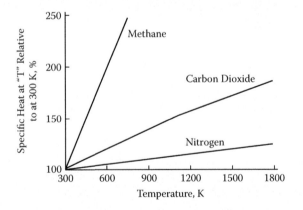

FIGURE 6.6 Variation of the specific heat with temperature relative to the corresponding values at 300 K for methane, carbon dioxide, and nitrogen.

combustion temperatures are approached. The net effect will be substantial reductions in the flame temperature and associated reaction rates through the presence of the diluents. In general, the dilution of methane with CO_2, N_2, or H_2O below the 15% range may be considered approximately, as the diluents are approximately interchangeable with an appropriate quantity of excess diluting air.

The heating value on a volume basis is an important characteristic of a fuel mixture since engine size, piping, and associated equipment needed depend on the volume of the fuel for a certain thermal loading. The levels of combustion and exhaust temperatures are also important because of their effects on the amount of heat that can be recovered in the case of turbocharging and cogeneration setups. With the lower wall combustion chamber temperatures associated with the combustion of methane–diluent mixtures, complete combustion requires longer retention times, which is favored more by slower-speed engines.

6.5   SOME OTHER NATURAL GAS MIXTURES

Shale gas: The gas trapped in low-permeability shale rock is described as shale gas, made up primarily of methane. The gas, which is not easy to extract from its reservoirs, has been receiving greater attention in recent years as a source of gas energy that is economic to extract in certain locations. When beds of sands or carbonates have porosities and permeabilities that are sufficiently low to inhibit the flow of the gas at commercial rates, they are known as tight formations. In order to produce shale gas, two technologies usually have to be coupled together: horizontal drilling and hydraulic fracturing. The fractures are induced along horizontal wells and in the shale rock by injecting pressurized water with added special sand and additives along the well bore at different intervals to increase the viscosity. The sand added to the water helps to prevent the induced fractures from closing under the great pressures of the rock. However, there are a number of environmental, social, health, and economic issues with shale gas production. Its exploitation requires substantial amounts of water, which results in the production of much wastewater. Other issues would include potential groundwater contamination, gas leakages, and migration into drinking water resources.

Coal bed methane (CBM): The emission of methane from coal beds, sometimes known as fire damp, is a serious source of greenhouse emissions, fire, and explosion hazards that have been associated with coal deposits. The rates of CBM production are the outcome of several major factors that vary from one formation to another and from source to source. Most of the methane is stored within the molecular structure of the coal and held by the overlaying rock or presence of water in the structure of the coal seam. Generally, the higher the pressure of the formation, the higher the gas content of the coal. The gas associated often with the presence of some carbon dioxide is released through desorption by lowering the pressure. Fractures in the coal provide pathways for gas to travel to the wellbore and ultimately guided it to the surface. In general, the older the coal, the higher its gas content. Coal bed methane is generally considered sweet gas since often it contains hardly any hydrogen sulfide.

The exploitation of CBM has improved considerably in recent years because of advances in drilling, including much better control of directional and horizontal drilling.

Fracture stimulation of the bed takes place through forcing a fluid under high pressure into a fracture in the coal seam to cause it to widen and create new passages for the migration and release of the gas, and thus improve gas productivity.

Methane hydrates: Some gases, such as methane, can have different combined forms of molecular structure with water at low temperatures and high pressures, which are described as hydrates. These products dissociate endothermically with an increase in temperature or a reduction in pressure. The molar methane content of hydrate is typically less than 16%. It is known that there are huge reserves of methane in the form of hydrates located at some sea floors and under permafrost of the arctic regions. However, it is unlikely that such a resource can be exploited effectively in the near future. Many industrial operations, such as in pipelines, try to prevent unwanted hydrate formation through the use of suitable inhibiting agents since it can impede fluid flow and undermine safety.

6.6   SOME COMMON NONNATURAL GAS–FUEL MIXTURES

Almost any gaseous fuel mixture may be employed to a varying degree of success in dual-fuel engines. This is largely by virtue of providing a substantial source of the required energy for ignition through the injection and subsequent ignition of an adequate amount of liquid diesel fuel to serve as a pilot. There is an increasing need to maximize the replacement of the diesel fuel by the usually cheaper available gaseous fuels while maintaining acceptable levels of emissions and engine performance.

Most of the gaseous fuels commonly available for consideration as an exploitable energy resource are fuel mixtures that are gaseous under normal ambient conditions. These include natural and liquid petroleum gases and a wide range of other fuel mixtures that are mostly products of industrial and natural processes, and that can vary widely in composition and properties. One broad classification of such fuel mixtures has been based on their heating value relative to that of natural gas/methane of approximately 1000 Btu/ft^3, i.e., 37.0 MJ/m^3. These are the following:

- High heating value gases: Having heating values of $\geq$500 Btu/ft^3 (18.6 MJ/m^3).
- Medium heating value gases: Having heating values of around 300–500 Btu/ft^3 (11.2 to 18.6 MJ/m^3).
- Low heating value gases: Having heating values of 100–300 Btu/ft^3 (3.7 to 11.2 MJ/m^3).

The exploitation of low heating value fuel mixtures in engines tends to be less attractive in applications where the compression of the fuel gas is required due to the excessive costs involved. Fewer combustion problems are usually encountered in the exploitation of high heating value gases, and to a lesser extent, even medium heating value gases, than of those of the low heating value variety. These fuel gases can represent an enormous potential energy resource, much of which may be still wasted due to technical, economic, and environmental reasons. The severity of the associated problems with their proper exploitation can vary widely and depends on the fuels' origin and their composition. These problems arise mainly from the difficulties associated with their combustion and the environmental and toxic effects associated with their exploitation and emissions.

Some of the main factors that need to be dealt with effectively when planning to exploit an insufficiently high heating value fuel resource and to employ it as an engine fuel are the following:

- What is its average composition and how does it vary with supplies and time? Are the supplies of the fuel adequate, uninterrupted, and continuing? What are the supply pressure and temperature values? Can the fuel be rendered sufficiently clean and noncorrosive?
- What are the combustion and other properties? Is the gas toxic? Does it need treatment before burning? Are its products environmentally unacceptable? Will the products of combustion need treatment or cause operational problems?
- Is a standby fuel required for either mixing with the fuel to enhance its properties or replacing it altogether, if needed, to ensure continued operation in case of supply interruption? Often natural gas supplies are used to supplement the fuel to improve its combustion properties.

There are many nonnatural gas–fuel mixtures that originate from a variety of sources and processes. These are increasing in importance and diversity because of their potential as a cheap energy resource, and to reduce the negative contribution of their release to the environment and minimize the associated potential fire and toxic hazards. As an example, it can be seen in Table 6.4 that most of the gases that

TABLE 6.4
Composition of Some Manufactured Gases from Coal

Composition (% by vol.)	Coal Gas	Coke Oven Gas	Water Gas	Producer Gas	Blast Furnace Gas
CO_2	4.0	2.0	5.0	4.0	11.0
O_2	0.4	0.4	—	—	—
C_nH_m	2.0	2.6	8.0	—	—
CO	18.0	7.4	33.5	29.0	27.0
H_2	49.4	54.0	39.5	12.0	2.0
CH_4	20.0	28.0	9.0	2.6	—
N_2	6.2	5.6	5.0	52.4	60.0

Source: Adapted from Foxwell, G.E., *The Efficient Use of Fuel*, British Ministry of Technology, HMSO, London, UK, 1958.

originate from the processing of coal contain methane with large concentrations of carbon dioxide, nitrogen, and water vapor. They reduce the available useful energy and maximum combustion temperature while undermining the combustion properties of the fuel mixture.

The following are examples of some of the most common gaseous fuel mixtures that are used in a variety of combustion devices, including engines of the dual-fuel type:

- Biogases, landfill gases, and sewage gases
- Coal mine gases and coal bed methane
- Blast furnace and coke oven gases
- Producer gases, pyrolysis gases, and wood gases

The presence of diluent gases such as carbon dioxide and nitrogen with methane primarily affects the thermodynamic and transport properties, and consequently the amount and mode of heat flow from the combustion zone. The peak temperature attained on combustion is reduced significantly with the increased presence of the diluent, as shown in Figure 6.7 for methane for a wide range of equivalence ratio values with different concentrations of carbon dioxide, nitrogen, and water vapor. Reaction kinetics can be very much affected by the lowering of combustion temperatures. Similarly, the burning velocity of the mixture is lowered significantly as the concentration of the diluent gas with the methane is increased. Again, this results mainly from the associated reduction in the flame temperature, reaction rates, and changes in the diffusivity and other transport properties. The flammable range, when established on the basis of whether an ignition source of adequate energy, such as an electric spark or pilot flame, is employed and can initiate a propagating flame, also narrows significantly.

Much attention has been given in recent years to the exploitation of low and medium heating value fuel mixtures, especially those derived from biological sources. This is supported by the following potential benefits:

- The gas to be consumed is often available as a by-product or waste essentially free of charge for substitution of increasingly expensive conventional fuels. It represents an environmentally friendly way to dispose of "problem" gas releases and a potential source for efficient power and heat generation in suitably modified wide ranges of combustion devices, including dual-fuel engines.
- By avoiding the uncontrolled releases of gases containing significant amounts of methane and carbon dioxide into the atmosphere, the associated greenhouse gas emissions are reduced. This is very important nowadays since the release of methane into the atmosphere is 21 times more environmentally harmful as a greenhouse gas than a similar amount of CO_2.
- Coke oven gas contains some hydrogen and carbon monoxide, while producer gas contains much nitrogen. The heating values of these fuel gases will vary with the specific composition of the gas and can be much lower than that of a typical natural gas. In engine applications, this will reflect negatively on the power that can be produced per engine cylinder volume in comparison to natural gas applications.

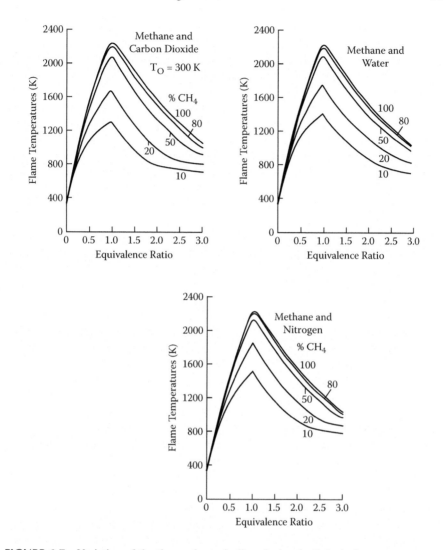

FIGURE 6.7 Variation of the thermodynamically calculated adiabatic flame temperature with changes in equivalence ratio for a range of variation of the molar concentration of the diluents carbon dioxide, water vapor, and nitrogen in methane at atmospheric pressure and initial ambient temperature. (From Karim, G.A., Hanafi, A.S., and Zhou, G., A Kinetic Investigation of the Oxidation of Low Heating Value Fuel Mixtures of Methane and Diluents, *Proceedings of the ASME/ETCE Emerging Energy Technology Conference*, Houston, TX, 1992, pp. 103–109.)

6.7 BIOGAS FUELS

It is both economically and environmentally beneficial to produce a combustible fuel gas through transforming organic materials often considered as waste into an economically useful commodity, while releasing energy relatively easily in conventional engine systems requiring a minimum of change. Clearly, this would also collectively represent a significant global environmental benefit. Additionally, the production of such fuel gases is often associated with the production of high-quality fertilizer material as a by-product, while providing some savings of natural resources, such as through reduction in the burning of wood.

Biogas has come to be a term used to describe gaseous products formed following the pyrolysis, gasification, or anaerobic digestion of organic matter originating from living sources of animal and plant origin described as biomass. The gasification of such material produces a mixture of gases that contain to various degrees many or all of the following: methane, hydrogen, carbon monoxide, carbon dioxide, nitrogen, water, and some simple hydrocarbons, mainly lower olefins. The occasional presence of some hydrogen tends to help compensate for the associated low burning rate in the fuel due to the presence of diluents.

Biogases are one of the most common examples of low to medium heating value fuel mixtures. In principle, all organic material can be used for biogas production. The gas is considered a renewable energy source derived from organic waste on a continual basis. It is widely used in developing countries for a wide range of applications, especially for cooking, and also as an engine fuel due to its availability at low cost. Unlike natural gas derived from petroleum sources, biogas production takes place over a relatively short time of days to weeks. The production of combustible gases from biomass is an old tried technology. However, as a result of the variable and dispersed nature of the biomass, it is often economically unattractive to accumulate substantial amounts in a single location. Accordingly, large-scale biogas generation projects tend to be rather uncommon or necessarily cheap. The gas produced from most biomass sources must be thoroughly cleaned up of any tarry materials, alkaline metal compounds, and dust before it can be used in general, and in engines in particular. When, the fuel gas is used for heating purposes in furnaces, the requirements for cleaning in comparison to engine applications tend to be of less importance.

The typical variation in the composition of biogases by volume is 50–80% methane, 15–45% carbon dioxide, water, traces of hydrogen sulfide, and traces of nitrogen gas. The relative amount of each gas varies depending on the maximum temperature, pressure, heating rate, and size and porosity of the biomass bed. Scrubbing away some of the CO_2 in water is needed for reducing this gas content and to permit satisfactory operation in some devices, such as engines. Figure 6.8 shows an example of the variety of processes and resulting products arising from the processing of biomass to yield a wide range of products, with most of the combustible products in the form of fuel gases.

In general, the rate of production and composition of biogases depend on a number of variables. These include temperature, available nutrients, retention time, pH value, solid contents, agitation, etc. Moderately increasing the average temperature tends to increase the rate of production of the gas. The external heat input required is often supplied through the burning of some of the product gas.

Due to the variation in biogas quality, flow conditions, and reaction kinetics, the modification of commercial devices (so as to be able to burn such fuels satisfactorily) tends to be heavily dependent on empirical approaches based mainly on operational experience and local conditions. The partial oxidation of biomass using

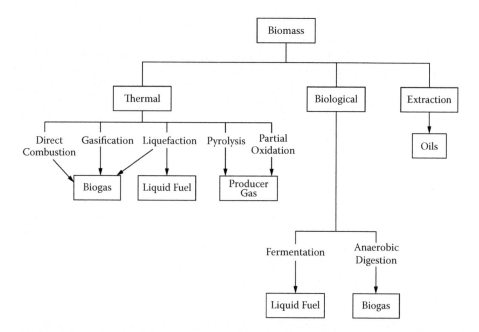

FIGURE 6.8   The various fuel–gas mixtures produced from the processing of biomass. (From Parik, P.P., Bhave, A.G., and Shash, I.K., Performance Evaluation of a Diesel Engine Dual Fuelled on Process Gas and Diesel, in *Proceedings of the National Conference on ICE and Combustion*, National Small Industries Corp. Ltd., New Delhi, India, 1987, pp. Af179–Af186.)

pure oxygen or oxygenated air is often relatively uneconomic and has tended to be uncommonly applied. Moreover, landfill gas, sewage gas, and marsh gas are established examples of biogases. It is generally known that for civic sewage processing plants, approximately 1 MJ of methane can be produced daily from the digestion of sewage per capita.

Fluidized beds are being considered increasingly suitable, in principle, for the production of biogas through the gasification of biomass, especially in small batches. They tend to be compact in size and construction, with high heat transfer and reaction rates because of the intimate and high level of mixing within the fluidized bed. However, their proper operation requires complex controls with some power consumption, such as for compression of the supplied air.

There are a number of key requirements for the successful employment of biogases for power production in engine applications. It must have an adequate heating value and should be free of dust and tars in order to produce continually sufficient power at a satisfactory efficiency and minimize engine wear, deposits, and maintenance problems while complying with the required exhaust emission controls. The maximum power that an engine can produce while running on biogas depends on the heating value of the gas, its composition, and specific engine characteristics. In comparison to natural gas operation, bigger size engines and supply lines would be required to compensate for the low specific energy content of the gas. Combustion devices, including engines, need to be suitably modified to permit the effective burning of biogases. These modifications, for example, include suitably increasing the size of the fuel piping and injector to permit handling the larger flow rates.

For effective usage of biogases as fuels, their methane and other combustible components may need to be enriched through measures such as water scrubbing, glycol absorption, application of molecular sieves, and membrane separation. These are employed in wastewater and sewage treatment plants, where the water does not need to be regenerated or recycled. The presence of any hydrogen sulfide in the gas has prejudicial effects on, for example, the lubricating oil, cylinder surfaces, valving, and exhaust gas treatment measures. It is also necessary when using natural gas and other gaseous fuels to ensure that they are sufficiently dry.

To operate dual-fuel engines on producer gas obtained from indigenous sources of agricultural materials and wastes represents an attractive approach for the production of power in relatively small units. For example, crop irrigation via engine-driven pumps is economically attractive. Of course, such engines are often started as a diesel, and then converted gradually and increasingly with load to dual-fuel operation.

Sewage gas and biogas frequently contain silicon and halogen vapor compounds in small concentrations, which require removal, such as through using activated carbon adsorber systems. These compounds can form undesirable deposits of silica in the engine, increasing component servicing requirements and wear and affecting the operation of oxidizing catalytic exhaust gas converters. Engines operating on biogases that may contain some siloxanes will tend to lose efficiency and require increased maintenance. Also, the silicon dioxide contaminates lubricating oil with increased accumulation of deposits.

6.8 LANDFILL GASES

A notable example of a low to medium heating value gaseous fuel is landfill gases. They are a form of biogas that is produced naturally through waste material decaying over time in an anaerobic oxygen-deficient environment. A typical landfill gas is made up largely of methane and carbon dioxide, with a small concentration of other gases, such as nitrogen and oxygen. Toxic components can also be found, but usually at very small concentrations, such as ammonia with hydrogen and carbon monoxide. Depending on the waste composition, low molecular weight intermediates such as organic acids and alcohols may also be found in the gas. Landfill gases pose a pollution problem by odor and a threat to the environment. They contaminate the soil and water table and emit powerful greenhouse gases. There are environmental and economical benefits in exploiting the gas for the production of power in engines or for heat in suitable furnaces since the strongly potent greenhouse gas methane is converted to relatively less pollutant products while simultaneously producing useful work and heat.

The generation of landfill gases from the biomass accumulation is affected largely by factors that include the availability of nutrients, temperature, moisture, pH value, atmospheric and climatic conditions, age, and variations in the water table. The amount of gas produced off a landfill site depends on factors such as its age, size, volume, type of waste, soil, and moisture content. The capture of gases produced by landfills requires pipes suitably embedded into the landfill mass to collect the gas and bring it to a collection point where it is processed. The gas is recovered by inducing a partial vacuum to drive the gases toward the collection well through the resulting pressure difference. The methane yield off landfills is usually optimum at a temperature range of 35–45°C. At temperatures below 10°C the production is significantly reduced, limiting production in winter months. A number of halogenated compounds (chemicals containing chlorine, fluorine, or bromine) are produced, and they may combine with hydrocarbons after combustion to produce toxic compounds. Such formation needs to be prevented by having the raw gas suitably refined before combustion. Such refinement often tends to be costly.

To use the gas in reciprocating engines requires its processing to remove various contaminants and reduce moisture. The troubling materials in gas applications in engines include chlorine-bearing compounds, sulfur compounds, organic acids, particles, and metal and silicone compounds. The concentration of CO_2 in gases derived from landfills varies widely, depending on the age and local conditions of the landfill, and can range from 40–55%.

6.9   SOME OTHER GASEOUS FUEL MIXTURES

The volumetric heating value of a fuel gas is an important indicator of its economic viability since the pipeline capacity and design of equipment depend on the volume of gas that needs to be handled for a certain thermal loading. The associated resulting temperature on combustion is also important because of its effect not only on the combustion rate, but also on the amount and quality of heat that can be released. The heating value of the fuel may be raised through supplementing the fuel gas with a high-grade fuel such as natural gas. Raising the temperature level through heat transfer employing heat exchangers tends not to be used due to cost, pressure losses, corrosion, and other material problems.

There is a wide range of other gaseous fuel mixtures suitable for employment in dual-fuel engines that are produced from a variety of materials having their own name while using different processing approaches. Some of these follow:

Bagasse: A gaseous fuel mixture produced from the processing of the remains of sugarcane after extracting the juice.

Blast furnace gas: A low heating value gas–fuel mixture produced as a by-product of the blast furnace during steelmaking.

Coal gases: For many decades the main domestic fuel gases used, including in engine applications, were derived from coal. Generally, coal gasification, which takes place at high temperatures and pressures, refers to the partial oxidation reactions of coal with air, oxygen, steam, carbon dioxide, or a mixture of these gases to yield a combustible gaseous product for use as a fuel. The principal gases produced are H_2, CO_2, steam, CO, and some methane. The composition of the product gas can be altered with the aid of catalysts and the proper control of operating variables such as to favor hydrogen and carbon monoxide production.

Producer gas: A fuel–gas mixture that is produced by the combustion of fuels such as coal with air and steam. The product gas contains hydrogen and carbon monoxide with nitrogen, but with hardly any oxygen. The gas is better suited to stationary engine applications. However, when used in mobile applications, the apparatus for generation of the gas tends to be bulky and often located on a trailer pulled by the vehicle.

Refinery gas: The noncondensable vapors resulting from the fractional distillation of crude oil. The gas is usually either flared or supplied for mixing with other fuel gases, such as propane and butane, mainly for domestic use and engine transport applications.

Synthesis gas: A manufactured fuel-gas mixture made from the reforming or partial oxidation of fossil fuels, such as coal, natural gas, or oil. The gas, which is made up mainly of hydrogen and carbon monoxide, serves as a raw material for the chemical synthesis of a wide range of fuels, e.g., alcohols and synthetic diesel fuel. This gas mixture is normally considered more valuable chemically, and normally employed less as an engine fuel. The production of alcohols and higher hydrocarbon fuels, including diesel fuel, is from the gas mixture while employing, as an example, the Fischer–Tropsch process.

Water gas: A manufactured gaseous fuel mixture consisting primarily of carbon monoxide and hydrogen made by the action of steam and hot carbon/coal. The product gas is used mainly to serve on combustion as a heating source.

REFERENCES AND RECOMMENDED READING

Ali, A.I., and Karim, G.A., Combustion, Knock and Emission Characteristics of a Natural Gas Fuelled Spark Ignition Engine with Particular Reference to Low Intake Temperature Conditions, in *Proceedings of the Institute of Mechanical Engineers*, London, 1975, vol. 189, pp. 139–147.

Anon., *SAE Handbook: Engine, Fuel, Lubricants, Emissions, and Noise*, vol. 3, Society of Automotive Engineers, Warrendale, PA, 1993.

Bartok, W., and Sarofim, A.F., *Fossil Fuel Combustion*, John Wiley & Sons, New York, 1991.

Bechtold, R.L., *Alternative Fuels Guidebook*, SAE, Warrendale, PA, 1997.

Bergman, H., and Busenthur, B., Facts Concerning the Utilization of Gaseous Fuels in Heavy Duty Vehicles, in *Proceedings of the Conference on Gaseous Fuels for Transportation*, August 1986, pp. 813–849.

Borman, G.I., and Ragland, K., *Combustion Engineering*, int. ed., McGraw Hill, New York, 1998.

Challen, B., and Barnescu, R., *Diesel Engine Handbook*, 2nd ed., SAE, Warrendale, PA, 1999.

Czerwinski, J., and Comte, P., *Influences of Gas Quality on Natural Gas Engine*, SAE Paper 2001-01-1194, 2001.

Diggins, D., *CNG Fuel Cylinder Storage, Efficiency and Economy in Fast Fill Operations*, SAE Paper 981398, 1998.

Eke, P., and Walker, J.H., Gas as an Engine Fuel, in *Proceedings of the Institute of Gas Engineers*, 1970, pp. 121–138.

Foxwell, G.E., *The Efficient Use of Fuel*, British Ministry of Technology, HMSO, London, 1958.

Gas Research Institute, *Technology Today*, Chicago, IL, 1991.

Golvoy, A., and Blais, E.J., *Natural Gas Storage on Activated Carbon*, SAE Paper 831678, 1983.

Heenan, J., and Gettel, L., *Dual Fueling Diesel/NGV Technology*, SAE Paper 881655, 1988.

Jeong, D.S., Suh, W.S., Oh, S., and Choi, K.N., *Development of a Mechanical CNG-Diesel Dual-Fuel Supply System*, SAE Paper 931947.54, 1993.

Karim, G.A., Some Aspects of the Utilization of LNG for the Production of Power in Internal Combustion Engines, presented at Proceedings of the International Conference on Liquefied Natural Gas, Institute of Mechanical Engineers, London, 1969.

Karim, G.A., The Combustion of Bio-Gases and Low Heating Value Gaseous Fuel Mixtures, *International Journal of Green Energy*, 8, 1–10, 2011.

Karim, G.A., *Fuels, Energy and the Environment*, CRC Press, Boca Raton, FL, 2012.

Karim, G.A., and Wierzba, I., *Methane–Carbon Dioxide Mixtures as a Fuel*, SAE Paper 921557, 1992.

Karim, G.A., Hanafi, A.S., and Zhou, G., A Kinetic Investigation of the Oxidation of Low Heating Value Fuel Mixtures of Methane and Diluents, Proceedings of the ASME/ETCE Emerging Energy Technology Conference, Houston, Texas, 1992, pp. 103–109.

Keller, E., *International Experience with Clean Fuels*, SAE Paper 931831, 1993.

Liss, W.E., and Thrasher, W.H., *Natural Gas as a Stationary Engine and Vehicular Fuel*, SAE Paper 912364, 1991.

Lom, W.L., *Liquefied Natural Gas*, Applied Science Publishers Ltd., London, 1974.

Parikh, P.P., Bhave, A.G., and Shash, I.K., Performance Evaluation of a Diesel Engine Dual Fuelled on Process Gas and Diesel, in *Proceedings of National Conference on ICE and Combustion*, National Small Industries Corp. Ltd., New Delhi, 1987, pp. Af179–Af186.

Pollock, E., ed., *NGV Resource Guide*, RP Publishing, Denver, CO, 1985.

Quigg, D., Pellegrin, V., and Rey, R., *Operational Experience of Compressed Natural Gas in Heavy Duty Transit Buses*, SAE Paper 931786, 1993.

Rain, R.R., and McFeatures, J.S., *New Zealand Experience with Natural Gas Fuelling of Heavy Transport Engines*, SAE Paper 892136, 1989.

Rose, J.W., and Cooper, J.R., eds., *Technical Data on Fuels*, 7th ed., British National Committee of World Energy Conference, London, 1977.

Tomita, E., Fukatani, N., Kawahara, N., and Maruyama, K., Combustion in a Supercharged Biomass Gas Engine with Micro Pilot Ignition Effects of Injection Pressure and Amount of Diesel Fuel, *Journal of Kones Powertrain and Transport*, 14(2), 513–520, 2007.

Turner, S.H., and Weaver, C.S., *Dual-Fuel Natural/Diesel Engines: Technology, Performance and Emissions*, No. GRI-94/0094, Topical Report Gas Research Institute, November 1994.

U.S. Energy Information Administration, U.S. Department of Energy, Washington, DC, 2014, http://www.eia.doe.gov/emeu/international/reserves.html.

Zabetakis, M., *Flammability Characteristics of Combustible Gases and Vapors*, Bureau of Mines Bulletin 627, U.S. Department of the Interior, Washington, DC, 1965.

7 Liquefied Petroleum Gases, Hydrogen, and Other Alternative Fuels

7.1 EVALUATING THE MERITS OF THE DIFFERENT GASEOUS FUELS FOR DUAL-FUEL OPERATION

Care is needed when evaluating the suitability of different fuels for dual-fuel engine applications. Different conclusions may be arrived at depending on the basis on which the evaluation is being made. For example, a comparison of fuels based on their heating values on a mass basis may produce a different ranking from one made similarly, but on a volume basis or on the basis of the energy release per unit of volume of fuel–air mixture. Similar yet different rankings may be obtained if the comparison were to be extended to include considering the energy expenditure in the fuel production and its dispensing. Also, in any fuel application, the use of the available energy should be optimized whether with respect to the direct chemical energy of the fuel or that arising also from its state, such as the compression work of compressed natural gas (CNG) or the cold in liquefied natural gas (LNG) applications. Moreover, in transport applications, the energy expenditure due to the increased mass of the vehicle arising from the additional weight of the fuel tanks needs to be considered. Similarly, the introduction into the engine of gaseous fuels such as methane or hydrogen displaces a significant portion of the intake air. Also, the enhancement of the volumetric efficiency in liquid fuel applications due to the evaporative cooling of the fuel will be missing whenever gaseous fuels are employed. These would lead to a reduction in the total energy that can be released in comparison to liquid fuels unless special additional remedial measures are taken to compensate for this reduction. In any case, the conversion of engines to operate on alternative fuels often tended to lack the degree of refinement and optimization of conventional diesel engines, which have benefited greatly from the continued support of much research and development over the years. Additionally, government policies of support and taxation of the use of the different fuels, together with some remaining technical challenges, have controlled and greatly distorted on occasion the merits of a wider employment of an alternative fuel. On this basis, the evaluation of the relative benefits of the conversion of a diesel engine to operate on an alternative fuel can be made largely on a tentative basis. The benefits and associated costs will vary widely with the types of fuel and engine, field of application, fuel availability, and infrastructure required, and whether significant design changes to the engine are needed or not.

7.2 LIQUEFIED PETROLEUM GAS (LPG)

Liquefied petroleum gas (LPG) is often identified loosely as simply propane. The gas is primarily made up of propane, with some butane and small concentrations of other constituent combustible vapors. Propane and butane are usually present in small concentrations in natural gases, and they are mostly removed from the gas before its high-pressure transmission through pipelines. Hence, the production of natural gas gives rise to the availability of relatively significant amounts of LPG.

Propane and butane are also produced from the refining processes of crude petroleum and will have with them varying quantities of other related fuel components, such as propylene, butylene, and isomers of the fuels. The presence of these components and fluctuations in their concentrations can give rise to potential complications as far as the utilization of the fuel in internal combustion engines by increasing the tendency to knock, and can change adversely the nature of exhaust emissions. On this basis, there can be a significant difference between the effective octane numbers of pure propane and commercial propane, which has a higher tendency to knock and produce smoke than pure propane. LPGs have been used as vehicular fuel for quite some time, with many of the applications related to the commercial and public sectors, such as taxis, trucks, and vans. Some of these are of the dual-fuel type.

Liquefied petroleum gas can exist in the liquid state at ambient temperatures when it is under moderate pressure. This makes it more easily portable for a variety of applications and on board vehicles than CNG. The constituents of LPG tend to be more complex than those of CNG, and tend to produce, on their presence in the atmosphere, more reactive and objectionable compounds.

LPGs are heavier than air, and the vapor can constitute a greater fire hazard than methane. The leakage of the fuel, for example, tends to concentrate the fuel vapor in the vicinity of the accident site, while following a leak of methane or hydrogen, the gas will disperse very rapidly and mainly vertically due to its highly buoyant nature. However, in this respect, a gasoline spill tends to represent a perhaps greater fire and explosion hazard than either methane or propane since the heavier than air flammable vapors, when released, can linger on for quite some time.

Figure 7.1 shows a schematic diagram of a typical LPG fuel container. Propane cylinders are required to be equipped with pressure relief valves to release any excessive pressure buildup, such as when cylinders are subjected to excessive heating. The cylinders need to be maintained in an upright position so that only gas is vented. When a cylinder is inverted or placed on its side so that the liquid fuel covers the outlet opening, then liquid propane can be discharged and will flash evaporate exceedingly quickly on its discharge while the pressure drops. In some limited conditions, the liquid fuel could pool or reach the point of use to create a serious fire and explosion hazards. Since LPG products are essentially odorless, an odorant, usually a mercaptan, is added in very small concentrations as a standard practice so that leakages can be detected readily. Propane is essentially nontoxic, although butane and higher hydrocarbons have a tendency to act as an anesthetic when vapors are breathed over a protracted time period. Figure 7.2 shows the valving and control arrangements for an LPG fuel tank.

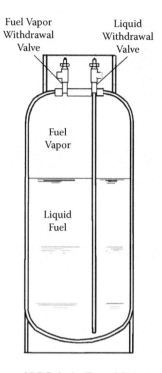

FIGURE 7.1 Typical container of LPG fuel. (From Matheson, *Guide to Safe Handling of Compressed Gases*, 2nd ed., Matheson Gas Products, New York, 1983.)

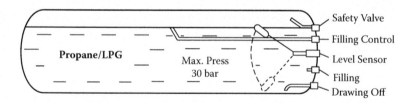

FIGURE 7.2 Typical representation of the valving and controls of an LPG fuel tank. (From Bergman, H., and Busenthur, B., Facts Concerning the Utilization of Gaseous Fuels in Heavy Duty Vehicles, in *Proceedings of the Conference on Gaseous Fuels for Transportation*, August 1986, pp. 813–849.)

Propane and its higher hydrocarbon content need to be limited to ensure no more than 1% condensation of the fuel at the lowest expected operational temperature. Also, odorization must ensure that gas leaks are clearly detectable at concentrations not over 1/5 of the normal flammability limit in air, i.e., for natural gas, <1%, and for propane, 0.4% by volume in air. Propane tanks must not be filled to more than 80% of volume capacity to allow the gas room for expansion as temperature changes. The pressure in propane tanks normally ranges from 0.6 to 1.4 MPa at 32°C, and 1 L of propane produces almost 40% less thermal energy than 1 L of diesel fuel.

7.3   OTHER ALTERNATIVE FUELS

The term *alternative fuels* usually indicates fuels that may be used in engine applications for transportation and power generation other than those conventional liquid fuels, such as gasoline and diesel fuels. Often, the term includes the following
groups of fuels:

- Natural wellhead and pipeline gases, methane, LNG, CNG, biogas, landfill
  gas, coal bed methane gas, shale gas, LPG, propane, and butane
- Industrially processed fuel-gas mixtures such as coal gas, coke oven gas,
  blast furnace gas, and synthesis gas
- Hydrogen, both liquid and gas
- A range of liquid fuels that include alcohols and bio-derived fuels

Some of the main advantages associated with the use of such alternative fuels
follow:

- There are potential environmental benefits in comparison to conventional
  liquid fuels, supplemented by some operational and performance benefits.
- Often, there may be cost advantages in relation to applications using
  cheaper and locally available fuels with lower associated operational and
  maintenance costs.
- There is diversification of the types of fuels and resources available, with
  increased relative independence of the national fuel supply and resources.
- They can be integrated with other fuel and energy requirements, such as
  with cogeneration applications and supplementing conventional fuels.
- They are efficient for power and heat generation in a wide range of power
  devices, including dual-fuel type engines. With some fuels it may enhance
  starting and cold-weather operation.
- Alternative fuels, especially those that are gases, may be available as waste
  products, are often virtually free of charge, and are suitable for substitution
  of conventional fuels with high greenhouse reduction potential, and include
  the disposal of "problem" gases.

Of course, there may also be some potential drawbacks associated with the operation of compression ignition engines on such fuels. Examples of these include:

- In transport applications, the fuels often have potentially inferior characteristics to those of conventional liquid fuels, bringing about increased costs,
  limitations in operational range, and increases in vehicular weight and bulk.
- The modifications needed in equipment and infrastructure result in
  increased costs. There may be a need to have available two fuels or more
  systems on board, which add to both cost and complexity of control.
- Performance of combustion engines may be adversely affected, depending on how well the conversion to alternative fuel operation has been

implemented. It may also bring about some deterioration in performance and the safety of operation.

- Alternative fuels invariably are mixtures of fuels where their composition may vary sufficiently with source and time to affect performance adversely when not suitably controlled and optimized.
- There are special problems with some fuels, such as those involving alcohols where there are some toxic, environmental, cold starting, safety, and materials compatibility problems. Also, alcohols suffer from the disadvantage of separating from a blend with diesel fuel in the presence of some water.

There are numerous points in favor as well as against the employment of alcohols as engine fuels, or even as supplementary additives to conventional fuels. For example, the production of ethanol from plant sources requires huge areas of farmland, which places the fuel in direct competition with food supplies. These cast doubt about ethanol as a long-term substantial fuel source globally, especially with the rapid increase in the world's population and the need to feed them.

7.4 HYDROGEN AS AN ENGINE FUEL

Hydrogen can be viewed as merely an energy carrier. It is not available naturally in its free state and requires energy to be produced from common sources such as water, but with much of the energy recovered on combustion. It is widely used by the petrochemical industry, for example, to manufacture a wide range of chemicals and for upgrading the quality of gasolines and other fossil fuels. The viability of hydrogen as a fuel, of course, is critically dependent on the effective and economic solution of a number of problems associated with its manufacture since it requires much energy and capital expenditure.

Hydrogen continues to have very limited use as an engine fuel at present. Much of it is manufactured from fossil fuels—natural gas, oil, and coal via their steam reforming or partial oxidation processes. Both catalytic and noncatalytic approaches are employed. Much of the bulk of hydrogen needed at present is produced in large-capacity units employing mainly natural gas as the fuel. The electrolysis of water, which in comparison is less widely employed, produces conveniently hydrogen in high purity while consuming a considerable amount of electrical energy. Much of this energy is generated mainly via the combustion of fossil fuels or hydroelectric and nuclear power. There is little prospect at present for the wide use of electrical energy from renewable sources for the production of hydrogen using wind, solar, or tidal energy. There are also very limited long-term prospects for hydrogen manufacture through bacterial action with solar energy and water, oxidation of metals, or special chemical cycles.

Hydrogen gas is commonly stored and transported as compressed gas, liquefied, in metal hydride form, or occasionally adsorbed in special but expensive alloys. The fuel gas has many attractive features as an energy resource carrier. For example, as a fuel, the following favorable characteristics can be assigned to hydrogen:

- It is a renewable fuel that can be manufactured from widely available sources, such as water, through the expenditure of energy.

- It is a clean fuel that produces much less exhaust emissions on combustion than other fuels. Its exhaust gas produces water and, on condensation, energy through cogeneration. However, its combustion in air produces oxides of nitrogen.
- Hydrogen is catalytically sensitive and has some very attractive combustion characteristics, such as clean combustion, fast burning rates, and a wide flammable mixture range. It requires low ignition energy, but with relatively high autoignition temperatures. It has a very high heating value on mass bases with high flame temperatures. It is highly buoyant, diffusive, and remains a gas down to extremely low cryogenic temperatures.

However, at present there are some limitations to its wide application as a fuel, especially in engines:

- It requires much useful energy for its manufacture and lacks at present the infrastructure for its wide distribution, resulting in its high cost on an energy basis in comparison to other fuel resources.
- It has potential problems relating to safety, material compatibility, portability, storage, handling, and transport. It is extremely difficult to liquefy and keep as a liquid, requiring much useful work.
- It has a very low heating value on volume and liquid bases, with its flames having very low luminosity.

The burning of the hydrogen-filled *Graf Zeppelin* in 1937 (Figure 7.3) remains a notable reminder of the hazards associated with the employment of hydrogen.

The vision of the "hydrogen economy" is for a future where most of the energy required can be made available cleanly and safely through hydrogen combustion, which is assumed to be produced from entirely renewable green energy sources. Electrical and transportation needs are then satisfied through reliance on hydrogen fuel cells. Such a vision of the future remains largely a concept. It is difficult to predict when and if it can be implemented.

FIGURE 7.3   The burning of the hydrogen-filled *Graf Zeppelin* in 1937. (From Lyons, P.R., *Fire in America*, National Fire Protection Association, Boston, MA, 1976.)

Current hydrogen pipelines are of relatively small diameter and of shorter length than those transporting natural gas. They do not operate at the very high pressures employed in natural gas pipelines. This is mainly due to the tendency of common steels to hydrogen embrittlement and low strengths.

What about liquid hydrogen (LH_2)? Where does it fit in the utilization of hydrogen as an engine fuel? Examination of the very limited experience in this respect leads to the conclusion that there is very little prospect for its wide application in this form in the near future. It is a cryogenic fluid with extremely low boiling temperature, which is only 20 K at atmospheric pressure. It has a very low liquid density (~70 g/L), which is around only 10% of that of gasoline, and for the same energy, it needs to have a few times the volume of liquid gasoline.

Also, the associated cold of liquid hydrogen is often wasted since it is often considered a nuisance and a hazard. The cryogenic nature of the fuel would require expensive and relatively bulky and heavy fuel tanks for its storage and transport. Moreover, there are numerous serious safety and dispensing problems associated with all aspects of the employment of hydrogen. There are some additional problems, such as increased ice formation, corrosion, materials compatibility, and storage time limitations. When liquid hydrogen, LH_2, is injected directly, heat pickup by the fuel is needed and a very cold gas is produced with a substantial increase in volume. The fuel has to be injected fast enough into the high-pressure combustion space and distributed effectively in a relatively short time. The big difference in density between the fuel and air, the considerably high flame speed, and the different turbulence characteristics would greatly influence the mixing and combustion processes in engine applications. Table 7.1 lists some comparative properties of hydrogen, methane, and iso-octane.

TABLE 7.1

Some Comparative Properties of Hydrogen, Methane, and Iso-Octane

Property	Hydrogen	Methane	Propane
Density at 1 atm and 300 K (kg/m³)	0.082	0.717	1.79
Stoichiometric composition in air (% by volume)	29.53	9.48	4.02
Stoichiometric fuel/air mass ratio	0.029	0.058	0.064
No. of moles of products to reactants	0.85	1.00	1.05
Heating values:			
HHV (MJ/kg)	141.7	52.68	50.40
LHV (MJ/kg)	119.7	46.72	46.37
HHV (MJ/m³)	12.10	37.71	94.01
LHV (MJ/m³)	10.22	33.95	86.49
Combustion energy per kg of stoichiometric mixture (MJ)	3.37	2.56	2.78
Diffusion coefficient into air at NTP (cm²/s)	0.61	0.189	0.160

Source: Adapted from Zabetakis, M., *Flammability Characteristics of Combustible Gases and Vapors*, Bureau of Mines Bulletin 627, U.S. Department of the Interior, Washington, DC, 1965; Rose, J.W., and Cooper, J.R., eds., *Technical Data on Fuels*, 7th ed., British National Committee of World Energy Conference, London, UK, 1977.

With suitably designed high-compression ratio engines using hydrogen as a fuel, its combustion characteristics, such as its very high flame propagation rates and extended lean mixture operational limits, permit somewhat better reduction of NOx emissions than other conventional hydrocarbon fuels. However, the main impediments for using hydrogen as an engine fuel, especially in transport applications, remain the size, weight, complexity, and cost of hydrogen storage on board. It is also noted that the molar concentration of hydrogen in the intake charge of a firing engine is much higher than for other gaseous fuels, which leads to a substantial displacement of some of the intake air, lowering the corresponding overall oxygen concentration. This will have a significant influence on the corresponding combustion processes and power output in hydrogen-fueled dual-fuel engines.

REFERENCES AND RECOMMENDED READING

Al-Garni, M., A Simple and Reliable Approach for the Direct Injection of Hydrogen in Internal Combustion Engines at Low and Medium Pressures, *International Journal of Hydrogen Energy*, 20(9), 723–726, 1995.

Aly, H., and Siemer, G., Experimental Investigation of Gaseous Hydrogen Utilization in a Dual-Fuel Engine for Stationary Power Plants, in *ASME-ICE 1993*, 1993, vol. 20, pp. 67–79.

Anon., *SAE Handbook: Engine, Fuel, Lubricants, Emissions, and Noise*, vol. 3, Society of Automotive Engineers, Warrendale, PA, 1993.

Bergman, H., and Busenthur, B., Facts Concerning the Utilization of Gaseous Fuels in Heavy Duty Vehicles, in *Proceedings of the Conference on Gaseous Fuels for Transportation*, August 1986, pp. 813–849.

Bols, R.E., and Tuve, G.L., eds., *Handbook of Tables for Applied Engineering Science*, CRC Press, Cleveland, OH, 1970.

Challen, B., and Barnescu, R., *Diesel Engine Handbook*, 2nd ed., SAE, Warrendale, PA, 1999.

Diggins, D., *CNG Fuel Cylinder Storage, Efficiency and Economy in Fast Fill Operations*, SAE Paper 981398.35, 1998.

Ecklund, E., Bechtold, R., Timbario, T., and McCallum, P., Alcohol Fuel Use in Diesel Transportation Vehicles, in *Institute Gas Technology 3rd Symposium on Nonpetroleum Vehicular Fuels*, October 1982, pp. 261–313.

Energy Mines and Resources, *Propane Carburation*, Government of Canada, Ministry of Supplies and Services, Ottawa, 1984.

Gopal, G., Rao, P.S., Gopalakrishnan, K.V., and Murthy, B.S., Use of Hydrogen in Dual-Fuel Engines, *International Journal of Hydrogen Energy*, 7(3), 267–272, 1982.

Goto, S., Furutani, H., and Delic, R., *Dual Fuel Diesel Engine Using Butane*, SAE Paper 920690, 1992.

Karim, G.A., *Fuels, Energy and the Environment*, CRC Press, Boca Raton, FL, 2012, p. 75.

Karim, G.A., and Rogers, A., Comparative Studies of Propane and Butane as Dual Fuel Engine Fuels, *Journal of the Institute of Fuel*, 40, 513–522, 1967.

Karim, G.A., and Wierzba, I., Comparative Studies of Methane and Propane as Fuels for Spark Ignition and Compression Ignition Engines, *SAE Transactions*, 92, 3677–3688, 1983.

Lyons, P.R., *Fire in America,* National Fire Protection Association, Boston, MA, 1976.

Matheson, *Guide to Safe Handling of Compressed Gases,* 2nd ed., Matheson Gas Products, New York, 1983.

Maxwell, T., and Jones, J., *Alternative Fuels: Emissions, Economics and Performance*, SAE, Warrendale, PA, 1995.

Oester, U., and Wallace, J.S., *Liquid Propane Injection for Diesel Engines*, SAE Paper 872095, 1987.

Rose, J.W., and Cooper, J.R., eds., *Technical Data on Fuels*, 7th ed., British National Committee of World Energy Conference, London, 1977.

Varde, K.S., *Propane Fumigation in a Direct Injection Type Diesel Engine*, SAE Paper 831354, 1983.

Varde, K.S., and Frame, G.A., Hydrogen Aspiration in a Direct Injection Type Diesel Engine— Its Effects on Smoke and Other Engine Performance Parameters, *International Journal of Hydrogen Energy*, 8, 549–555, 1983.

Zabetakis, M., *Flammability Characteristics of Combustible Gases and Vapors*, Bureau of Mines Bulletin 627, U.S. Department of the Interior, Washington, DC, 1965.

8 Safety Considerations

8.1 FUEL SAFETY REQUIREMENTS

One of the most important concerns associated with the production, transport, storage, or utilization of any fuel is how to guard against the potential toxic and environmental negative effects as well as the associated risks of fire and explosion. An enormous amount of information exists in the public domain that includes guidelines and directives for ensuring safe operation and handling of fuels of all types at all levels and applications. These include numerous rules, regulations, codes, good practice statements, instructions, etc., originating from all levels of government, industry, and professional, international, and trade associations. There are detailed reports in the open literature of case studies that may have involved life and property losses with important lessons to learn highlighted. These need to be studied, familiarized with, and followed by all those who work in the energy, engine, and fuel fields. Moreover, there is a need for a thorough knowledge of the potential fire and explosion hazards associated with the operation of engines on gaseous fuels in general, and for the planning to guard against the possibility of any damage that may arise from a hazardous situation. This is especially needed when research type testing and investigations are considered, such as those within a university laboratory setting, where inexperienced students may be present or are involved and where novel, unconventional, or untried procedures may be attempted. Efforts should always be first directed at eliminating the generation of any potentially combustible mixture rather than the mere elimination of potential ignition sources.

Some of the main hazards associated with gaseous fuel application, in addition to the health and environmental hazards, are fuel leakages, the dissipation of the fuel and its mixing with the surrounding air, the ease of its ignitability and associated flammability limits, deflagration speed, thermal radiation, and health hazards. These are mainly functions of the fuel properties, such as viscosity, density, volatility, flash point, flammability limits, ignition energy, flame speed, flame emissivity, heat release rate, autoignition temperature, and toxicity. Some overall related characteristics that need to be considered when evaluating the safety aspects of the usage of various fuels in engine applications are the following:

- Storage and portability characteristics of the fuel and other related potentially hazardous material that may be used, such as oxygen and catalysts
- The tendency to form combustible mixtures following accidental discharge, such as gas leakages
- Ignition, explosion, flame spread, and detonation characteristics

- Effective and prompt control measures provided for the fuel spread and its combustion
- Environmental and toxic consequences of fuel employment and discharges

Relative to conventional liquid fuels such as gasoline and diesel fuels, gaseous fuels usually require only small amounts of energy for ignition, especially in the case of hydrogen. Many relatively low ignition energy sources can ignite such a fuel. These can include sources such as sparks, sufficiently hot surfaces, open flames, and static electricity discharges, including those generated by the human body and its apparel. Therefore, the value of the autoignition temperature of a fuel, its heating value or flammability limit, does not alone sufficiently indicate the relative explosion hazards of a fuel.

The detonation characteristics of any fuel in air are of far-reaching importance as safety is concerned. The detonatable mixture range of a fuel in air is narrower than the corresponding flammability mixture limits under the same conditions. The maxim burning and detonation velocities are different, and sometimes by orders of magnitude. It is very difficult for methane to undergo detonation, while hydrogen-air mixtures are much easier to detonate.

Sufficient care must be taken to ensure that the fuel gas is of an acceptable purity and does not contain significant amounts of material, such as sulfur compounds, or is excessively wet, to avoid corrosion problems within the fuel container interior, which may undermine the strength of the container and transfer fine debris to the engine and its working and control parts.

Fuel storage containers must meet all the appropriate safety regulations and must be tested at regular intervals. All fuel tubing and fittings, pressure switches, indicators, regulators, and valves must also be of appropriate design and comply with the required regulations. A typical gaseous fuel, when leaking from a container into the surrounding environment, can only form a certain maximum volume of combustible mixture. Methane, by virtue of its relatively narrow range of flammability limits, can generate a maximum combustible mixture that is small relative to other common fuels. For example, the discharge of one unit volume of methane would generate a maximum volume of combustible mixture that is around 40% of that following the release of an equivalent volume of propane, and 20% the volume of the mixture formed following the release of a similar amount of gasoline vapor.

The rate of discharge of a gaseous fuel following an accident will depend on the nature of the leak and the associated area of discharge. The discharge from a high-pressure cylinder or a line into the atmosphere will be proportional to the size of the discharge area and inversely proportional to the gas density. Accordingly, hydrogen will leak much faster than methane or propane. However, the extent of hazard will depend largely on the tendency to form a combustible mixture, its resulting volume, and the length of time for it to linger in the vicinity of the discharge location. The mixture may get ignited either from a potential source or by feeding a fire that is already engulfing the fuel source.

The tendency for a compressed fuel gas at high pressure to disperse into the surroundings out of a leak is governed by the resulting gaseous turbulent jet mixing in the surrounding air. Buoyancy and diffusional effects are dominant in the dispersion of fuels such as methane and hydrogen, unlike the much heavier than air vapors of gasoline or propane. These heavy vapors tend to linger on following their discharge and disperse slowly. Thus, they retain the hazard for a longer time and spread the potential for fire and explosion away from the region of the leak.

8.2 METHANE AND COMPRESSED NATURAL GAS (CNG) OPERATIONAL SAFETY

Methane remains a flammable gas down to low temperatures. It is colorless and can act as a simple asphyxiant when it displaces the air needed for breathing. All handling and transfer operations of methane are to be conducted in well-ventilated areas, and all the equipment and accessories are to be grounded in case there may be an escaping gas that can form flammable mixtures with air. Methane has a relatively high value of the lean flammability limit, high buoyancy, and high diffusional properties, which make it somewhat less likely to form readily explosive fuel–air mixtures under ambient conditions in open air than other common pure gaseous fuels.

The most common gaseous fuel used in engine applications is processed natural gas made up mainly of methane. An odorant such as a mercaptan is commonly added for ease of detection in very low concentrations in case of a leakage. It is, of course, well recognized that there are potential safety issues arising from the fact that fuels such as compressed natural gas are usually fed from very high-pressure containers. However, these are cylinders of exceptional strength with safety fittings. The conductive nature of the metallic tanks, for example, tends to quickly even out the effects of external heating, such as due to a fire nearby. This is in contrast to the conventional tanks of liquid fuels that are relatively fragile, often poorly conductive, with a tendency to develop hot spots when externally exposed to overheating. This would lead to serious safety problems and eventual rupture, spilling the contents, which tend to stay in pools nearby rather than vaporize and disperse rapidly, as the case with gaseous fuels, especially those of the lighter than air methane and hydrogen.

In the case of a high-pressure leak, the gas will get discharged as a very high velocity jet. The minimum amount of energy needed to effect ignition of the resulting fuel–air mixtures within the flammable range is rather large in comparison to that of hydrogen or propane. Moreover, the quenching of methane flames by cold surfaces, such as in flame traps, tends to be more successful than in the case of fires of propane or gasoline.

Figure 6.1 shows a typical CNG cylindrical container as fitted to supply fuel to an engine showing two stages of regulation and valving for reducing the high supply pressure to the much lower pressure for mixing with the intake engine air.

8.3 LNG OPERATIONAL SAFETY

The transportation, fueling, and storage of LNG represent a potentially serious safety hazard due to the cryogenic liquid nature of the fuel retained at extremely low temperature (111K at atmospheric pressure), with the potential for fire and explosion following a spillage in accidents. The hazards associated with an LNG tank rupture and fuel spillage in principle far exceed those associated with CNG or other liquid

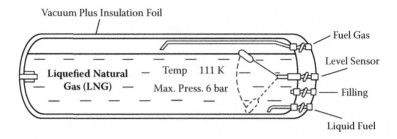

FIGURE 8.1 A representation of a typical small-capacity LNG fuel tank. (From Bergman, H., and Busenthur, B., Facts Concerning the Utilization of Gaseous Fuels in Heavy Duty Vehicles, in *Proceedings of the Conference on Gaseous Fuels for Transportation*, August 1986, pp. 813–849.)

fuels, such as gasoline. Tanks storing LNG, which are very well insulated, as schematically shown in Figure 8.1, are normally double-walled. The LNG is usually kept at pressures not much higher than atmospheric. Heat transfer from the environment into the well-insulated LNG tank will boil off increasingly more fuel, raising the pressure within the tank. Modern tank designs will not release the gas through insulated relief valves until a significant buildup of pressure to a prescribed value over a protracted period of time, typically such as 2 weeks elapsing. On arrival in marine terminals, LNG is stored in specialized tanks and in underground caverns. It is then vaporized when needed and usually transported by pipelines as a gas to augment conventional natural gas supplies.

The exploitation of natural gas sources and the reduction of its flaring and consequent emissions have been made possible by advances in the technology of the liquefaction and transportation of the gas. Nevertheless, ensuring safe operation with LNG remains a paramount concern to the industry and public.

The conversion of natural gas to the liquid state somewhat changes its composition and properties, including its heating value. All the sour component gases and vapors, such as H_2S and CO_2, and any water vapor are removed. Some of the larger molecular hydrocarbons may remain, but in very small concentrations. The LNG is gasified shortly after arrival by marine tankers using typically seawater as the heating medium for transport as pipeline gas.

The LNG tank wall is normally constructed of aluminum alloy, with heat leakage into the tank minimized by having vacuum sidewall insulation. On land, a dike is constructed around the tank to protect the surroundings in case of a spillage, such as following an accident. Reliquefication equipment is usually installed on large units to reliquefy the boil-off gas and avoid releasing it into the open atmosphere. Figure 8.2 shows a typical arrangement for vaporizing the low-temperature LNG to a gas state, as fed to an engine while using the hot water jacket, rather than the exhaust gases. A similar procedure may be needed for LPG applications.

LNG has a limited shelf life. If an engine is left idle for several days, then the pressure within the tank starts to increase and the LNG begins to boil. The tank begins to vent once a prescribed maximum pressure is reached. Much energy is required for an empty tank that was kept idle for some time to be cooled down to the cryogenic

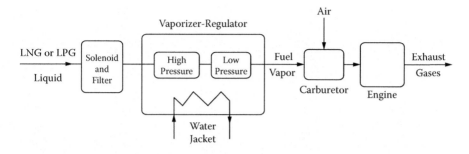

FIGURE 8.2 An arrangement of an LNG feed to an engine.

temperature of LNG. When supplies of LNG are readily available, the time needed for the safe filling of fuel tanks becomes important. A fast-fill LNG supply station is usually much less expensive than comparable CNG stations. The LNG filling speed may even be comparable to that of diesel fuel. However, because of the cryogenic nature of LNG, anyone fueling a vehicle must take special safety measures, such as the wearing of safety goggles and protective gloves. Some of the procedures to deal with LNG-associated hazards follow:

- Provision of fire protection and foam-producing equipment in tanks and other firefighting facilities, including the provision of explosion-proof equipment such as pumps, electrical equipment, and fittings.
- Provision of gas detectors and alarm systems at strategic locations to ensure adequate protection against low-temperature embrittlement of steels that may be in contact with the cryogenic liquid or its cold vapor.
- Provision in large LNG marine engines of detection of gas leakage into the crankcase, with an inert gas flushing of the crankcase to avoid the onset of crankcase explosion.
- Protection against the accidental rupture of tanks and avoiding resonance in LNG tanks through sloshing of the liquid, especially during its transportation in tankers.
- Preventing hydrates and ice formation that may plug lines by ensuring the gas is dehydrated with only a minimum practical amount of water.
- Protection provided against accidental interruption of the working of the various associated components, such as due to a vapor-lock forming in a line.

8.4 OPERATIONAL SAFETY OF LPG GASES

Any accidental releases of the heavier than air LPG gases will tend to flow to low levels and will persist undispersed much longer than with similar releases of the buoyant natural gas. Propane cylinders, shown schematically in Figures 7.1 and 7.2, are equipped with pressure relief valves to release any excessive pressure buildup, such as when cylinders and contents are subjected to excessive heating. The cylinders need to be kept in an upright position so that only gas fuel is discharged. When a cylinder is inverted or placed on its side so that the liquid fuel covers the outlet opening, then liquid propane could be discharged. This liquid will flash evaporate exceedingly quickly into gas on its discharge while the pressure drops through the control equipment. In extreme conditions, the liquid fuel could pool or reach the point of use to create a very serious fire and explosion hazard.

Propane cylinders should not be transported without proper caps covering the valve assembly to protect against breakage in case of an accident. Since LPG products are essentially odorless, an odorant, usually a mercaptan, is added in very small concentrations as a standard practice so that even small leakages can be detected readily.

8.5 SOME ENGINE SAFETY MEASURES

The conventional diesel engine and all aspects of its operation are usually considered to be relatively safe in comparison to the spark ignition gas or gasoline engines. The conversion of diesel engines to gas fuel operation is usually made without major changes in the design and operational features of the engine, including its safety guards. Many of such engine conversions tend to be associated, especially in recent years, with engines where both the engine and operating conditions are strictly controlled and usually monitored very closely by extensive rapid response sensors and control equipment. Accordingly, it is not to be assumed that because an engine is operating on a gaseous fuel, its probability to be involved in an incident is necessarily much higher than that of engines that operate on conventional liquid fuels. Statistics of gas-fueled engine incidents so far tend to confirm this.

In applications where the high-pressure supply of the fuel gas has to expand down to the low intake pressure conditions, significant cooling of the fuel would result, which may require warming through the supply of some external heating. The heating needed by the fuel may become excessive and needs to be avoided. Alternatively, in such cases, it may become advisable to provide some heating to the air and then introduce to it the cool fuel stream to obtain the desired mixture temperature at engine intake.

The occurrence of backfiring into the intake manifold of engines needs to be avoided through measures that reduce to a minimum the volume of the air that can become mixed with the fuel gas in any proportions outside the engine cylinder and within the intake manifold. Also, some of the gas injected during the engine valve overlap period may pass through the cylinder and onto the exhaust manifold to contribute to not only fuel wastages and increased unburned gas emissions, but also the potential for encountering uncontrolled explosions. It is also strongly advisable to remove surge tanks that are usually fitted on the intake side for metering the air consumption or damping some of the intensity of intake pressure pulsations. This is since, in the likelihood of any backflow of some of the inducted fuel gas, an explosive mixture may be formed, which in the event of a backfire can get ignited and develop into an explosion.

In some engine applications, especially those of very large size, it may be necessary to guard against the possibility of an explosion in the engine sump involving lubricating oil and fuel vapors. An inert atmosphere, such as of carbon dioxide or nitrogen, may need to be retained in the oil sump. In sufficiently large size engines, measures are taken for the relief of any potential buildup of pressure following ignition within the crankcase, combined with the fitting of suitable flame traps to quench any burning gases discharging into the engine room and surroundings.

Typically, for most large stationary dual-fuel engines, and especially for large applications before the engine is shut down, it is first changed over to diesel operation so that the stopping procedure commences. Such an approach for safety and emissions reasons ensures there will be no gas remaining in either the engine or the exhaust system following shutdown. In some engine applications, in case of an emergency, the unmetered discharge of a diluent gas such as nitrogen or carbon dioxide is made into the intake system and cylinder, together with cutting off the supply of the gaseous fuel. It then renders harmless any gaseous fuel that could have been admitted or still remains in the system.

In most recent engine applications where very high fuel gas supply pressures are involved, the gas pipes are double-walled for added protection. The space between the supply pipe walls is continuously vented, and fitted gas sensors promptly detect any gas leakage.

REFERENCES AND RECOMMENDED READING

Anon., *SAE Handbook: Engine, Fuel, Lubricants, Emissions, and Noise, vol. 3, Society of Automotive Engineers*, Warrendale, PA, 1993.

Badr, O., El-Sayed, N., and Karim, G.A., An Investigation of the Lean Operational Limits of Gas Fuelled Spark Ignition Engines, *ASME Journal of Energy Resources Technology*, 118(2), 159–163, 1996.

Bergman, H., and Busenthur, B., Facts Concerning the Utilization of Gaseous Fuels in Heavy Duty Vehicles, in *Proceedings of the Conference on Gaseous Fuels for Transportation*, August 1986, pp. 813–849.

Challen, B., and Barnescu, R., *Diesel Engine Handbook*, 2nd ed., SAE, Warrendale, PA, 1999.

Coward, H.F., and Jones, G.W., *Limits of Flammability of Gases and Vapors*, Bulletin 503, Bureau of Mines, 1952.

Eke, P., and Walker, J.H., Gas as an Engine Fuel, in *Proceedings of the Institute of Gas Engineers*, 1970, pp. 121–138.

Karim, G.A., *Some Considerations of the Safety of Methane (CNG) as an Automotive Fuel— Comparison with Gasoline, Propane and Hydrogen Operation*, SAE Paper 830267, 1983.

Karim, G.A., *Dual Fuel Engines of the Compression Ignition Type—Prospects, Problems and Solutions—A Review*, SAE Paper 830173, 1983.

Karim, G.A., *Fuels, Energy and the Environment*, CRC Press, Boca Raton, FL, 2012.

Karim, G.A., and Wierzba, I., Comparative Studies of Methane and Propane as Fuels for Spark Ignition and Compression Ignition Engines, *SAE Transactions*, 92, 3677–3688, 1983.

Karim, G.A., and Wierzba, I., Safety Measures Associated with the Operation of Engines on Various Alternative Fuels, *Reliability Engineering and Systems Safety Journal*, 37, 93–98, 1993.

Lee, J.T., Kim, Y.Y., Lee, C.W., and Caton, J.A., An Investigation of a Cause of Backfire and Its Control Due to Crevice Volumes in a Hydrogen Fueled Engine, *ASME Journal of Engineering for Gas Turbines and Power*, 123, 204–210, 2001.

Matheson, *Guide to Safe Handling of Compressed Gases*, 2nd ed., Matheson Gas Products, New York, 1983.

Pounder, C.C., ed., *Diesel Engine Principles and Practice*, George Newens, London, 1955.

Rose, J.W., and Cooper, J.R., eds., *Technical Data on Fuels*, 7th ed., British National Committee of World Energy Conference, London, 1977.

Zabetakis, M., *Flammability Characteristics of Combustible Gases and Vapors*, Bureau of Mines Bulletin 627, U.S. Department of the Interior, Washington, DC, 1965.

9 Combustion of Fuel Gases

9.1 SOME COMBUSTION FUNDAMENTALS

Combustion is an increasingly important topic of much economic and social importance. It addresses key issues such as the diversification and efficient utilization of fuel resources, while curbing environmental pollution and enhancing safety. By its nature, it requires sufficient knowledge of a range of fields within science and engineering. The recent advances in combustion diagnostics and measurement have made experimentally based combustion research, though very effective, increasingly more demanding, with often very expensive and specialized methods and expertise needed.

To bring about substantial improvements to the performance of internal combustion engines, there is a need to understand the nature of the combustion processes. These, by definition, involve chemical reactions that are exothermic and relatively fast and are influenced by physical processes relating to turbulent flow mixing and heat transfer. The proper understanding or accounting of such processes needs to specify what type and sequence of changes take place and consider their consequent temporal changes in composition and associated properties of the reactive system. Often, this is no simple task. Figure 9.1 shows a typical schematic representation of the temperature rise with time during the course of a combustion reaction of an adiabatic process. Initially, due to the low temperature at the commencement of the reaction, there is only a little temperature rise for quite some time, which represents a delay period. Often for convenience, this time is assumed to be represented arbitrarily by that taken to produce 5–10% of the total temperature rise or of the energy released. Following this delay period, the temperature begins to increase more rapidly as a result of the exponential dependence of the reaction rate, and hence the energy release rate on temperature. This initial rapid acceleration of the combustion reaction rate is associated with the onset of ignition. Toward the end of the combustion process the rise in the rate begins to slow down despite the prevailing high temperature levels due to the depletion of the reactants and the buildup of products. Under ideal and adiabatic conditions, the temperature will gradually converge toward a value approaching the ideal calculated value of the adiabatic flame temperature. In practice, due to a variety of factors, such as the inevitable heat loss to the surroundings from the high-temperature reaction zone, the possible lack of sufficient time, and possible incompleteness in the combustion reactions, the associated temperature rise can be significantly lower than that for the ideal adiabatic case. For the same reacting fuel–air mixture, the temperature rise-time development is accelerated considerably as higher initial temperatures are employed. The exponential

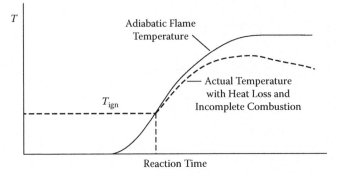

FIGURE 9.1 Typical temperature changes of a reacting mixture with time for both adiabatic and nonadiabatic combustion. The ignition temperature is shown. (From Karim, G.A., *Fuels, Energy, and the Environment*, CRC Press, Boca Raton, FL, 2012.)

nature of the dependence on absolute temperature of the reaction rate, and hence the associated energy release rate, for sufficiently low initial temperature values can slow down the temperature rise, and may even bring it to a virtual halt, thus quenching the combustion process altogether.

A variety of forms of representations may be employed to describe the rate of combustion of fuels. These may include the temporal rate of change of the consumption of the fuel, oxidant, or the appearance of the products. It can also be represented by the rates of change in temperature, the associated energy release, or the pressure rise. These forms of parametric variables are readily interrelated. Figure 9.2 shows a schematic representation of the course of the combustion process of a homogeneous fuel–air mixture displaying the variation of the reaction rate, and hence the energy release rate, with the extent of completion of the reaction.

This extent can be represented in terms of the fractional temperature rise, (e.g., $\sim[T - T_o / T_f - T_o]$) or relative mass of fuel consumed or combustion products

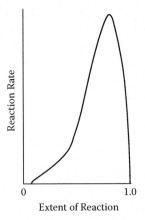

FIGURE 9.2 Schematic representation of the variation of reaction rate, and hence the energy release rate, with the extent of reaction in terms of fractional fuel mass burned.

formed. It is evident that the maximum rate of energy release or reaction rate is encountered well below the peak temperature reached in combustion. The property that has the most influence on the overall performance of a combustion system and acceptance of a fuel is the associated combustion reaction rate. It controls key performance parameters, such as the flame propagation rate, energy release rate, and nature of the exhaust products. These in turn influence notable performance characteristics, such as the mixture combustion limits, autoignition, and combustion efficiency, among others.

The combustion of fuel–air mixtures does not proceed via a single-step reaction where the reactants are converted directly to products. Instead, fuel combustion involves many simultaneous reaction steps where numerous transient species are generated and consumed in the course of the oxidation reaction. Of these elementary reaction steps, each has different reacting species and reaction rates. The net effect of these reactions while converting the reactants to products is to establish the key parameters of the combustion process, such as the rates of fuel and oxygen consumption, the energy release rate, and the production rates of the different reacting species, both transient and stable, that eventually appear in the exhaust gas.

Historically, a simplifying and approximate approach was to consider the reaction rate on an overall global basis to involve only the fuel and oxygen, while ignoring all details of the changing reaction activity of the mixture and the numerous transient and eventual product species. The results of corresponding experimental observations were then employed to produce an optimized fitted relationship for an apparent assumed single-step overall reaction for the observed conversion rate of the fuel and air to products by combustion. The relevant fitted key kinetic data in this simplistic approach are obtained through a best fit of the observed experimental data available. An example of a formulation for the oxidation of common fuels such as methane in air described as global combustion reaction rate is represented as

$$d[n_{CH4}]/dt = k[n_{CH4}]^a \, [n_{O2}]^b \, [n_{Diluent}]^c \, e^{-E/RT}$$

where $[n]$ is the molar concentrations, often for simplicity referred to, somewhat erroneously, as those of the initial mixture rather than the transient values, and k, a, b, c, and E are constants for the specific reaction. However, the actual combustion reaction activity cannot strictly be represented universally sufficiently reliably by such a relatively simple formulation, and these constants can vary widely depending on the conditions under which the experimental data were obtained. For the presence of some common diluents with methane, such as carbon dioxide in biogases, the index c of the above equation tends to be generally much smaller than unity. Thus, the reaction rate under *isothermal conditions* will be affected relatively little by the diluent addition. It is the energy released by the reaction and the consequent temperature rise with time that will be significantly affected by the increased presence of the diluents in the mixture. These factors are mainly responsible for bringing about the substantial reduction normally observed in the oxidation rates in combustion processes of fuels containing diluents. Similarly, this is also reflected in the slow reaction rate activity of lean fuel–air mixtures where the excess air lowers the resulting temperature and reduces the reactivity of such mixtures.

The temporal progress of the reaction and the associated kinetic ignition delay time are primarily dictated by the time required to build a sufficient pool of the highly reactive transient unstable species known as radicals, e.g., OH, H, O, and HO_2. The carbon in the fuel initially gets oxidized through a number of steps to carbon monoxide, which does not react significantly with the available oxygen to produce carbon dioxide, until much of the hydrocarbon molecules are sufficiently depleted. Accordingly, the typical chain reaction of the oxidation of methane in air is where the carbon in the fuel, for example, goes sequentially to the ultimate final product of carbon dioxide via formaldehyde and carbon monoxide. By employing a fairly comprehensive detailed representative scheme for the possible important elementary reaction steps in the path of the oxidation of fuels, the course of the oxidation reactions can be followed computationally right to the ultimate end product. The exponential dependence of the reaction rates on absolute temperature rapidly accelerates them as the temperature increases. The time needed to complete the combustion process will be appreciably reduced, and almost exponentially with high temperatures. The concentrations of the products eventually, when sufficient time is provided, will be essentially those expected in equilibrium at the high final temperatures reached.

Extensively detailed and comprehensive information can be obtained about the chemical processes in the combustion of more common complex fuels than methane, such as, for example, n-heptane. During the course of the reaction, the concentrations of various species change significantly before approaching their expected eventual values of concentration. The dominant role played by the changing mixture temperature is evident throughout. It is also evident that by interrupting the full course of the sequence of combustion reactions suddenly, such as through quenching, some intermediate products may survive to the exhaust stage. This is of much importance in relation to the reduction of exhaust emission components in internal combustion engines and securing high combustion efficiency. It reflects on the important roles played of the resulting temperature, time, and initial composition of the reactive mixture.

The time needed for self-ignition of a homogeneous fuel–air mixture is usually logarithmically dependent on temperature. Accordingly, high temperatures are needed to effect rapid reaction to ignition and completion of fuel conversion within the available residence time in an engine, which is approximately inversely proportional to engine speed. Moreover, the time needed for ignition under isothermal conditions, when they can be ensured, is essentially linearly related to the equivalence ratio, as shown in Figure 9.3 for lean methane–air mixtures. This time increases rapidly as leaner mixtures or increased diluents are involved.

A stoichiometric mixture contains the theoretical amount of air that will permit combustion of the fuel. However, in practice there is no assurance that simply by providing the correct fuel-to-air ratio, combustion will be completed or even initiated. There are many possible contributory factors. These include having too low temperature and pressure, excessive heat loss, insufficient time given for completing the reaction, insufficient ignition energy provided, presence of diluents, too excessive turbulence, and the fuel and air not being thoroughly mixed. Remedial action can address such influences and may contribute toward correcting them accordingly.

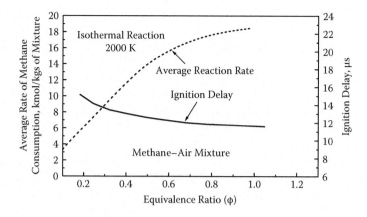

FIGURE 9.3 Calculated variations in the average reaction rate and the corresponding ignition delay with changes in equivalence ratio for lean mixtures of methane air under assumed isothermal conditions of 2000 K temperature. (From Karim, G.A., *International Journal of Green Energy*, 8, 1–10, 2011.)

9.2 COMBUSTION, FLAMES, AND IGNITION PROCESSES

The propagation of the flame within the fuel–air mixture is produced through the energy and active species released by the chemical reaction transported via heat and mass transfer to the adjoining fresh mixture yet to burn. Thereby in turn, its temperature is raised and its reaction activity enhanced to begin releasing its own energy. This way, the flame propagation characteristics depend on the chemical nature of the mixture as well as the nature and rate of transport processes of heat and mass from the reaction zone. Because of the associated vigorous transport processes associated with turbulence, turbulent flames are much faster and can release a much greater amount of energy per unit volume and time than its laminar counterpart. Flames within engines are highly turbulent so as to produce a rapid energy release in a relatively limited compact space and short time. An important parameter in fuel combustion is the maximum amount of energy that can be released by the chemical reaction per unit volume and time. This maximum is normally associated with stoichiometric mixtures.

Ignition may be defined as the initiation of combustion through accelerating the exothermic reactive processes beyond the prevailing rate of dissipation of energy out of the reaction zone to the surroundings. The initiation of ignition requires a minimum amount of energy that depends on the fuel used and other operating conditions. This energy is supplied from an external source such as a spark or through mixing with sufficiently hot gases, such as the hot products from a pilot flame. Stoichiometric mixtures normally require the least minimum amount of energy for ignition. In practical combustion systems, much higher ignition energies and temperatures than the minimum are supplied to ensure prompt ignition and continued combustion under sometimes adverse local conditions, such as an engine starting under cold temperature. Autoignition or self-ignition takes place as a result of self-heating from the acceleration of the exothermic chemical reactive processes without the need for an external energy source of ignition. The autoignition characteristics of fuel–air mixtures are of paramount importance from the point of performance, safety, and suitability of a fuel for spark ignition or compression ignition engine applications. Fuels for spark ignition automobile engines require a high resistance to autoignition so as to resist the tendency for the undesirable uncontrolled autoignition of part of the cylinder mixture, described as knock. However, for diesel engine operation, the fuel requires only low ignition temperatures since combustion needs to proceed compression ignition. Higher molecular weight hydrocarbons in general autoignite in engines more readily and earlier than lighter fuels. Increasing the initial mixture temperature or pressure lowers the autoignition temperature. Fuels in atmospheres containing higher concentrations of oxygen than in air autoignite more readily. For hydrocarbon fuels, the larger the number of carbon atoms, the lower the autoignition temperature. Also, fuels of the isomer variety have higher ignition temperatures than their corresponding straight-chain compounds.

9.3   DIFFUSION VS. PREMIXED COMBUSTION

There are basically two major types of combustion and flames. These are of the unpremixed fuel and air diffusion type and those of the premixed type. In diffusion combustion, the fuel and oxidants are initially either separate or not thoroughly premixed. The oxidation reactions and consequent flame propagation are then governed largely by the processes of interdiffusion and the resulting extent of mixing between the fuel and air. Commonly, the relative slowness of these mixing processes controls the rate of fuel burning and associated energy releases. This is because the physically controlled processes are approximately related linearly to local temperature, and they are much slower than those of the chemical reaction processes, which are essentially logarithmically dependent on temperature. In practice, diffusion flames, in comparison to the premixed type, have a relatively wide region over which the local composition changes significantly. Thus, they are more adaptable and tolerant of changes in the quality and type of fuel than the premixed type. Examples of diffusion type combustion can be seen in the combustion of fuel jets and droplets. Figure 9.4 schematically shows a jet of fuel entraining air and burning after discharging from the orifice of a burner. The concentration of the fuel as it travels to regions away from the orifice gradually decreases and becomes diluted through mixing with air entrained from the surroundings. After a certain distance from the point of discharge, a stoichiometric mixture envelope becomes located. Before this location, fuel-rich mixtures are located, while beyond it, increasingly lean mixture envelopes are developed. Hence, when the fuel jet is ignited, a diffusion flame will be formed and gets anchored while burning more readily around the most reactive fuel–air mixture region, which is available around the stoichiometric mixture envelope. The resulting local temperature varies both radially and axially with its peak values located around the stoichiometric mixture envelope. Also, at any horizontal plane of the jet there are corresponding radial variations in the concentrations of the fuel, oxygen, and products of combustion. Basically, a similar set of processes takes

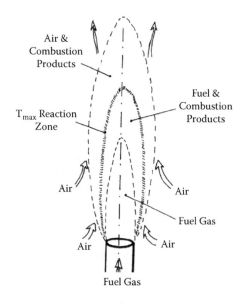

FIGURE 9.4 Fuel jet diffusion flame. (From Karim, G.A., *Fuels, Energy, and the Environment*, CRC Press, Boca Raton, FL, 2012.)

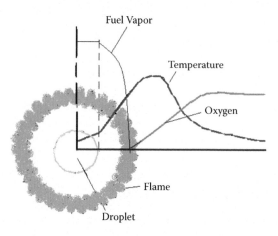

FIGURE 9.5 Schematic of diffusion combustion around a burning liquid fuel droplet showing the typical variations in temperature and the concentrations of fuel vapor and oxygen under quiescent flow conditions. (From Karim, G.A., *Fuels, Energy, and the Environment*, CRC Press, Boca Raton, FL, 2012.)

place around the burning of an evaporating liquid fuel droplet, as shown schematically in Figure 9.5. The liquid fuel, as it evaporates, diffuses outwardly to meet the surrounding air, forming a wide range of fuel vapor–air mixtures. On ignition, the resulting flame will be located around the stoichiometric mixture shell, with the fuel vapor on the inside and the air and combustion products outside.

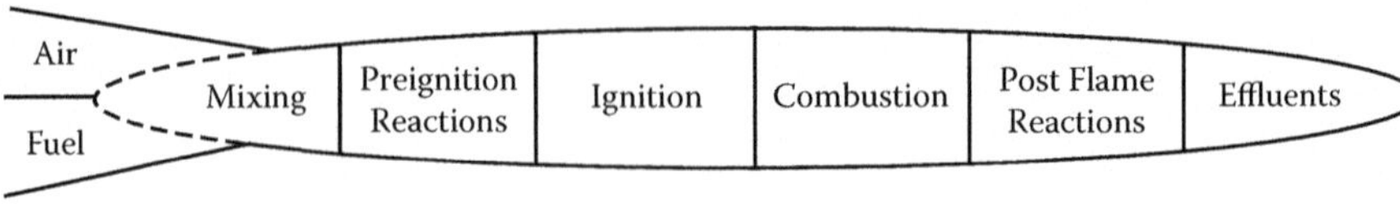

FIGURE 9.6 Representation of the main stages in the combustion of fuel gas in air.

In premixed combustion, where the fuel and oxidant have been mixed together before combustion, such as in the case of the common spark ignition gas engine, the associated chemical reaction rates are quite significant and controlling since the fuel and air are already mixed. Hence, the type of fuel and its relative concentrations in the air are exceedingly important, although the local flame speed is still strongly influenced by the local heat and mass transfer from the flame front region to the adjoining fresh reactant mixtures. For regions that are only partially premixed, additional fuel or oxidant may be required to initiate and undergo combustion. These examples demonstrate the complex interplay between the multitude of physical and chemical processes that bring about the combustion of the fuel, and reflect on the complexity of the combustion process in engines in general and the dual-fuel engine in particular.

In general, the chemical processes are significantly controlling in situations where they are comparatively slow compared to the physical mixing processes, such as at low temperatures. At sufficiently high temperatures, the oxidation reactions that are exponentially dependent on temperature tend to proceed quickly in comparison to the other physical processes. Occasionally, for the sake of simplicity, its influence on the rate of combustion may then be deemphasized or even ignored.

The length of a diffusion flame needs to be controlled to ensure that no excessive surface impingement and heat transfer will result in the buildup of unacceptably high temperature surfaces. Figure 9.6 shows a schematic representation of the typical stages in the combustion of fuel and air, showing their initial mixing followed by preignition reactions, which lead to ignition and flame propagation. Some post-flame reactions may continue while the products of combustion are of sufficiently high temperature.

9.4 DIESEL ENGINE COMBUSTION

The pilot jet flame of burning diesel fuel supplies energy and active oxidation products to the flowing mixtures of the fuel–air mixtures. The spread of the flame into adjoining mixtures produces sufficient thickening of the local jet flame to allow a flame to propagate, setting the whole flowing surroundings alight. When pilot flames are used, though they may be small in size, they are sufficient to provide the energy needed for combusting the surroundings. In dual-fuel engines, the contribution of the pilot to the combustion processes of the gaseous fuel will depend on many factors, which include its size, type, and quality of its fuel, such as in terms of its cetane number, injection timing, and spray characteristics.

The early stages of combustion process in the dual-fuel engine proceed in a heterogeneous manner despite the premixed nature of the bulk of the charge. Initially, much of the combustion takes place preferentially within the central regions of the cylinder, away from the walls. In the vicinity of the fuel jet boundaries, suitable conditions prevail in terms of the local fuel-to-air ratio and temperature combustion proceeds to completion usually in a relatively short time, aided by the turbulent movement of the premixed fuel and air. In regions of locally rich fuel–air mixtures, the oxygen gets rapidly consumed initially. It can then be followed by high-temperature endothermic reforming reactions that would produce a lowering of temperature while consuming the unconverted fuel and producing partial oxidation products. The heterogeneous nature of the mixture region in the neighborhood of the pilot flame is reminiscent of the stratified rich-lean mixture combustion approach that has been employed in some combustion devices to reduce the localized production of oxides of nitrogen. This effect contributes to the lowering of the overall production of oxides of nitrogen in comparison to straight diesel engine combustion. In general, since stoichiometric combustion is present in extended regions of the pilot with the premixed charge, there is a need to optimize the mixing and combustion processes so as to achieve superior performance and a lowering in NOx and particulate emissions levels.

To help in producing the right conditions for mixing and propagating the combustion process, entry conditions into the cylinder are made to produce an induction swirl, especially in the open chamber type diesel engines. This is accomplished, for example, through proper design of the air intake passages or occasionally by shrouding the intake valve. The extent of such measures is limited by reducing the amount of restriction suffered by the airflow and the resulting intake pressure drop. There is also a secondary contribution to the mixture motion within the cylinder due to squish-radial airflow toward the end of compression, especially aided by the proper shaping of the piston bowl.

The development of fuel spray following commencement of injection would be controlled primarily by physical factors, such as the breakup of the liquid fuel jet into small fragments and droplets with parcels of vapor, which is usually described as atomization, followed by its penetration, air entrainment, and vaporization. The growth of the spray is dependent on the fuel–air interaction through momentum exchange, with the mixing processes enhanced by the high-pressure fuel injection and increased turbulence of the swirling flow through the proper design of the geometry of the intake port and combustion chamber. The overall magnitude of the swirl velocities is some 10–15 times the engine speed.

The combustion process in engines of the regular diesel type may be viewed as proceeding regionally through the following main stages: ignition delay, followed by rapid combustion arising mainly from regions where there has been some mixing of the fuel vapor and air, often described as premixed type combustion. This is then followed by a more gradual combustion of the diffusion type and followed later by after-burning. Soot formation during the early part of the combustion process in diesel engines is a common occurrence, resulting from the heterogeneous nature of the high hydrocarbon fuel and the high localized combustion temperatures. However, most of the carbon produced is consumed during the later stages of combustion. Only when relatively large amounts of liquid fuel are injected, such as near full load, may some of this carbon survive in significant proportion to the exhaust stage.

The performance of diesel engines of recent design is closely controlled by numerous key design and operational variables. Some of these are as follows:

- Fuel type and its physical and chemical properties, fuel injection characteristics, and timing
- Various features of engine design, such as the compression ratio, turbocharging, exhaust gas recirculation (EGR), cogeneration, any throttling employed, valve timings, engine size, number of cylinders, speed, load ranges, and combustion chamber fluid flow and turbulence characteristics
- Specific operating conditions, such as overall equivalence ratio, percentage of total load, speed, and water jacket temperature

Much research and development have been successfully directed at reducing diesel engine emissions. These measures include improvements to combustion, the quality of the fuels used, and the injection characteristics. As an example, they include variable high fuel injection pressure and timing. Measures such as these allowed the diesel engine to make significant inroads in the relatively light duty and transport applications, including the automotive field.

In the direct injection diesel engine of the open chamber type, the combustion process involves mostly the interaction of the liquid fuel jet with the bulk air motion within the cylinder. This motion is normally brought by suitably sited and shaped inlet ports directing the incoming air in a swirling and circulating motion. This is subsequently modified through the motion of the piston, changes in the geometry of the cylinder, and heat transfer effects. Throughout compression and combustion there are no significant differences in pressure between the different regions of the chamber.

In the indirect injection or divided chamber engines, the fuel is injected in a separate chamber open and attached to the main cylinder. Some of the liquid jet breakup, atomization, and vaporization take place in the prechamber. A highly heterogeneous fuel vapor–air jet is then ejected into the main chamber. Air is forced into the constant volume prechamber, while the main chamber is decreasing markedly in volume during compression. Engine designs aim to have a smooth communicating passage of the right shape, orientation, and size so as to reduce potential work losses due to increased frictional and aerodynamic losses. After ignition, the rich mixture, products of combustion, partial oxidation products, and unconverted fuel within the prechamber are forced as a burning jet into the much leaner surroundings and air of the main chamber, permitting burning to proceed and continue more vigorously. Since mixing in comparison to the direct injection engines is less dependent on the momentum of the fuel jet, lower injection pressures may be used. Moreover, lower maximum pressures, rates of pressure rise, and noise are associated with the indirect injection engines, but they tend to have slightly lower efficiencies, higher thermal stresses, later burning, and are harder to start. These are mainly a consequence of the high heat transfer and work dissipation, with the flow of gases through the restriction between the prechamber and main chamber. Largely due to the mode of combustion where essentially rich mixture combustion is followed by lean combustion, emissions may be lower.

9.5   HEAT RELEASE ANALYSIS

The complex combustion processes of a typical diesel engine have been analyzed through a variety of approaches that vary widely in their complexity and usefulness. A widely used approximate approach is obtained from a simplified processing of the cylinder pressure ~ time record to produce an effective equivalent combustion energy release variation with time. The analysis infers that a gross characterization of the thermal consequences of the combustion process is possible by considering the observed pressure change to have been the outcome of the energy released via combustion and that of the heat transfer to the surroundings. This approach assumes that the equivalent combustion energy is released throughout the charge and may be considered to be akin to an equivalent transient heat supplied to the contents of the cylinder while accounting for the corresponding changes in their properties. Such an analysis is clearly simplistic, but represents a useful approach that interprets the overall features of the combustion process in compression ignition engines. In its simplest form, the equivalent rate of heat release is given in terms of the observed change in pressure and volume and the corresponding changes in properties of the system. However, further improvements to this simple analysis, such as when applied to consider the combustion process in a dual-fuel engine, may need some of the assumptions appropriately modified to consider a two-zone system separately representing those of the pilot combustion zone and those of the bulk gaseous fuel–air system, with a consideration of the interaction between them. Figure 9.7 shows an

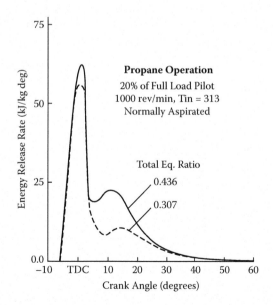

FIGURE 9.7   A typical energy release diagram for a propane-fueled engine operating on a constant value pilot of 20% of the corresponding full-load diesel operation as the total equivalence ratio increased from 0.307 to 0.436. (From Karim, G.A., and Rogers, A., *Journal of the Institute of Fuel*, 40, 513–522, 1967.)

example of the temporal development of the energy release rate for a propane-fueled dual-fuel engine. A relatively large pilot is employed with two values of the total equivalence ratio. An increase in the premixed portion of the heat release is evident.

The most significant sources of errors in the calculated rate of heat release in diesel engines are due to cylinder pressure measurements and associated errors caused by factors such as incorrect phasing between the pressure and crank angle signals, absolute pressure referencing offset, and electronic noise. Additionally, there are the fundamental errors associated with the heat release model due to deficiencies such as not accounting for the spatial variations in the gas properties and charge to wall heat transfer. These are especially important in considering the combustion process in dual-fuel engines. Nevertheless, some useful insight into the combustion process can be derived when using such a simplified approach.

A typical temporal variation in the values of the combustion heat release of a diesel engine is shown in Figure 9.8. It may be considered to be made up of the rapid energy releases due to ignition of the partially premixed fuel vapor followed by subsequent incremental diffusional burning, as further fuel segments are injected, ignited rapidly, and oxidized. But these injected fuel segments take increasingly longer time to be consumed due to the gradual depletion of oxygen, prevaporized and mixed fuel portions, and increased combustion products. Accordingly, the rate of heat release variation in diesel engines has been viewed to be reflecting grossly two different modes of fuel combustion (Figure 9.9). The first is of a predominantly autoignition premixed nature, while the second is the cumulative energy release due to progressive diffusional burning. The so-called premixed energy release part will vary in intensity and relative size, depending on the injection and length of the delay, while the diffusional part will be changing with the amount and rate of fuel injection and the associated mixing and combustion processes.

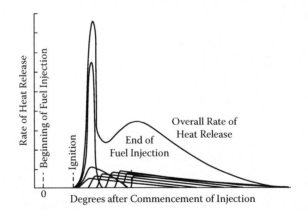

FIGURE 9.8 Schematic representation of the components of combustion heat release in a diesel engine. (From Gee, D., and Karim, G.A., *The Engineer*, 222, 473–479, 1966.)

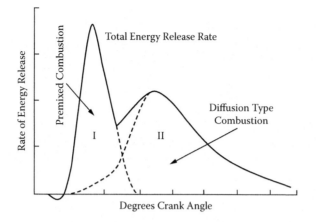

FIGURE 9.9 A representation of the premixed and diffusion type combustion as visualized through the heat release diagram of a diesel engine.

9.6 FLAMMABILITY AND COMBUSTION LIMITS OF GAS FUEL–AIR MIXTURES

The terms *lean*, *weak*, or *lower flammability limit* indicates the minimum concentration, usually by volume, of the fuel vapor needed in air that will support continued flame propagation from an adequate ignition source. This flame propagation is not considered to be within quiescent homogeneous fuel–air mixtures under a specified set of operating conditions. The corresponding maximum fuel concentration is referred to as the *rich*, *upper*, or *higher flammability limit*. Accordingly, the value of the flammability limit of a fuel is strictly restricted to apply only to a specific set of operating conditions that are far removed from what is available in engines during ignition. Similarly, the term *flame spread limit* relates to limit mixture condition for a flame to begin spreading under different prevailing conditions from an ignition source through the mixture. An increase in temperature widens the limits almost linearly, permitting leaner and richer mixtures to support flame propagation. Also, an increase in pressure widens the limits, while the presence of diluents with a fuel narrows the limits. Figure 9.10 shows a schematic representation of the typical lean and rich limits of flame propagation within a homogeneous mixture of fuel and air. The narrower mixture zone associated with autoignition is also shown.

Mixtures of fuel gases and air at the corresponding composition of the lean flammability limits and various initial mixture temperatures tend to be associated with an approximately constant value of flame temperature. The presence of a diluent with the fuel in lean limit mixtures can be viewed as a replacement of some of the excess air to bring the resulting mixture closer to the stoichiometric value. The maximum amount of diluent that can be tolerated and support flame propagation will be approximately that which will be at a stoichiometric fuel–air ratio within the total mixture. Accordingly, the flammable mixture region will narrow gradually

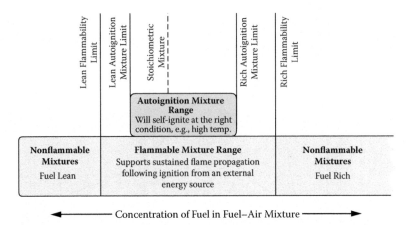

FIGURE 9.10 A representation of the range of flammable regions in a fuel–air mixture. (From Karim, G.A., *Fuels, Energy, and the Environment*, CRC Press, Boca Raton, FL, 2012.)

with the addition of increasing amounts of the diluent to the fuel. Preheating of mixtures would permit combustion with higher concentrations of diluents in the fuel. Moreover, sufficiently less lean mixtures may fail to propagate a flame when excessive heat loss is present as takes place in flame quenching.

The laminar burning velocity of a fuel in air is lowered as the concentration of a diluent added to the fuel is increased. This is mainly a consequence of the combined effects of the reduction in flame temperature, reaction rates, diffusivity, and transport properties of the mixture. For sufficiently fast turbulent combustible streams, flame propagation rates can be reduced since the rate of dissipation of the energy release by the flame can be much too fast to permit adequate time for the release of sufficient chemical energy so as to propagate the reaction further. Throughout, the stoichiometric mixture remains the most capable of supporting flame propagation. Moreover, the extent of random variation in the length of the flame initiation phase, as well as the subsequent flame propagation within the combustible flowing streams, increases in intensity with leaner mixture.

The employment of catalysts is a major common approach to enhancing the reaction rates in combustion processes. However, their use with methane as the fuel, in comparison to other common fuels, tends to be relatively limited. The combustion period, which is the time taken to complete the combustion process in an engine and can be considered a function of the inverse of the mean flame speed, increases as the concentration of the fuel is reduced or that of the diluent is increased. This is also reflected by a continued increase in exhaust temperature. Carbon dioxide would bring an earlier flame propagation failure than similar concentrations of nitrogen under the same operating conditions. There is some limited information about the effective lower operational mixture limits in engines, particularly for fuel-gas mixtures under various operating conditions. However, there are approximate approaches that can offer some guidance to these limits as a function of a number of key operational variables.

9.7   METHANE OXIDATION WITHIN FUEL MIXTURES

The course of the combustion of fuel–air mixtures such as those of methane–air involves the production and consumption of numerous transient species generated during the oxidization reaction, together with the interaction of numerous reaction steps, each with its own rate and reactive species. The net effect is to establish features of the combustion process, including the production and consumption rates of the different reacting species, both stable and unstable.

The time needed to autoignition of fuel–air mixtures is reduced logarithmically with mixture temperature. High temperatures are needed to effect rapid reactions to ignition and subsequent completion of the fuel conversion to products. This is particularly needed when substantial concentrations of diluents or combustion products are present. Examination of details of the set of reactions that go on during the oxidization of methane indicates that the presence of diluents with methane does not markedly affect the initiation reactions of its oxidation. Their presence will lead to a slowing down of the overall conversion of the fuel and the associated energy release rates to an extent that will depend on variables, such as the diluent involved, its concentrations, the temperature, and to a lesser extent, pressure. The interrupting of the full course of the combustion reactions can produce higher concentrations of carbon monoxide and lower concentrations of carbon dioxide, where the peak concentrations of carbon monoxide are not necessarily associated with fuel mixtures that contain very high concentrations of carbon dioxide with the methane. This reflects the important role of the resulting temperature level of the reactive mixture.

The rates of oxidation reactions of methane, as well as the control of eventual combustion products, may be enhanced through the homogeneous mixing with the reactants' small proportions of combustion products. Such mixing not only leads to a slight preheating of the reactive mixture, hence somewhat speeding up the reactions, but also provides small amounts of active species and radicals, leading to enhancement of the rates of the early stages of the reactions.

The presence of a small amount of higher hydrocarbon vapor, such as those vapors produced from the diesel fuel pilot with methane, reduces the ignition delay significantly, even without necessarily making a substantial change in temperature, the energy released, or the composition of the products. As shown, for example, in Figure 9.11, the presence with methane of small concentrations of n-heptane vapor, e.g., to an extent of a fraction of 1%, can bring about a substantial acceleration of the oxidation reaction of methane in air. Moreover, as shown, for example, in Figure 9.12, the nonmethane higher hydrocarbon fuel components in a typical natural gas react with the air far more rapidly and well ahead of the methane to release energy and radicals earlier in the course of the reaction. These lead to speeding up the overall reaction and reducing the ignition delay. These trends indicate a potential approach for improving the reactivity of fuel mixtures of methane even in the presence of diluents. It can be noted that the reaction initiation time represents the bulk of the whole reaction time since much of the early part of the ignition delay time is associated with relatively low temperature levels. The dominant role played by a change in mixture temperature is evident throughout.

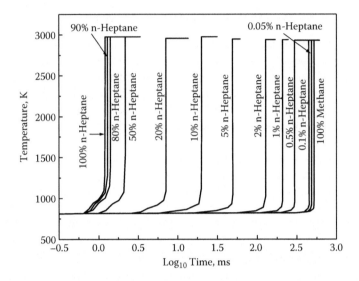

FIGURE 9.11 Calculated temperature rise with time under constant pressure for a range of fuel mixtures of varying concentrations of methane and n-heptane in stoichiometric air at an initial temperature of 760 K. (From Khalil, E., Modeling the Chemical Kinetics of Combustion of Higher Hydrocarbon Fuels in Air, PhD dissertation, University of Calgary, Canada, 1998.)

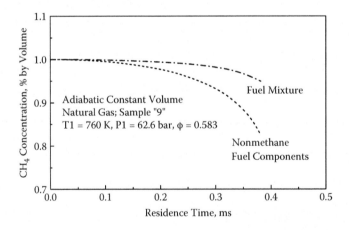

FIGURE 9.12 Variation of the normalized concentration of methane and the nonmethane components in a natural gas with time for an equivalence ratio of 0.583 and initial temperature of 760 K, at a constant pressure of 62.6 bar. (From Khalil, E., Modeling the Chemical Kinetics of Combustion of Higher Hydrocarbon Fuels in Air, PhD dissertation, University of Calgary, Canada, 1998.)

In general, the presence of some hydrogen or higher hydrocarbons with methane can increase the reaction rate and tolerance to the presence of diluents in the fuel. Moreover, with the presence of relatively small concentrations of diluents with the methane, little or sometimes even no changes are usually needed to equipment design and operation. However, with high concentrations, various changes to the engine and associated equipment may become necessary. These can include a need for dedicated injectors and the provision of an auxiliary supply of good quality fuel, either for starting and low-load applications or for supplementing the fuel mixture through direct mixing. Larger pilots for ignition will be also needed. On this basis, it is a common practice to have operation on the low heating value fuel-gas mixtures reserved primarily for relatively high and steady load operation. Additional measures may, when proven economical, include preheating the intake gas charge through mainly heat circulation from the exhaust gases.

REFERENCES AND RECOMMENDED READING

Aisho, Y., Yaeo, T., Koseki, T., Saito, T., and Kihara, R., *Combustion and Exhaust Emissions in a Direct Injection Diesel Engine Dual-Fueled with Natural Gas*, SAE Paper 950465, 1995.

Alcock, J.F., and Scott, W.M., Some More Light on Diesel Combustion, in *Proceedings of the Institute Mechanical Engineers (Auto Division)*, 1962, pp. 179–191.

Bade Shrestha, O.M., and Karim, G.A., The Operational Mixture Limits in Engines Fuelled with Alternative Gaseous Fuels, *ASME Journal of Energy Resources Technology*, 128, 223–228, 2006.

Bade Shrestha, S.O., Wierzba, I., and Karim, G.A., An Approach for Predicting the Flammability Limits of Fuel-Diluent Mixtures in Air, *Journal of the Institute of Energy*, 122–130, 1998.

Badr, O., El-Sayed, N., and Karim, G.A., An Investigation of the Lean Operational Limits of Gas Fuelled Spark Ignition Engines, *ASME Journal of Energy Resources Technology*, 118(2), 159–163, 1996.

Bartok, W., and Sarofim, A.F., *Fossil Fuel Combustion*, John Wiley & Sons, New York, 1991.

Bergman, H., and Busenthur, B., Facts Concerning the Utilization of Gaseous Fuels in Heavy Duty Vehicles, in *Proceedings of the Conference on Gaseous Fuels for Transportation*, August 1986, pp. 813–849.

Bols, R.E., and Tuve, G.L., eds., *Handbook of Tables for Applied Engineering Science*, CRC Press, Cleveland, OH, 1970.

Borman, G.I., and Ragland, K., *Combustion Engineering*, int. ed., McGraw Hill, New York, 1998.

Challen, B., and Barnescu, R., *Diesel Engine Handbook*, 2nd ed., SAE, Warrendale, PA, 1999.

Danyluk, P.R., Development of a High-Output Dual-Fuel Engine, *ASME Journal of Engineering for Gas Turbines and Power*, 115, 728–733, 1993.

Gee, D., and Karim, G.A., Heat Release in a Compression-Ignition Engine, *The Engineer*, 222, 473–479, 1966.

Golvoy, A., and Blais, E.J., *Natural Gas Storage on Activated Carbon*, SAE Paper 831678, 1983.

Gunea, C., Razavi, M.R., and Karim, G.A., *The Effects of Pilot Fuel Quality on Dual Fuel Engine Ignition Delay*, SAE Paper 982453, 1998.

Heywood, J., *Internal Combustion Engines*, McGraw Hill, New York, 1988.

Karim, G.A., Some Aspects of the Utilization of LNG for the Production of Power in Internal Combustion Engines, presented at Proceedings of the International Conference on Liquefied Natural Gas, Institute of Mechanical Engineers, 1969.

Karim, G.A., Combustion in Gas Fueled Compression Ignition Engines of the Dual Fuel Type, *ASME Journal of Engineering for Gas Turbines and Power*, 125, 827–836, 2003.

Karim, G.A., The Combustion of Bio-Gases and Low-Heating Value Gaseous Fuel Mixtures, *International Journal of Green Energy*, 8, 1–10, 2011.

Karim, G.A., *Fuels, Energy and the Environment*, CRC Press, Boca Raton, FL, 2012.

Karim, G.A., and Khan, M.O., Examination of Effective Rates of Combustion Heat Release in a Dual Fuel Engine, *Journal of Mechanical Engineering Science*, 10, 13–23, 1968.

Karim, G.A., and Khan, M.O., *Examination of Some of the Errors Normally Associated with the Calculation of Apparent Rates of Combustion Heat Release in Engines*, SAE Paper 710135, 1971.

Karim, G.A., and Rogers, A., Comparative Studies of Propane and Butane as Dual Fuel Engine Fuels, *Journal of the Institute of Fuel*, 40, 513–522, 1967.

Karim, G.A., and Ward, S., The Examination of the Combustion Processes in a Compression-Ignition Engine by Changing the Partial Pressure of Oxygen in the Intake Charge, *SAE Transactions*, 77, 3008–3016, 1968.

Karim, G.A., and Wierzba, I., Comparative Studies of Methane and Propane as Fuels for Spark Ignition and Compression Ignition Engines, *SAE Transactions*, 92, 3677–3688, 1983.

Khalil, E., Modeling the Chemical Kinetics of Combustion of Higher Hydrocarbon Fuels in Air, PhD dissertation, University of Calgary, Canada, 1998.

Khalil, E., and Karim, G.A., A Kinetic Investigation of the Role of Changes in the Composition of Natural Gas in Engine Applications, *ASME Journal of Engineering for Gas Turbines and Power*, 124, 404–411, 2002.

Lyn, W.T., Calculation of the Effect of the Rate of Heat Release on the Shape of Cylinder Pressure Diagram and Cycle Efficiency, in *Proceedings of the Institute of Mechanical Engineers*, London, 1961, pp. 34–37.

Obert, E.E., *Internal Combustion Engines and Air Pollution*, Harper & Row, New York, 1973.

Rose, J.W., and Cooper, J.R., eds., *Technical Data on Fuels*, 7th ed., British National Committee of World Energy Conference, London, 1977.

Samuel, P., and Karim, G.A., *An Analysis of Fuel Droplets Ignition and Combustion within Homogeneous Mixtures of Fuel and Air*, SAE Paper 940901, 1994.

Schiffgens, H.J., Brandt, D., Dier, L., and Glauber, R., Development of the New Man B&W 32/40 Dual Fuel Engine, in *ASME-ICE Division*, 1996, vol. 27-3, pp. 33–45.

Schiffgens, H.J., Brandt, D., Rieck, K., and Heider, G., Low-NOx Gas Engines from MAN B&W, in *CIMAC Congress*, Copenhagen, 1998, pp. 1399–1414.

Turner, S.H., and Weaver, C.S., *Dual-Fuel Natural/Diesel Engines: Technology, Performance and Emissions*, Topical Report, No. GRI-94/0094, Gas Research Institute, Chicago, November 1994.

Zabetakis, M., *Flammability Characteristics of Combustible Gases and Vapors*, Bureau of Mines Bulletin 627, U.S. Department of the Interior, Washington, DC, 1965.

10 The Conversion of Diesel Engines to Dual-Fuel Operation

10.1 ECONOMICS OF CONVERSION

An estimate of the costs of converting a diesel engine to dual-fuel operation can only be made on a somewhat tentative basis. There are a wide variety of diesel engines of different makes, sizes, and ages on the market and in operation. These have numerous design and operational features that make universal generic retrofit equipment hard to devise when premium performance with cost savings is targeted. The costs are also a function of the number of engines to be converted and the type, availability, and cost of the fuels that are to be used with them. The potential benefits that arise from the employment of gaseous fuels that can permit, in principle, some design change reductions in the specific energy consumption can help to contribute toward a reduction in overall costs. For example, the liquid fuel tank and other support equipment and facilities may be reduced significantly from those of regular engines, which operate exclusively on liquid fuels. However, for example, the variable cost differential between the two types of engines in transport applications would have to include the costs of additional items, such as those of the fuel containers, maintenance, and insurance, and the associated cost of fuel gas compression. Against these costs is the value of the benefits of having positive features that are associated with the operation of properly converted engines, such as those of the invariably cheaper gaseous fuel and the improved environmental features. The economy of operating engines, whether on natural gas or other gaseous fuels, is greatly influenced by other factors that may include unit load and plant load, hours operated per year, fuel cost and its variations, capital and operational costs, and the prevailing and future prices of liquid fuels and electric power.

Reported experience with a converted diesel engine to dual-fuel operation engines may not necessarily be representative of the potential performance that can be obtained with other types. Until relatively recently, modifications to engine design and operational features, when made so as to use cheaper alternative fuels, were not optimized commonly or sufficiently. Consequently, the merits of dual-fuel engine operation could not be demonstrated or assessed properly. This contrasts with the commonly employed liquid-fired diesel engines that have benefited immensely from the research and development efforts and the long operational experience accumulated over the years.

In gaseous fuel engine applications, the use of all forms of energy available needs to be maximized. This is whether it is in the form of the direct chemical energy of the fuel or the indirect energy arising from the state of the fuel, such as the compression work inherent in CNG applications or the low temperatures of LNG or LH_2. Also, in any evaluation of engine applications, the total energy consumption for producing the alternative fuel, such as that needed to compress the fuel gas or liquefy it, needs to be taken into account. Moreover, an important consideration in dual-fuel engine design and expectations of performance is the type of gaseous fuel used and the potential variations in its quality. Substantial differences may exist between the possible gaseous fuels and their mixtures that may become available.

The volumetric efficiency of the engine at wide-open throttle tends to determine the maximum power that can be developed at any set of operating conditions. Gaseous fuels such as methane or hydrogen, when introduced by fumigation into the engine intake manifold, displace some of the air that otherwise would be inducted by the engine. Also, the enhancement of the volumetric efficiency due to the vaporization cooling of the intake charge in liquid fuel applications is missing when gaseous fuels are employed, requiring remedial measures. Similarly, in the case of the conversion of two-stroke engines to dual-fuel operation, timed injection of the gas into the intake manifold or its injection directly into the cylinder needs to be made, rather than fumigating the fuel gas into the intake air system. Also, with turbocharged engines, an effective after-cooler may need to be used, with the compression ratio having to be reduced a little also.

10.2   THE REQUIREMENTS FOR CONVERSION
TO DUAL-FUEL OPERATION

In most dual-fuel engine applications, proper diesel engine performance needs to remain available so that the engine can be switched from dual-fuel operation back to diesel promptly and smoothly when required. Of paramount importance throughout is to keep the cost of conversion down, such as by avoiding as much as possible the making of relatively major modifications to the engine or its instillation, while maximizing the benefits arising from operating in the dual-fuel engine mode. This often represents a challenge, especially in transport applications, where there is a need to achieve superior economics combined with improved performance and lower emissions. The levels of durability and reliability of the converted engines need to be at least comparable to and may be superior to those of the corresponding diesel type. The diesel engine idle and overspeed governing equipment are normally maintained to ensure engine safe operation throughout in either mode. Many influencing factors need to be addressed effectively for the conversion of any diesel engine to dual-fuel operation, such as the following:

- The driving characteristics required, such as those of torque ~ speed, acceleration, frequent stopping and starting, and smooth operation with acceptable levels of the exhaust gas temperatures

- The quality and storage capacity of fuels required, the number and weight of containers, fuel and capital equipment costs, and fuel supply outlet availability
- Requirements for the associated controls of exhaust gas emissions, enhanced safety, reductions in noise, vibration, and wear with water requirements, consumption, and capacity
- Availability and costs of a suitable engine, the extent of its usage and availability of service and fueling centers, and lubrication requirements

Sufficiently wide availability of the fuel gas supplies is an important consideration. In countries where the natural gas supply infrastructure is well established, the fuel gas is delivered via the pipe supply system to the point of consumption, and with the fuel quality controlled and consistent. However, for standby power generation applications, the liquid fuel needs to be supplied and stored in sufficiently large tanks, which may be inconvenient and contributes to increased costs. Also, the liquid fuel supply may not get used for long periods of time, which can lead to fuel quality degradation. Other deteriorating effects may develop, such as valve seat recession requiring remedial measures to be taken, such as providing seat inserts and hardened cast iron cylinder heads. Also, the use of appropriate lubricating oils of high pH value may be needed.

The rate of pressure rise in the engine cylinder is a function of numerous factors. These include the type of gas used, equivalence ratio, size of pilot, injection timing, and turbulent state of the charge. Since most converted engines are required to continue to satisfy all the requirements as a diesel, these factors may not get optimized sufficiently for dual-fuel operation. For example, in transport applications it would be expected that the injection characteristics of the pilot needed when operating as a dual fuel may not necessarily be the same as those optimized for operation as a diesel. This would also be dependent on the nature and intensity of the turbulent flow of the charge for the two types of operation. The engine controls need to be capable of operating satisfactorily on a wide range of pilot-to-gas ratios. These are increasingly being achieved through computer controls to increasingly ensure satisfactory operation throughout.

Modern conversion kits are increasingly using electronic processors to sense and regulate the gas and diesel fuel flows. They can be programmed to promptly vary the ratio of the gaseous to liquid fuel flows, depending on operating conditions and fuel requirements. A typical example of a conversion kit may include an electronic programmable processor that continuously computes the rate of diesel fuel the engine would have used if it had been running on diesel only. The processor may then divert a certain portion of the liquid fuel and replace it with an equivalent amount of the gaseous fuel. Accordingly, to convert diesel engines to dual-fuel operation requires very careful consideration of numerous economic, technical, environmental, social, and operational key influencing factors. Some of these considerations, which may include the following, can be sufficiently overriding that the lowest cost solution occasionally may not be chosen.

- Total costs of equipment and associated installation, their expected durability and working life, maintenance frequency and its costs, fuel costs, availability, quality required and its potential flexibility current supply prices and future trends, recycling and secondhand value of kit equipment, desirable uniformity with other installations, and availability and relative cost of alternative competing sources

- Retrofitting possibilities and costs, associated design costs, ease of replacement of parts and equipment, avoiding increased complexity while maintaining the flexibility of equipment and operation, whether there is a need and the cost of special materials and lubricants, ensuring ease of control, and whether remote control of installations is feasible

- Standby, startability, and ease of shutdown, safety of operation, insurance costs, taxes, subsidies and incentives, manpower requirements and their quality, possibility of the sale of by-product heat and exhaust gas

- Environmental impact, current legislation nationally and internationally and possible future changes in regulations, noise and vibrations, corrosion and wear problems, choice of innovative vs. well-tried design and social acceptance

- The need and type of auxiliary power generation or supply, availability of water of the right quality and quantity

Some of the retrofit approaches offered in the past did not necessarily take full advantage of the potential benefits of dual-fuel operation in relation to normal diesel or sometimes even comparable spark ignition operation. The convenience, cost reductions, and other benefits offered, such as those of the significant reductions in NOx and particulate emissions, while using plentiful and cheaper fuel gas resources, were considered to be sufficiently worthwhile to forgo proper optimization of the converted engine. In general, the conversion to dual-fuel operation may often be viewed to be sufficiently attractive beyond merely the potential savings and economy in fuel costs.

There is a continuing need for advances in developing dual-fuel engine applications, such as in relation to fuel metering and injection systems, engine controls and exhaust gas reductions and after-treatment, on-board fueling systems, fuel storage vessels and delivery technologies, and other fuel gas compression and refueling systems.

10.3 SOME FEATURES OF CONVERSION TO DUAL-FUEL OPERATION

Because of the wide diversity of common diesel engine types and sizes, and to cut development times and costs, it has been desirable that the engine type chosen to be converted is representative of engines most commonly found in the transportation and power sectors. The cost of conversion needs to be kept low, and it is necessary to cut operational costs throughout. The conversion should avoid requiring significant modifications to the engine, and its operational life should not be undermined, and whenever possible, it should be extended beyond that associated with the conventional diesel engine. Any modifications to the engine need to be as simple and as few as possible since often there is insufficient space for installing voluminous devices. Efforts should also be made to have the overall specific energy/fuel consumption kept lower, and at least not in excess of that associated with normal diesel operation.

Also, the extent and mode of exhaust gas emissions should not be inferior and kept even somewhat superior to those encountered with normal diesel operation.

The gaseous fuel intake delivery system in the premixed dual-fuel engine type should not produce unacceptably high pressure drop or excessive cooling in the intake manifold. The volumetric efficiency of the engine should not be undermined by the conversion of the engine to gaseous fuel operation. Turbocharged operation retained with dual-fuel engine operation is to provide adequate and effective power supplement over the whole operational range while ensuring effective safety controls. Throughout, it is desirable to have the conversion capable of controlled variations in the relative size of the pilot, its injection timing, and its injection characteristics to ensure optimum performance, while tolerating some variation in the composition of the gaseous fuel supplied.

It was often expedient to have much of the conversions of diesel engines undertaken using multipoint natural gas injection at each intake port, while using the existing diesel engine system to inject a small quantity of diesel fuel to function as the ignition source. At idling and low loads, the engine runs on diesel fuel only. At high loads, the quantity of diesel fuel injected is increasingly reduced and replaced by natural gas under control of the engine computer. The natural gas injection system was installed in the intake manifold, and the electronic controls added to the fuel pump. Usually a minimum modification to the original form of the diesel engine was aimed at. This way it was possible that simultaneous improvements to both power and emissions could be attained. There is a continuing need for a more effective, less costly, and trouble-free conversion of the recently much improved diesel engines. It pays to remember that the conventional internal combustion engine in general and diesel engines in particular have taken many decades of research, development, and experience to reach the present state of reliability and superior quality of performance.

10.4 ENGINES DEDICATED WHOLLY TO DUAL-FUEL OPERATION

In dual-fuel engine applications, it is generally assumed that the capacity of the engine to operate in its diesel mode must not only be retained but also must not be undermined in any way by conversion. This would include its prompt availability as a diesel power plant whenever needed during dual-fuel applications. It is also desirable to ensure optimum dual-fuel engine performance, which would include the capacity to operate on gaseous fuels at as large a percentage of the fueling as possible, and use as little pilot diesel fuel as possible. The engine also needs to retain the ability to switch between dual-fuel and diesel operations freely without interruption to either the output or its quality. The requirement to maintain the capacity to perform as a diesel engine that abides by the numerous accepted limiting requirements can represent a serious barrier to achieving the full potential of dual-fuel operation. Should an engine be suitably designed and operated optimally as a dual-fuel engine right from the start, then the requirement of its capacity to change to diesel operation may need to be relaxed and confined to only very short durations or merely as an emergency measure. On this basis, a properly optimized engine for dual-fuel operation can yield superior performance characteristics in many respects compared to its diesel version. Of course, all aspects of dual-fuel operation must be rendered absolutely safe throughout and retain the capacity for effective and prompt engine control in an emergency.

It is becoming increasingly evident that in order to make best use of the gaseous fuel resource, engines that are fully dedicated to gas operation need to be developed and optimized in design and operation primarily for dual-fuel application. This needs to be made even if the facility to convert engine operation promptly to diesel operation is sacrificed or its corresponding diesel engine performance not ensured fully under all operating conditions. To achieve this would require further improvements in the understanding and management of the combustion processes under dual-fuel operation by engine designers, manufacturers, and operators. An engine suitably designed and exclusively operated as a dual fuel should be associated with the following positive features:

- A freedom from the constraints associated with diesel operation would permit various degrees of effective optimization. As an example, if a vehicle is dedicated to methane operation, then there will be some cost cutting and savings arising from the simpler controls and the removal of a large liquid fuel tank, simplifying the fuel injection system and part of the emission control facilities. The performance may be especially suitably dedicated to limited operating conditions, such as those in electric power generation or hybrid operation.
- Once the dual-fuel engine is free from having to retain the capacity to operate satisfactorily as a diesel, then the full potential advantage of the high resistance to knock of methane can be exploited. In any case, it should be possible to have with the dedicated engine to dual-fuel operation a much simpler engine design and controls leading to lower capital and operational costs.

It is suggested that with the high-compression ratio reciprocating internal combustion engines of current design, when fueled with natural gas and properly operated as a dedicated dual-fuel engine, few if any contemporary power sources, including those of the diesel engine type, can successfully compete with them on the basis of economic returns per capital invested. The economics of the combination of a cheap fuel gas and an efficient diesel engine should often prove to be a very attractive investment opportunity while offering environmental and resource clear advantages.

REFERENCES AND RECOMMENDED READING

Addy, J.M., Bining, A., Norton, P., Peterson, E., Campbell, K., and Bevillaqua, O., *Demonstration of Caterpillar C10 Dual Fuel Natural Gas Engines in Commuter Buses*, SAE Paper 2000-01-1386, 2000.

Barbour, T.R., Crouse, M.E., and Lestz, S.S., *Gaseous Fuel Utilization in a Light Duty Diesel Engine*, SAE Paper 860070, 1986.

Beck, J., Karim, G.A., Mirosh, E., and Pronin, E., Bus Fuel Efficiency Local Emissions and Impact on Global Emissions, in *NGV 96 Conference and Exhibition*, Kuala Lumpur, Malaysia, October 1996.

Beck, N., Johnson, W., George, A., Peterson, P., vander Lee, B., and Klopp, G., *Electronic Fuel Injection for Dual Fuel Diesel Methane*, SAE Paper 891652, 1989.

Bell, S., *Natural Gas as a Transportation Fuel*, SAE Paper 931829, 1993.

Beppu, O., Fukuda, T., Komoda, T., Miyake, S., and Tanaka, T., Service Experience of Mitsui Gas Injection Diesel Engines, in *Proceedings of CIMAC Congress*, Copenhagen, 1998, pp. 187–202.

Bergman, H., and Busenthur, B., Facts Concerning the Utilization of Gaseous Fuels in Heavy Duty Vehicles, in *Proceedings of the Conference on Gaseous Fuels for Transportation*, August 1986, pp. 813–849.

Blizzard, D., Schaub, F.S., and Smith, J., Development of the Cooper-Bessmer Clean Burn Gas-Diesel (Dual Fuel) Engine, in *ASME-ICE Division*, 1991, vol. 15, pp. 89–97.

Blyth, N., Development of the Fairbanks Morse Enviro-Design Opposed Piston Dual Fuel Engine, presented at ASME-ICE Division, 1994, Paper 100375.

Brogan, T.R., Graboski, M.S., Macomber, J.R., Helmich, M.J., and Schaub, F.S., Operation of a Large Bore Medium Speed Turbocharged Dual Fuel Engine on Low BTU Wood Gas, in *ASME-ICE Division*, 1993, vol. 20, pp. 51–66.

Callahan, T., Survey of Gas Engine Performance and Future Trends, presented at Proceedings of ASME Conference, ICE Division, Salzburg, Austria, 2003, Paper ICES2003-628.

Challen, B., and Barnescu, R., *Diesel Engine Handbook*, 2nd ed., SAE, Warrendale, PA, 1999.

Checkel, M., Newman, P., van der Lee, B., and Pollak, I., *Performance and Emissions of a Converted RABA 2356 Bus Engine in Diesel and Dual Fuel Diesel/Natural Gas Operation*, SAE Paper 931823, 1993.

Danyluk, P.R., Development of a High Output Duel Fuel Engine, *ASME Journal of Engineering for Gas Turbines and Power*, 115, 728–733, 1993.

Dardalis, D., Matthews, R.D., Lewis, D., and Davis, K., *The Texas Project, Part 5—Economic Analysis: CNG and LPG Conversions of Light-Duty Vehicle Fleets*, SAE Paper 982447, 1998.

Ding, X., and Hill, P.G., *Emissions and Fuel Economy of a Prechamber Diesel Engine with Natural Gas Dual Fuelling*, SAE Paper 860069, 1988.

Ericson, R., Campbell, K., and Morgan, D., Application of Dual Fuel Engine Technology for On-Highway Vehicles, presented at Proceedings of ASME-ICE Division, Salzburg, Austria, 2003, Paper ICES2003-586.

Gettel, L., and Perry, G.C., *Natural Gas Conversion Systems for Heavy Duty Truck Engines*, SAE Paper 911663, 1991.

Grosshans, G., Development of a 1200 kW/Cyl. Low Pressure Dual Fuel Engine for LNG Carriers, in *Proceedings of CIMAC*, Copenhagen, 1998, pp. 1417–1428.

Heenan, J., and Gettel, L., *Dual Fueling Diesel/NGV Technology*, SAE Paper 881655, 1988.

Herdin, G.R., Gruber, F., Plohberger, D., and Wagner, M., Experience with Gas Engines Optimized for H2-Rich Fuels, presented at ASME-ICE Division, 2003, Paper ICES2003-596.

Ishida, M., Chen, Z.L., Luo, G.F., and Ueki, H., *The Effect of Pilot Injection on Combustion in a Turbocharged D.I. Diesel Engine*, SAE Paper 841692, 1994.

Ishyama, T., Shioji, M., Mitani, S., Shibata, H., and Ikegami, M., *Improvement of Performance and Exhaust Emissions in a Converted Dual Fuel Natural Gas Engine*, SAE Paper 2000-01-1866, 2000.

Maxwell, T., and Jones, J., *Alternative Fuels: Emissions, Economics and Performance*, SAE, Warrendale, PA, 1995.

Mayer, R., Meyers, D., Shahed, S.M., and Duggal, V.K., *Development of a Heavy Duty On-Highway Natural Gas Fueled Engine*, SAE Paper 922362, 1992.

Ogden, J.M., Williams, R.H., and Larson, E.D., Societal Lifecycle Costs of Cars with Alternative Fuels/Engines, *Energy Policy*, 32, 7–27, 2004.

Quigg, D., Pellegrin, V., and Rey, R., *Operational Experience of Compressed Natural Gas in Heavy Duty Transit Buses*, SAE Paper 931786, 1993.

Rain, R.R., and McFeatures, J.S., *New Zealand Experience with Natural Gas Fuelling of Heavy Transport Engines*, SAE Paper 892136, 1989.

Schiffgens, H.J., Brandt, D., Dier, L., and Glauber, R., Development of the New Man B&W 32/40 Dual Fuel Engine, in *ASME-ICE Technical Conference*, 1996, vol. 27-3, pp. 33–45.

Schiffgens, H.J., Brandt, D., Rieck, K., and Heider, G., Low-NOx Gas Engines from MAN B&W, in *CIMAC Congress*, Copenhagen, 1998, pp. 1399–1414.

Sinclair, M.S., and Haddon, J.J., *Operation of a Class 8 Truck on Natural Gas/Diesel*, SAE Paper 911666, 1991.

Turner, S.H., and Weaver, C.S., *Dual-Fuel Natural/Diesel Engines: Technology, Performance and Emissions*, No. GRI-94/0094, Topical Report Gas Research Institute, November 1994.

Ursu, B., and Perry, C., *Natural Gas Powered Heavy Duty Truck Demonstration*, SAE Paper 961669, 1996.

Zaidi, K., Andrews, G., and Greenhough, J., *Diesel Fumigation Partial Mixing for Reducing Ignition Delay and Amplitude of Pressure Fluctuations*, SAE Paper 980535, 1998.

11 The Diesel Fuel Pilot

11.1 THE FUNCTION OF THE PILOT

Mainly for economic and environmental reasons, the size of the diesel fuel pilot needs to be minimized relative to that of the cheaper fuel gas. This is to be done while maintaining optimum combustion and engine performance characteristics. When the regular diesel engine injection system is retained and employed for pilot delivery in dual-fuel operation, the desirable size of the pilot may be too small and outside the normal operating ranges of the injector and fuel pump systems. The pilot quantity often may be much below the amount of fuel needed to keep the diesel engine idling at the same speed, which leads to off-design operation and poor injection system performance. The employment of such small quantities can contribute to increased cyclic variations accentuated by variations in the mass of fuel injected per cycle, with the need to improve heat transfer, avoid overheating of the injector tip, and guard against occasional backfire. Reducing the relative size of the pilot will also be associated with conditions where the injection of the fuel may be completed well within the ignition delay period. With further increases in the pilot fuel size, longer injection times will be involved, which then tend to produce only secondary performance effects. With the employment of relatively large pilots (e.g., greater than 30% of normal diesel operation), as shown, for example, in Figure 11.1 for a dual-fueled engine operating on propane, engine performance with respect to power and specific energy consumption can become superior even to those corresponding values of normal high load diesel performance.

Figure 11.2 also shows a dual-fuel engine operating on a gaseous fuel mixture of methane and nitrogen. Increasing the size of the pilot reduces the length of the delay period over the fuel mixture range. Improvements in the fuel ignition characteristics, such as through employing a higher-cetane-number liquid fuel, can permit economy in the size of the pilot employed for the same engine performance. For dedicated dual-fuel engine applications, in order to obtain the desired performance, a special pilot fuel injection system needs to be specially developed and installed, in addition to the injection system of the engine for conventional diesel operation. This would ensure proper fuel injection characteristics while minimizing the amount of pilot fuel relative to the fuel gas component. The capacity to operate as a diesel engine is retained with adequate cooling of injectors and reductions in exhaust emissions. Figure 11.3 shows a dual-fuel application with the main fuel nozzle of engine as a diesel is retained, while a second injector exclusively provides the pilot. Two suitably independent fuel pumps are provided.

The proper functioning of the pilot is a key requirement for achieving the potential attractive characteristics of the dual-fuel engine. The effects of changes in a number

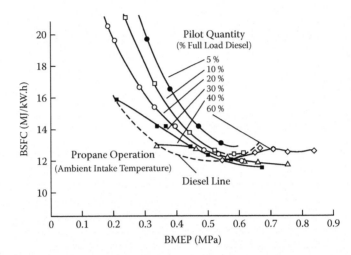

FIGURE 11.1 An example of the variation with pilot size in the brake specific fuel consumption (BSFC) with brake mean effective pressure (BMEP) of a dual-fuel engine operating at constant speed on propane at atmospheric conditions. (From Karim, G.A., and Rogers, A., *Journal of the Institute of Fuel*, 40, 513–522, 1967.)

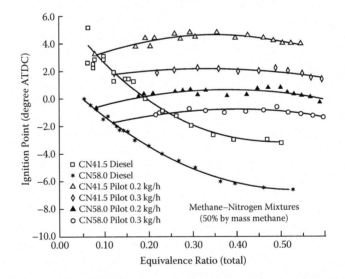

FIGURE 11.2 Variations in the ignition point with total equivalence ratio for a dual-fuel engine operating on methane–nitrogen mixtures at constant speed and ambient operating intake conditions for a number of cases of different size pilots and different pilot cetane number quality. (Adapted from Gunea, C., Examination of the Pilot Quality on the Performance of Gas Fueled Diesel Engine, MSc thesis, University of Calgary, Canada, 1997.)

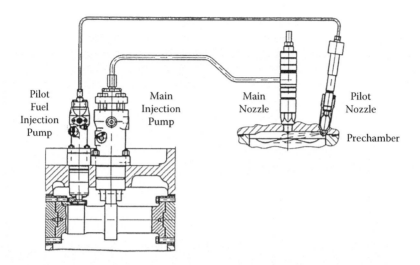

FIGURE 11.3 An example of a dual-fuel engine having a dedicated pilot fuel injector in addition to the central main regular diesel injector, with two separate fuel pumps. (From Schiffgens, H.J., Brandt, D., Rieck, K., and Heider, G., Low-NOx Gas Engines from MAN B&W, in *CIMAC Congress*, Copenhagen, Denmark, 1998, pp. 1399–1414.)

of influencing factors must be considered and attended to effectively. These will include the pilot injection characteristics and the liquid fuel physical and chemical properties, including its cetane number. The design configurations of the pilot injector are also important, such as the location of the injector into the cylinder head, number of jets and their orientation angles, orifice size employed, fixed or variable injection timing, length of injection period, single or multiple injections within the cycle, cooling and overheating characteristics of the injector, and whether in line, jerk pump, or common rail injection systems are employed. Of course, these major variables of the fuel injection process control the mixing processes prior to and subsequent to ignition, the length of the ignition delay, inflammation, energy release rates, and exhaust emissions. Over the range of pilot quantities that may be used in diesel engine conversion to dual-fuel engine applications, the rates of preignition processes appear to increase almost proportionally with the size of the pilot. This is brought about by the increase in the number of potential ignition centers produced and the associated improvement in the injection characteristics, such as fuel jet penetration and its atomization and distribution. These lead to improvements in the mixing of the pilot fuel vapor and the cylinder mixture, producing higher pressures and temperatures. There is also proportionally less pilot fuel vapor dissipation, down to concentrations much too low for ignition and flame propagation. The associated bigger energy also releases lead to hotter surfaces that are increasingly produced through higher charge temperatures and the possibility of some fuel burning off surfaces. Hotter residuals are also produced, with a greater production of active species that are also to be found within the residual gas.

Large pilots will be associated with longer injection periods that would provide more time for ignition to take place. A slightly earlier pilot injection advance may be employed in dual-fuel engines operating on methane, in comparison to those operating on other gaseous fuels, such as propane and hydrogen. Methane reactions tend to be somewhat slower and require more time to ignition, while being more resistant to autoignition and knock. Usually, the size of the pilot that can be effectively employed at high loads tends to be reduced to avoid the incidence of knock within the high-compression-ratio cylinder charge. Also, it serves to prevent excessive emissions, especially those relating to oxides of nitrogen, unburned hydrocarbons, particulates, and carbon monoxide.

11.2 PILOT SIZE CONSIDERATIONS

Some of the main motivating factors for reducing the size of the pilot relative to that of the fuel gas in dual-fuel engines applications are

- The reason is almost always primarily economical. The price of the liquid fuel tends to be greater than that of the fuel gas component, making operation as a dual fuel more economically attractive than when operating exclusively as a diesel engine with liquid fuels.
- Environmental factors are becoming increasingly important since lower undesirable exhaust emissions can be obtained with dual-fuel operation, compared to diesel or even operation with a large pilot. There is a very substantial reduction in the emissions of carbon and particulates, with a significant reduction in the production of oxides of nitrogen. This is mainly since the bulk of the energy release is produced by the combustion of very lean gaseous fuel–air mixtures and the associated reduction in the size of the very hot combustion zone, which mainly involves the liquid diesel pilot fuel.
- With most common fuel gases there will be a reduction in the overall carbon-to-hydrogen ratio of the fuel system in comparison to diesel operation, together with a potential for lower specific fuel consumption. These will lead to relatively less emissions of carbon dioxide, a greenhouse gas. There is additionally the potential for improved energy efficiency through the employment of cogeneration, aided by the exhaust gas of dual-fuel engines tending to be cleaner yet relatively a little hotter.
- Reducing the pilot size can lead to improvements in engine durability, reduced maintenance costs, and quieter running. There will also be less heat transfer, lubricant degradation, and surface deposition of the liquid fuel, leading to its surface ignition and combustion.

At the same time, there are a number of potentially limiting features associated with the employment of very small pilots in premixed gas-fueled dual-fuel engines that need remedial action:

- There is the potential for increased concentrations of the unburned fuel gas methane and carbon monoxide at light load as a result of operation on very lean gas fuel–air mixtures. However, although methane as a pollutant is less reactive and does not contribute actively to the production of photochemical smog, which is a desirable feature, it is a potent greenhouse gas. Its release into the atmosphere without sufficient control will be increasingly objectionable as a potential contributor to global warming.

- The low reactivity of the unconverted methane present in the exhaust gas emissions makes it less amenable to catalytic oxidation than higher hydrocarbons by conventional catalytic converters. However, its low flame propagation rates, especially with lean mixtures, produce longer combustion periods, which tend to lead to relatively hotter exhaust gas, which can assist in speeding up catalytic oxidation reactions in general.

- It is difficult to maintain good performance at very light load, as it will result in gas-fumigated or port-injected engines having to operate on very lean mixtures. This will lead to lower effective energy efficiencies and power outputs. The control of the fuel injection and combustion processes becomes more demanding, and engine performance in the diesel mode may suffer.

- The use of a sufficiently small amount of liquid fuel pilot injection may require a redesign of the combustion chamber with added costs and increased complexity of control. Hence, very small pilots tend to be more manageable in large-bore stationary engines with long periods of steady running at high loads, which can produce substantial savings in fuel costs.

11.3 REDUCTION OF THE RELATIVE SIZE OF THE PILOT

An important question that needs to be answered is: What is the minimum size of the pilot needed for the ignition of a lean fuel–air mixture in dual-fuel engines? Also, why does the energy of the pilot often appear to be orders of magnitude larger than that needed in the electric spark commonly employed in gas engines? Most probably, it is due to the ultra-high temperature produced within the spark region, coupled with the plasma-like electrically generated radical composition, localized quiescent heat release, and ultra-fast localized energy releases. These are to be compared to the ignition process in diesel and dual-fuel engines of much slower but quite large energy releases that take place over a much bigger mass of the mixture. The penetration of the injected liquid fuel jet helps to promote good fuel–air mixing, while its overpenetration can lead to surface impingement with adverse engine performance and increased exhaust emissions. The development of the fuel spray and its breakup into small fragments following the commencement of injection penetration, entrainments, and vaporization can be controlled through appropriate changes mainly to physical factors, such as through the fuel–air momentum exchange.

The simplest approach to achieve the desired reduction in the relative size of the pilot is through using the diesel engine injector and fuel pump assembly, except this will restrict the reduction of the size of the pilot. Such an approach will likely lead, especially at light load, to increased cyclic variations, long ignition delays, increased emissions of methane and carbon monoxide, and a drop in power, adversely affecting the performance of the engine as a diesel. However, remedial measures may be considered, which would include some changes in the design and geometry of the fuel injector. Examples of these can include changes in the injection pressure and timing, the number of holes of the injector and its orientation, and ensuring adequate cooling and injector cleanliness without wall deposition of the fuel. Some other measures can include paying greater attention to the cleanliness and quality of the liquid fuel employed, including the use of a higher-cetane-number fuel or suitable additives. It is necessary to ensure that the fuel injection takes place into the hottest region of the cylinder, and to consider a form of skip firing by varying the size of the pilot between cylinders and periodically over a number of cycles. A major approach is to use a specially designed smaller injector located either separately, in the same body as the regular injector, or in a specially manufactured sleeve body with the same dimension to facilitate its housing in the original space within the engine cylinder head. Yet another approach is to install a small prechamber that can accommodate a small injector such that the injection of the small pilot will lead to ignition within this chamber. The products of its combustion can then exit rapidly into the main chamber to ignite the premixed lean gas–air mixture. Figure 11.4 shows a typical prechamber diesel engine converted to dual-fuel operation with pilot injection, while Figure 11.5 shows an example of a combined fuel injector body for both the pilot and main liquid fuel injection for diesel operation.

Special attention needs to be given to ensure that there is adequate penetration of the liquid fuel jet while preserving its integrity, without premature overdispersion down to concentration levels where the resulting mixtures become too lean to ignite regularly or satisfactorily.

The recent development and widespread availability of common rail liquid fuel systems in modern diesel engines of various types and sizes has helped to improve

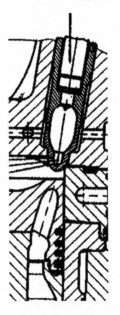

FIGURE 11.4 An example of a prechamber diesel engine converted to dual-fuel operation with pilot injection. (From Schiffgens, H.J., Brandt, D., Rieck, K., and Heider, G., Low-NOx Gas Engines from MAN B&W, in *CIMAC Congress*, Copenhagen, Denmark, 1998, pp. 1399–1414.)

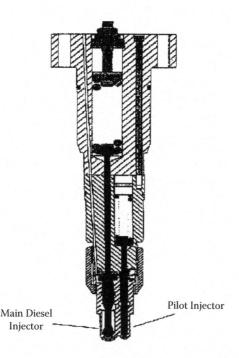

Main Diesel Injector   Pilot Injector

FIGURE 11.5   A combined fuel injector for both the dual-fuel engine pilot and the liquid fuel for diesel operation. (From Challen, B., and Barnescu, R., *Diesel Engine Handbook*, 2nd ed., SAE, Warrendale, PA, 1999.)

the functioning of dual-fuel engines. For example, these injection systems permit much wider variations in the fuel quantity injected, as well as the start and rate of its injection. Very high injection pressures (e.g., 150–1400 bar) with multiple injections can be employed. Such flexibility and quality in liquid fuel injection makes it possible to achieve significant improvements to the corresponding duel-fuel operation, especially for transport engine applications, compared to that with conventional liquid fuel jerk pump injection.

With the application of common rail liquid fuel injection, it is becoming increasingly possible to inject very small fuel quantities in the range of 1–5% of the diesel engine full-load quantity, in most engine sizes and designs.

11.4   PILOT FUEL DELIVERY SYSTEMS

The injection pressure of cam-driven liquid fuel systems for in-line pumps, electronically controlled unit pumps, or unit injectors is a function of engine speed and not an independent variable. Common rail systems of recent design, which consist of a low-pressure fuel supply pump, high-pressure pipes connected to the injectors in each engine cylinder, and an electronic control unit system, are capable of higher injection pressures and producing suitably shaped injection rates combined with the facility of variable injection timing. These measures have helped to reduce the extent of the emissions of NOx and particulates. In approaches where the whole system is managed by an electronic control unit, the rail functions as a pressure accumulator as well as a damping volume for pressure oscillations within the system. In engines that are wholly dedicated to dual-fuel operation, it is expected that the injector and its system will be optimized for such operation, which can lead to the development of injectors that display notable differences from those developed for conventional diesel operation.

The prime function of the pilot liquid fuel spray in dual-fuel engines is the reliable and deliberate control of the ignition of the gaseous fuel–air mixture. It is not as employed in regular diesel engine operation where the continued injection of the liquid fuel merely supplies it for continued combustion and providing the remaining part of the energy release. Accordingly, the optimum control of the key parameters of liquid fuel pilot injection often may not necessarily be the same as those for conventional diesel engine operation. This is often overlooked in practice since the engine is normally required to remain available to operate wholly in the diesel mode whenever required. For example, unlike most conventional diesel engine operations, where relatively fixed injection timing is usually employed, changes in the injection timing of the pilot in dual-fuel operation can be made beneficial and have the potential for bringing about further improvements in engine performance. This is becoming increasingly possible, especially with the latest development in diesel injection systems employing the common rail system, albeit adding further complexity and cost to proper engine control and associated hardware.

Since most dual-fuel engines have been developed on the basis that they need to have the capacity to revert promptly, whenever required, to the regular diesel engine mode of operation, then from a practical point, conventional diesel fuel injection systems are usually employed. This tends to impose a limit to how small a pilot needs to be for satisfactory engine performance. On this basis, the minimum pilot fuel quantities have been largely limited to well over 5% of the corresponding full-load liquid fuel quantity when operating on diesel alone. However, often, in many dual-fuel engine applications, especially where frequent variations in operating conditions are involved, such as in transport applications, the diesel pilot quantity tends not to be reduced below 10 or 15% of the total fuel energy input. The employment of such relatively large pilots can have some merits, such as lowering unburned hydrocarbon emissions and avoiding excessive nozzle injector overheating. But, it is also associated with limitations, such as leading to early onset of knock, overheating, and increased NOx emissions.

The using of two independent injectors produces the most favorable performance as a dual-fuel engine with very high diesel fuel replacement. It also ensures adequate cooling of the injector and contributes to the reduction of exhaust emissions. However, this approach tends to add to capital and operational costs and would require suitable modifications to engine controls and the combustion chamber to accommodate the second injector and the addition of a suitable fuel pump. When a different type of liquid fuel may be used as the pilot, then further complexities are introduced that make such a practice relatively less common. Accordingly, this mode of conversion of a typical diesel engine can be tedious and add to costs.

11.5 MICRO-PILOT APPLICATIONS

In order to substantially reduce the relative size of the pilot and implement additional improvements to dual-fuel combustion, the so-called micro-pilot injection is employed. This is associated with the injection of the pilot made directly into a small prechamber attached to the cylinder head, as shown in Figure 11.4. The use of such a system greatly maximizes the use of the gaseous fuel relative to the liquid fuel, while maintaining engine speed and load ranges similar to those of conventional diesel engines. Such arrangements can also lead to higher efficiency and lower pollutant values relative to those of the corresponding diesel engine operation.

In comparison, the employment of a relatively large liquid fuel injection with a much smaller proportion of gaseous fuels is less widespread. This is mainly a reflection of the common desire for fuel cost reduction reasons to maximize the exploitation of the gaseous fuel resource while economizing liquid fuel usage. However, there are occasions when dual-fuel engines may operate in this mode, such as under transient operating conditions or when operating on very lean gaseous fuel–air mixtures, leading to a reduction in the concentrations of unconverted fuel and particulates into the exhaust gas while maintaining high energy efficiency.

Specially developed micro-pilot injectors can have the pilot quantity reduced down to around 1% of the energy supplied at full load. Usually, the reduction in the size of the pilot that can be safely employed may become limited so as to avoid the incidence of knock within the cylinder charge when operating with some gaseous fuels and highly turbocharged engines, or at light load produce excessive emissions, especially those relating to unburned hydrocarbons and carbon monoxide.

There are more recent injection systems that employ a combined injector body for the injection of the pilot diesel, soon followed by the injection of high-pressure gaseous fuel. Hydraulically actuated electronically controlled unit injection (HEUI) systems have also been developed and applied for dual-fuel applications. There are obvious advantages to having two fuel injection systems integrated in one injector body of the same size and shape as the original diesel injector. Figure 11.6 shows an example of a combined injector body for the separate injection of liquid and gaseous fuels in dual-fuel applications. With the injection of small liquid fuel pilot, the mixing rate tends to be of a lower level, produced partly by the effect of the small amount of liquid fuel jet injected, but enhanced through the high-pressure fuel injection. Increased turbulence and swirling flows within the combustion chamber are produced mainly through intake port and combustion chamber geometry. Recent development with proper programming showed these systems can follow complex changes in fuel injection timing and pilot quantity.

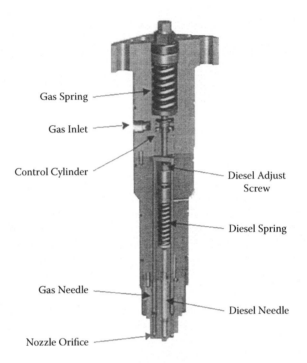

Gas Spring

Gas Inlet

Control Cylinder

Diesel Adjust
Screw

Diesel Spring

Gas Needle

Diesel Needle

Nozzle Orifice

FIGURE 11.6 A combined injector body for the separate injection of liquid and gaseous fuels in dual-fuel engine applications. (From Nylund, I., Gas Engine Development at Wartsila NSD, paper no. 2000-ICE-330ICE, vol. 35-2, 2000, pp. 131–137.)

REFERENCES AND RECOMMENDED READING

Abd Alla, G.H., Soliman, H.A., Badr, O.A., and Abd Rabbo, M.F., *Effect of Pilot Quantity on the Performance of a Dual Fuel Engine*, SAE Paper 1999-01-3597, 1999.

Beck, N., Johnson, W., George, A., Peterson, P., vander Lee, B., and Klopp, G., *Electronic Fuel Injection for Dual Fuel Diesel Methane*, SAE Paper 891652, 1989.

Beppu, O., Fukuda, T., Komoda, T., Miyake, S., and Tanaka, T., Service Experience of Mitsui Gas Injection Diesel Engines, in *Proceedings of CIMAC Congress*, 1998, pp. 187–202.

Carlucci, A.P., Ficarella, A., and Laforgia, D., Control of the Combustion Behavior in a Diesel Engine Using Early Injection and Gas Addition, *Applied Thermal Engineering*, 26, 2279–2286, 2006.

Challen, B., and Barnescu, R., *Diesel Engine Handbook*, 2nd ed., SAE, Warrendale, PA, 1999.

Chrisman, B., Callaham, T., and Chiu, J., *Investigation of Macro Pilot Combustion in Stationary Gas Engine*, ASME Paper 98-ICE-106, 1998.

Danyluk, P.R., Development of a High Output Dual Fuel Engine, *ASME Journal of Engineering for Gas Turbines and Power*, 115, 728–733, 1993.

Dumitrescu, S., Hill, P.G., Li, G.G., and Ouellette, P., *Effects of Injection Changes on Efficiency and Emissions of a Diesel Engine Fuelled by Direct Injection of Natural Gas*, SAE Paper 2000-01-1805, 2000.

Goto, S., Nishi, Y., and Nakayama, S., High Density Gas Engine with Micro Pilot Compression Ignition Method, paper presented at Proceedings of ASME Conference, ICE Division, Salzburg, Austria, 2003, Paper ICES2003-679.

Gunea, C., Examination of the Pilot Quality on the Performance of Gas Fueled Diesel Engine, MSc thesis, University of Calgary, Canada, 1997.

Gunea, C., Razavi, M.R., and Karim, G.A., *The Effects of Pilot Fuel Quality on Dual Fuel Engine Ignition Delay*, SAE Paper 982453, 1998.

Harrington, J., Munashi, S., Nedelcu, C., Ouellette, P., Thompson, J., and Whitfield, S., *Direct Injection of Natural Gas in a Heavy-Duty Engine*, SAE Paper 2002-01-1630, 2002.

Hlousek, J., Common Rail Fuel Injection System for High Speed Large Diesel Engines, in *ASME-ICE Division*, 1998, vol. 31-1, p. 50, Paper 98-ICE-122.

Hodgins, K.B., Gunawan, H., and Hill, P.G., *Intensifier-Injector for Natural Gas Fuelling Diesel Engines*, SAE Paper 921553, 1992, p. 51.

Imitrescu, S., and Hill, P.G., *Effects of Injection Changes on Efficiency and Emissions of a Diesel Engine Fuelled by Direct Injection of Natural Gas*, SAE Paper 2000-01-1805, 2000.

Ishida, M., Chen, Z.L., Luo, G.F., and Ueki, H., *The Effect of Pilot Injection on Combustion in a Turbocharged D.I. Diesel Engine*, SAE Paper 841692, 1994.

Johnson, W.P., Beck, N.J., Van der Lee, A., Koshkin, V.K., Lovkov, D., and Platov, I.S., *An Electronic Dual Fuel Injection System for the Belarus D144 Diesel Engine*, SAE Paper 901502, 1990.

Karim, G.A., and Rogers, A., Comparative Studies of Propane and Butane as Dual Fuel Engine Fuels, *Journal of the Institute of Fuel,* 40, 513–522, 1967.

Krepec, T., Giannacopoulos, T., and Miele, D., New Electronically Controlled Hydrogen Gas Injector Development and Testing, *International Journal of Hydrogen Energy*, 12, 855–861, 1987.

Kurtz, E.M., Mather, D.K., and Foster, D.E., *Parameters That Affect the Impact of Auxiliary Gas Injection in a DI Diesel Engine*, SAE Paper 2000-010233, 2000.

Li, G., Ouellette, P., Dumitrescu, S., and Hill, P., *Optimization Study of Pilot Ignited Natural Gas Direct Injection in Diesel Engines*, SAE Paper 1999-01-3556, 1999.

Lin, Z., and Su, W., *A Study on the Determination of the Amount of Pilot Injection and Lean and Rich Boundaries of the Premixed CNG-Air Mixtures for a CNG/Diesel Dual Fuel Engine*, SAE Paper 2003-01-0765, 2003.

MacCarley, C.A., and Vorst, W.D.V., Electronic Fuel Injection Techniques for Hydrogen Powered I.C. Engines, *International Journal of Hydrogen Energy*, 5, 179–203, 1980.

Nylund, I., *Gas Engine Development at Wartsila NSD*, in *ASME-ICE Division*, 2000, vol. 35-2, pp. 131–137, Paper 2000-ICE-330ICE.

Park, T., Traver, M.L., Atkinson, R., Clark, N., and Atkinson, C.M., *Operation of a Compression Ignition Engine with a HEUI Injection System on Natural Gas with Diesel Pilot Injection*, SAE Paper 1999-01-3522, 1999.

Saito, H., Sakurai, T., Sakaoji, T., Hirashima, T., and Karnno, K., *Study on Lean Burn Gas Engine Using Pilot Oil as the Ignition Source*, SAE Paper 2001-01-0143, 2001.

Schiffgens, H.J., Brandt, D., Rieck, K., and Heider, G., Low-NOx Gas Engines from MAN B&W, in *CIMAC Congress,* Copenhagen, Denmark, 1998, pp. 1399–1414.

Shioji, M., Ishiyama, T., and Ikegami, M., Approaches to High Thermal-Efficiency in High Compression-Ratio Natural-Gas Engines, in *Proceedings of 7th International Conference on Natural Gas Vehicles*, Yokohama, Japan, 2000, pp. 13–21.

Sierens, R., and Verhelst, S., Influence of the Injection Parameters on the Efficiency and Power Output of a Hydrogen Fueled Engine, *ASME Journal of Engineering for Gas Turbines and Power*, 125, 444–449, 2003.

Singal, S.K., Pundit, B.P., and Mehta, P.S., *Fuel Spray–Air Motion Interaction in DI Diesel Engine: A Review*, SAE Paper 930604, 1993.

Singh, S., Krishnan, S.R., Srinivasan, K.K., Midkiff, K.C., and Bell, S.R., Effect of Pilot Injection Timing, Pilot Quantity and Intake Charge Conditions on Performance and Emissions for an Advanced Low-Pilot-Ignited Natural Gas Engine, *International Journal of Engine Research*, 5(4), 329–348, 2004.

Stumpp, G., and Ricco, M., *Common Rail—An Attractive Fuel Injection System for Passenger Car DI Diesel Engines*, SAE Paper 960870, 1996.

Turner, S.H., and Weaver, C.S., *Dual-Fuel Natural/Diesel Engines: Technology, Performance and Emissions*, No. GRI-94/0094, Topical Report Gas Research Institute, November 1994.

Umierski, M., and Stommel, P., *Fuel Efficient Natural Gas Engine with Common-Rail Micro-Pilot Injection*, SAE Paper 2000-01-3080, 2000.

Wong, W.Y., Midkiff, K.C., and Bell, S.R., *Performance and Emissions of a Natural Gas Fuelled Indirect Injected Diesel Engine*, SAE Paper 911766, 1991.

Yonetani, H., Hara, K., and Fukatani, I., *Hybrid Combustion Engine with Premixed Gasoline Homogeneous Charge and Ignition by Injected Diesel Fuel–Exhaust Emission Characteristics*, SAE Paper 940268, 1994.

Yoshida, K., Shoji, H., and Tanaka, H., *Study on Combustion and Exhaust Emission Characteristics of Lean Gasoline-Air Mixture Ignited by Diesel Fuel Direct Injection*, SAE Paper 982482, 1998.

12 Gaseous Fuel Admission

12.1 MODES OF FUEL GAS ADMISSION

The modes of gaseous fuel introduction into dual-fuel engines vary widely in complexity of design and operation. These modes are of much importance and can control the progress of the combustion process and associated engine performance. In general, it is required that the admission of the gaseous fuel into the engine cylinder and its mixing with the required air should produce a homogeneous mixture with uniform properties at the time of pilot injection. This may not be an easy task to accomplish fully, especially when relatively simple approaches are adopted, while the alternative option for resorting to diesel engine operation is retained.

It has been the practice for some time to opt for simple and cheap conversion of high-compression-ratio diesel engines to dual-fuel operation, relying heavily on the established practice of carbureting, as implemented in spark ignition engines with gasoline as the fuel. Such relatively simple approaches were invariably associated with serious limitations to dual-fuel engine performance. The controlled admission of the gas directly into individual cylinders through independently controlled valving offers improvements in engine performance, albeit with increased complexity and cost. This has increasingly become the practice in a variety of forms in recent years.

The main methods that have been employed for gas introduction into gas-fueled diesel engines in the order of their increased complexity and cost are as follows:

1. The simple introduction of the gaseous fuel with a relatively low supply pressure into the engine intake manifold so as to mix with the incoming air. This simple approach is often described as fumigation. The carburetors and mixers employed tend to be similar to the variety of devices used in gasoline engine applications. They attempt to provide homogeneously mixed fuel and air at a constant overall fuel/air ratio for any engine load while producing very low pressure drop within the intake system. Figure 12.1 typically shows a simple nozzle arrangement used for admission of a fuel gas into the intake airstream of an engine. However, invariably with many gaseous fuels, there is a need for processing and filtering of the fuel gas before its introduction into the engine at around atmospheric pressure. Figure 12.2 shows a typical arrangement of an engine plant combined with a biomass processing system to produce biogas that is fed directly to the engine and consumed by it. Such arrangements are especially suited to stationary field applications.
2. Continuous steady-flow fuel gas injection at low pressures into the intake manifold independently from the valve action of the different cylinders.

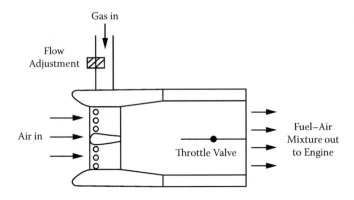

FIGURE 12.1 A simple nozzle arrangement for admission of fuel gases into the intake airstream of an engine.

3. Controlled timed intermittent fuel gas injection either into the intake manifold of a multicylinder engine or individually at the inlet of each cylinder synchronized with the opening and closing of the respective valves.
Figure 12.3 shows an example of an engine design that introduces the fuel
gas independently controlled into the incoming air just outside the cylinder,
while Figure 12.4 shows an arrangement for gas fuel introduction independently to each cylinder in a multicylinder engine.
4. Timed high-pressure fuel gas injection into the individual cylinders either
during the intake stroke after exhaust valve closure or at different times
within the relatively early stages of the compression process.
5. Timed sufficiently high-pressure gas fuel injection into the cylinder. This
may be performed either later in the compression stroke or after the commencement of pilot liquid fuel injection or its ignition.

The timed injection of the fuel gas, rather than its fumigation, is an effective
method for a better controlled fuel introduction reducing the escape of any unconsumed fuel gas into the exhaust. The overall efficiency and emissions of the engine
are expected to be improved, especially at light loads. When the fuel gas is injected
into the engine intake manifold, it is usually timed individually for each cylinder
immediately ahead of the intake valve and well before its closure timing, when this
is performed properly; then it can aid in bringing about regular and early ignition.
It may also help in producing proper mixture stratification that may be employed
to improve combustion and reduce emissions. Figure 12.5 shows an example of the
details of a high-pressure gas injector employed in dual-fuel engine operation.

In turbocharged engine applications, often the fuel gas, especially when available
in sufficiently elevated supply pressure, is injected beyond the compressor outlet into
the cylinder and not at entry into the intake manifold. The scavenging of common
two-stroke engines and highly turbocharged four-stroke engines can represent an
obstacle to the fumigation of gaseous fuels when employing relatively simple fuel
introduction methods. However, direct injection of the fuel gas into the cylinder,
when controlled properly, can produce low fuel gas loss to the exhaust manifold,

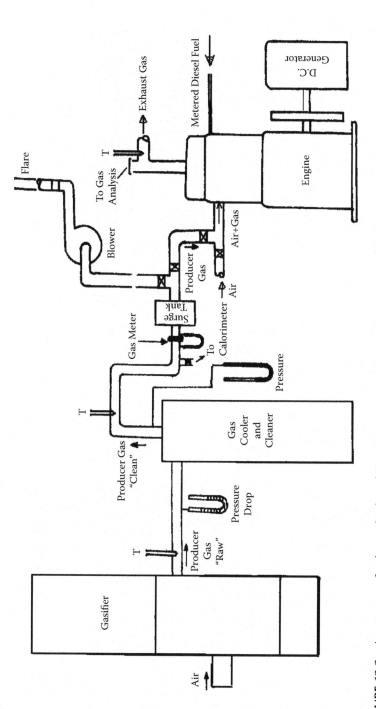

FIGURE 12.2 Arrangement for the production of biogas and its feed arrangement to a single-cylinder dual-fuel engine. (From Parikh, P.P., Bhave, A.G., and Shash, I.K., Performance Evaluation of a Diesel Engine Dual Fuelled on Process Gas and Diesel, in *Proceedings of the National Conference on ICE and Combustion*, National Small Industries Corp. Ltd., New Delhi, India, 1987, pp. Af179–Af186.)

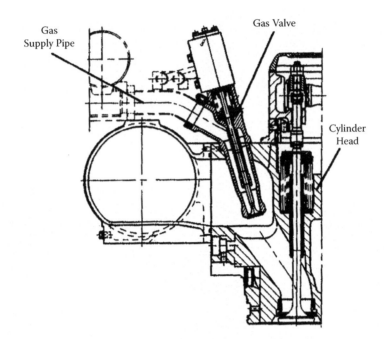

FIGURE 12.3   An example of an engine design that introduces the fuel gas independently into the incoming air just outside the cylinder. (From Schiffgens, H.J., Brandt, D., Dier, L., and Glauber, R., Development of the New Man B&W 32/40 Dual Fuel Engine, in *ASME-ICE Division Technical Conference,* 1996, vol. 27-3, pp. 33–45.)

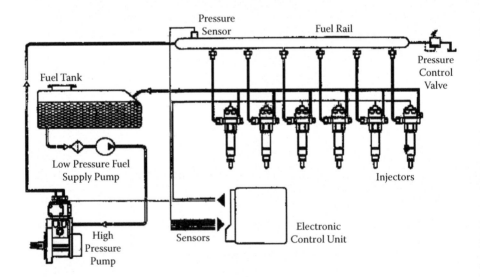

FIGURE 12.4   A typical arrangement of fuel introduction independently to each cylinder in a multicylinder engine. (From Stumpp, G., and Ricco, M., *Common Rail—An Attractive Fuel Injection System for Passenger Car DI Diesel Engines,* SAE Paper 960870, 1996.)

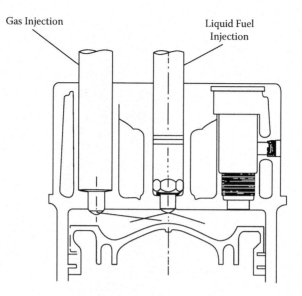

Gas Injection   Liquid Fuel Injection

FIGURE 12.5   An example of a dual-fuel engine where the fuel gas is injected separately during the compression and before pilot injection. (From Challen, B., and Barnescu, R., *Diesel Engine Handbook*, 2nd ed., SAE, Warrendale, PA, 1999.)

particularly in two-stroke engines. Figure 12.5 shows an example of a dual-fuel engine where the fuel gas is injected during the compression and well before pilot injection. Figure 12.6 also shows a two-stroke dual-fuel engine where the fuel gas is injected independently directly into the cylinder, avoiding the fuel passing unconverted into the exhaust gas with the scavenging air.

The gas injection into the individual cylinders of engines well after intake and exhaust valve closure requires a sufficiently high gas supply pressure and a suitable gas injection valve (see Figure 12.7). This approach offers the possibility of increasing the load and reducing emissions while curtailing the tendency of the engine to backfire or knock for a range of gaseous fuels. However, there is a need for more advanced control methods, leading to increased initial capital and running costs, such as having a specially designed fuel gas injector, and often with modifications made to the cylinder head with a compressor to provide the required supply of high-pressure gas. More elaborate safety devices and adequate systems for reducing NOx and particulate emissions will be also required. Generally, changes in the gas injection timing can have a significant effect on the mixing processes of the fuel gas and air, affecting the values of the ignition delay, peak pressure, emissions, and other engine performance parameters. In comparison, relatively small changes in pilot injection timing tend to produce only minor effects. The injection of a sufficiently high pressure gas just before, during, or after the pilot fuel injection period requires very careful synchronization. It can produce sufficient complications that would lead to deterioration in the control of the combustion process altogether, and hence adversely affect engine performance and emissions.

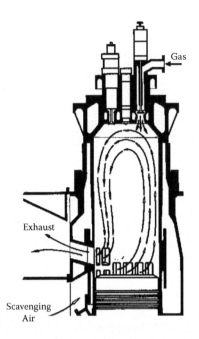

FIGURE 12.6 A two-stroke dual-fuel engine where the fuel gas is independently injected into the cylinder. (From Steiger, A., *Sulzer Technical Review*, 3, 1–8, 1970.)

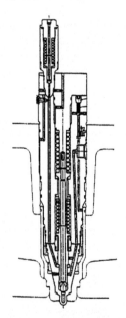

FIGURE 12.7 An example of the details of a high-pressure gas injector employed in dual-fuel engine operation. (From Beppu, O., Fukuda, T., Komoda, T., Miyake, S., and Tanaka, T., Service Experience of Mitsui Gas Injection Diesel Engines, in *Proceedings of CIMAC Congress*, Copenhagen, Denmark, 1998, pp. 187–202.)

The direct fumigation of some diesel fuel into the inlet air, which is not easy to accomplish properly, mainly due to the low volatility of the diesel fuel, may produce reductions in the ignition that are capable of achieving reductions in the ignition delay, rates of pressure rise, and particulate emissions. However, some low heat releases can take place during the course of the compression stroke, which may affect the overall specific fuel consumption and heat transfer processes. To apply some of the injected fuel directly to an externally heated glow plug, or partially reacting the fuel, such as through its passing before introduction into the cylinder in a suitably externally heated reactor controlled electrically, can produce improvements in performance over a limited range of engine operation.

Some of the improvements in engine performance produced by the proper synchronization of gas injection with the opening and closing of the intake valves may include the prevention of some fuel escaping directly into the exhaust duct, reduction of cyclic variation, and the incidence of backfiring, which results mainly from valve overlap. In the naturally aspirated large-bore engines of relatively low speeds, the gas is admitted more easily directly into the individual cylinder airstream. Other approaches may allow the gas to go into the airstream only when the inlet valve is open. However, this procedure may adversely affect safety, emissions, mixing, and the intake pressure characteristics.

12.2 SOME OF THE LIMITATIONS OF FUEL GAS FUMIGATION

In a typical diesel engine the amount of air available is governed mainly by the size of the cylinder, intake mixture density, and engine speed. The composition of the cylinder charge during compression is simply that of air with the small amount of residual gas that becomes mixed with it. Accordingly, the average value of the prevailing temperature at the time of liquid fuel injection can be established while accounting for the relatively small effects of heat transfer. However, a closer examination of details of the events taking place within the cylinder during compression would show a gradual buildup in stratification of the properties of the cylinder contents to produce some nonuniformities in temperature distribution, with higher temperatures appearing within the central regions of the cylinder. This can be quite significant, especially in high compression ratios with low swirl engines, which can affect the injection, preignition, and subsequent combustion processes of the liquid fuel jet, favoring the central regions of the cylinder.

In the homogeneous premixed gas-fueled dual-fuel engine the introduction of the fuel gas with the intake air produces changes in the physical and transport properties of the mixture, such as those in the specific heat ratio and, to a lesser extent, in heat transfer parameters. For example, these would markedly affect the peak temperature at TDC with the type of gaseous fuel used and its concentration in the cylinder charge. As shown in Figure 12.8, for a constant-compression-ratio, unthrottled, gas-fueled fumigated engine operating at a fixed intake temperature and in the absence of pilot fuel injection, the addition of hydrogen to the engine intake affects the maximum cylinder charge temperature at the end of compression only a little with the increase of equivalence ratio. This is because the thermodynamic properties of hydrogen are close to those of air. However, the corresponding values of temperature

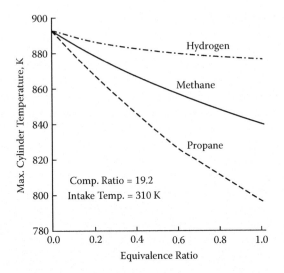

FIGURE 12.8 Calculated changes in the maximum temperature at the end of compression with equivalence ratios for hydrogen, methane, and propane in air. (Adapted from Liu, Z., An Examination of the Combustion Characteristics in Compression Ignition Engines Fueled with Gaseous Fuels, PhD thesis, University of Calgary, Canada, 1995.)

with the addition of methane or propane decrease more markedly since the properties of these gases are much different from those of air. With the addition of propane for the case shown, a drop in temperature of around 100 K can be observed for the stoichiometric mixture. This would show that the effective compression ratio of the engine would be lowered substantially by the fumigation of the fuel gas in comparison to the geometric compression ratio. Diesel engine compression with air will not suffer from such an important reduction that undermines the combustion processes of the dual-fuel engine and its performance.

The temperature at the end of compression is also markedly affected by the extent of heat transfer during the compression stroke, which also depends on the type of fuel fumigated and its concentration. For example, as shown in Figure 12.9 for a typical set of operating conditions, the corresponding calculated maximum heat transfer rates decreased only a little with the increased admission of propane or methane, but the rates increased significantly with the increased admission of hydrogen.

During compression the exposure of a premixed charge of a fuel and air to a rapidly increasing temperature, but over a relatively extended time, will result in some reaction activity that influences the rate of temperature change and the extent of any stratification in the properties of the mixture. The contributions of the pre-ignition reaction activity and the associated energy release to the development of temperature are also strongly dependent on the gaseous fuel used and the presence of residuals. An example of the extent of enhancing the calculated values of the pre-ignition energy release rates of the charge using detailed kinetics with the increased admission of hydrogen is shown in Figure 12.10. However, with the corresponding admission of methane or propane, the peak values of the energy release rates display

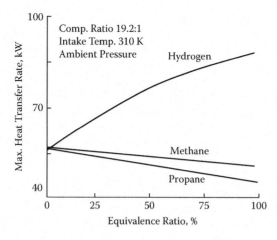

FIGURE 12.9 Variation of the calculated maximum heat transfer rate with equivalence ratios during compression of mixtures of hydrogen, methane, and propane in air. (Adapted from Liu, Z., An Examination of the Combustion Characteristics in Compression Ignition Engines Fueled with Gaseous Fuels, PhD thesis, University of Calgary, Canada, 1995.)

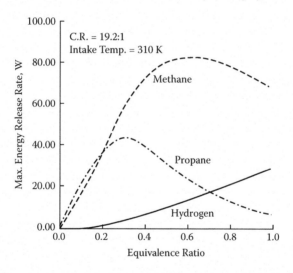

FIGURE 12.10 Variations in the calculated values of the maximum preignition energy release rates with equivalence ratios during compression for mixtures of hydrogen, methane, and propane in air at ambient intake pressure. (Adapted from Liu, Z., An Examination of the Combustion Characteristics in Compression Ignition Engines Fueled with Gaseous Fuels, PhD thesis, University of Calgary, Canada, 1995.)

maximum values that are significantly on the lean side of the stoichiometric. For any equivalence ratio, the rates associated with methane admission are higher than those with propane or hydrogen. This is somewhat unexpected in view of the relatively slower reaction rates of methane. However, it is more a reflection of the changes increasingly brought about to the mean charge temperature with fuel gas admission.

The peak value of the release rate for propane operation is obtained with a leaner mixture than that with methane, since the temperature with propane is lowered more greatly than with methane. For the typical case shown, the two corresponding equivalence ratios are around 0.30 and 0.60 for propane and methane operations, respectively. The changes with hydrogen admission mainly reflect the effect of increasing the equivalence ratio. Hence, the increased admission of either methane or propane to an engine may not necessarily enhance the preignition reactivity of the gaseous fuel–air mixture under constant compression ratio engine conditions. This is an important consideration in dual-fuel performance and its differences from that of diesel operation.

There are changes in the intake partial pressure of oxygen due to the displacement of some of the air by the fuel gas introduced. These changes, which depend on the gas equivalence ratio, can be very substantial with some of the common gaseous fuel. For example, in the case of a homogeneous stoichiometric mixture of hydrogen and air, the volumetric displacement can be around 30%, and for methane–air, it can be around 10%. Of course, in the case of operation on gasoline vapor, it is negligibly small and is below 1%. This reduction in the available amount of oxygen represents a serious burden on the capacity of the engine to produce sufficient power unless remedial action is taken, such as employing a slightly larger engine, higher compression ratios, or increased turbocharging. In dual-fuel operation that retains the capacity to revert promptly to diesel operation, this would be difficult to implement.

There are other important considerations arising from the employment of fumigation that have a significant bearing on the behavior of dual-fuel engines and their widespread acceptability as a means for producing power while using gaseous fuels. For example, the fact that partial load control is done via changes in the equivalence ratio of the admitted gas, while often keeping the pilot quantity constant, will result in idling or light load in very lean fuel gas–air mixtures employed. This would result in the flame propagation within the charge not being fully completed within the time available, which results in incomplete combustion of the fuel with poor output efficiency unless sufficiently large pilots are employed that can help improve the consumption of the gaseous fuel. Similarly, at very high loads and especially with some fuel mixtures, the operation on homogeneous gas fuel mixtures will be conducive to uncontrolled autoignition reactions, leading to the onset of knock.

The introduction of fuel gas into the airstream of the engine can bring about changes to the intake mixture temperature. It can improve the volumetric efficiency and power output, but may contribute to the excessive lowering of the mixture temperature, which can contribute negatively to increased ignition delay as well as increased undesirable emissions. For example, in the case of operation on liquid LPG or LNG, the rapid vaporization of the liquid fuel will be at the expense of heat removal out of the intake manifold and the air charge. Furthermore, in the case of operation on CNG, the cooling following the expansion of the fuel gas to the intake pressure will also result in cooling of the intake charge. Of course, following fuel fumigation into the incoming air, the presence of some of the premixed fuel–air charge outside the cylinder represents a potential for much increased safety hazards in case of a backfire, as well as increased exhaust emissions. Furthermore, the fumigation of a fuel gas in turbocharged engines has additional potential problems, arising from issues of poor vaporization and maldistribution of mixtures that can lead to charger malfunction. In the case of the fumigation of some liquid fuels such as alcohols, it must be reduced at low loads to prevent flame quenching, misfire, and increased emissions, while at high loads to prevent preignition and knock.

In spite of these limitations associated with resorting to fuel gas fumigation in dual-fuel engines, it remains widely used. This is mainly due to its simplicity and cheapness of conversion of engines to dual-fuel engines. Moreover, it is both very convenient and economical to exploit fuel gas resources that have low thermal value per volume admitted, as well as fuel gas supplies that become available at relatively low supply pressures.

To reduce or eliminate the shortcomings of fumigated systems, measures such as accurate time duration pulse injection of fuel gas at each intake port, combined with appropriate algorithms in the fuel management microprocessor, need to be used.

12.3   IN-CYLINDER DIRECT INJECTION OF ULTRA–HIGH-PRESSURE FUEL GAS

There are systems developed where the fuel gas is injected at very high pressure directly into the engine cylinder. This is, for example, done at the end of the compression stroke and after the start of ignition of pilot fuel injection. Such applications vary widely in their design and operational arrangements, depending on factors such as those of the field of application, size of engine, relative fraction of the pilot used, and whether the engine remains capable of operation as a diesel to deliver its full power or not. There are also a number of approaches that employ designs with combined injector body for independently injecting the pilot diesel from that of the fuel gas. Some arrangements that mainly target the road transport sector have the injected liquid fuel proceeds first, followed soon after by the injection of the gaseous fuel under very high pressure, with hydraulically actuated and electronically controlled unit injection systems. Such approaches require complex controls and an effective understanding of the complex physical and chemical processes that take place in the combustion and phasing of the consecutive high-pressure gas injections, and that of the pilot, under different loads and speed. An important consideration is the reduction of capital and operational costs while rendering the system more flexible.

Sufficiently high pressure gas supply (e.g., 350 bar) needs to be of a value very much in excess of that of the high-cylinder pressure prevailing at around TDC of the high-compression-ratio engine injected just after pilot ignition. Such very high fuel gas supply pressures, which are usually produced by a multistage compressor driven by the engine, are required so that the fuel gas can be injected fast enough into the cylinder where very high pressures and temperatures already exist. The associated mixing processes of the burning liquid fuel with the injected fuel gas are sufficiently complex and require careful control, staging, and understanding. The high-pressure injection of the fuel gas needs to be completed fast enough to allow for its ignition, combustion, and contribution to the combustion energy release that takes place mostly just after TDC, so as to produce the rated power output at high efficiencies. Figure 12.11 shows a schematic presentation of a combined injector of liquid pilot and fuel gas at high pressure

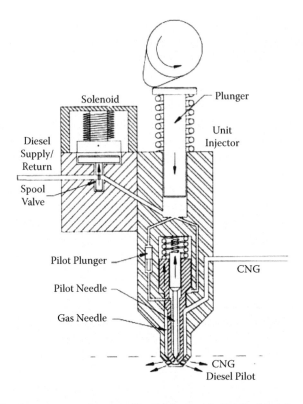

FIGURE 12.11 Arrangements for a combined injector of pilot liquid and gas fuels at high pressure employing a combined injector body. (From Dumitrescu, S., Hill, P.G., Li, G.G., and Ouellette, P., *Effects of Injection Changes on Efficiency and Emissions of a Diesel Engine Fuelled by Direct Injection of Natural Gas*, SAE Paper 2000-01-1805.36, 2000; Douville, B., Ouellette, P., Touchette, A., and Ursu, B., *Performance and Emissions of a Two-Stroke Engine Fueled Using High Pressure Direct Injection of Natural Gas*, SAE Paper 981160, 1998.)

directly into the cylinder toward the end of compression, while Figure 12.12 shows such an injector as developed by Westport Innovations.

Such approaches represent added complexity and costs to the already complex and limited operational flexibility of the control systems of the dual fueling of modern diesel engines. They also tend to be more restrictive in their capacity to vary the key controlling operational variables. In recent engine applications, the fuel supply systems are increasingly of the common rail variety, with the gas injection valve controlled by another system. This can be a standard liquid fuel injection pump that supplies controllably high pressure liquid fuel to the gas valve and keeps it open while injecting the required amount of gas. For transport applications, the needed fuel gas compressor would represent added complexity and cost, requiring the fuel to be supplied and carried as CNG, which has been compressed initially partly toward the required injection pressure value. This would be associated with the burden of extra weight and the consumption of a sizable amount of energy taken out of the useful engine output. Moreover, in some of these applications, to reduce the complexity

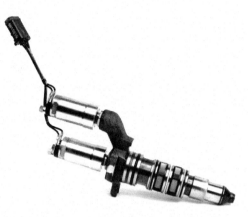

FIGURE 12.12   Westport high-pressure direct injection injector.

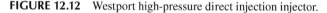

of operation, they opt to dispense with the capacity of producing the full-rated output of the engine as a diesel in the absence of fuel gas injection.

In the operation of dual-fuel engines employing high-pressure gas injection, the full safety of all aspects of the operation is of paramount importance and needs to be fully assured. For example, the gas pipes must be capable of withstanding the very high internal gas pressure and any of its fluctuations. It is made of double-walled high-strength alloyed steel tubing, often with a provision of connection to a nitrogen gas purging system. Very small leakages of gaseous fuel, should they occur, are detected, such as through special gas detectors or through the increases in the exhaust gas temperature and in the levels of unburned hydrocarbon emissions. In the case of engines for transport applications, the associated garages and maintenance shops would require additional venting, ample methane detectors, and alarms, with the electrical system meeting the necessary safety codes. Moreover, a gaseous fueled vehicle engine would normally not be parked inside a closed facility overnight unless attended with the heating system for the shop also suitably upgraded for safety.

There are other different arrangements for providing pilot fuel injection, as well as the fuel gas. For example, the gas may be introduced via a special injector into the cylinder sometime during the compression stroke, and then followed by the injection of the pilot either from a special injector or from the main diesel injector. Alternatively, the two fuels are introduced from a centrally located injection valve with separate double circles of injection holes. One set is for the pilot, and the other for the fuel gas. Both have their own separate actuators and controls. The controls would ensure that the gas is injected promptly through the choked orifices in accordance with the load requirements. In these applications, the engine is usually started on liquid fuel and switched over later to dual-fuel operation. The liquid fuel is then reduced to the appropriate pilot quantity rate as the gas supply rate increases proportionally. A similar procedure is normally followed in reverse during shutdown, to be ended sometimes with nitrogen purging of the gas fuel system. In the case of an emergency, the fuel supply is cut off promptly, followed by nitrogen gas purging.

With the injected liquid fuel, after a short ignition delay, burning commences virtually simultaneously at a number of points. Flames then begin to spread from these ignition points to the local adjacent regions of fuel vapor–gaseous fuel–air mixtures. A flame starting from such points would not have to travel far before it meets flames from similar neighboring areas. Accordingly, there is less likelihood for knock despite the high compression ratios employed. In a way, this would be akin to having an ignition system of too many multiple sparks. However, under certain operating conditions, such as with large pilots or less lean gaseous fuel–air mixtures, very fast rates of pressure rise could be encountered due to the rapid energy release rates arising from diesel fuel autoignition, as well as the multiple fast flame propagation fronts formed.

When the liquid fuel is injected first, it is allowed to ignite in the mode of a conventional diesel engine when the gaseous fuel is injected suitably into the burning liquid fuel. It is evident that these arrangements require great care in controlling the timing of the injection of the two fuels, choice of their respective sizes, and the associated mixing processes. These will be influenced by many key controlling variables, including the specific aspects of the design of the engine installation, operating conditions of the engine, such as speed, load, and turbocharging, and the composition of the gaseous fuel.

In the case of the high-pressure gas injection, the temperature, composition, and velocity distribution will be markedly different from those in the case of the diesel engine or the premixed version of the dual-fuel engine. The introduction of a high-velocity and very cold gas jet into the high-temperature environment can markedly distort the temperature distribution, and thus heat transfer, often in an uncertain manner.

12.4 LNG, LPG, AND HYDROGEN ADMISSIONS

Most engine applications of LNG supply the fuel in suitable cryogenic tanks as liquid, but it is vaporized as it is introduced into the engine as a very cold gas to mix with the relatively warm air. The vaporization set up employed may be as shown in Figure 12.13.

The engine coolant is usually used to supply the heat needed to vaporize and warm the fuel sufficiently for application into the engine. The pressure is further regulated before feeding the gas to the engine, where it is metered and mixed with the air. The use of LNG, which requires strict protective safety measures, is rather

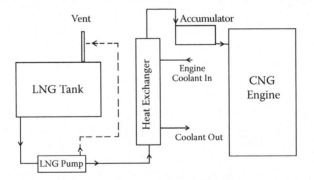

FIGURE 12.13 A schematic of a typical installation for vaporizing the stored LNG before being fed as a high-pressure gas to the engine.

attractive because much more energy is stored per unit volume container than CNG. In transport engine applications, LNG compares favorably to the use of CNG. It represents a lower fuel system weight, shorter refueling times, and a more uniform and a higher quality fuel. Moreover, the liquid nature of LNG permits its pressurization to high levels with relatively simple cryogenic pumps.

There are some major potential challenges associated with the use of hydrogen as an engine fuel in general and dual-fuel applications in particular, especially in transport applications. The introduction of hydrogen into engines is usually controlled on a volume basis. Hydrogen as a fuel suffers from the fact that it has an extremely low heating value on a volume basis. Moreover, per volume of fuel, it requires a very high amount of air for combustion. For example, its stoichiometric mixture in air represents nearly 30% of the mixture. This would lead to very significant derating to the engine. Other features relate to the increased fire and explosion hazards associated with its operation, handling, and storage. The very large difference in the heating value of hydrogen needs careful treatment and control. For example, the gas supply should be fitted with a suitable automatic cutoff device. It is also desirable to admit the gas into the air during the induction stroke only and to avoid backfire into the intake manifold. It is preferable to delay the mixing of the gas with the air until inside the cylinder with the gas admission period timed carefully to suit. The velocity of the gas on its admission is usually quite high, which helps in the mixing process during compression.

The admission of hydrogen to engines via fumigation has been frequently tried using procedures essentially similar to those for CNG. However, much care is needed to guard against the risk of backfire and knock. Also, the injection of hydrogen at sufficiently high pressure directly into the engine with pilot ignition represents an equally complex system, especially in the need to compress the hydrogen to sufficiently high pressures and the large volumes of gas that need to be introduced sufficiently fast.

In the limited experience with employing LH_2 as a fuel in engines, a very cold gas is quickly formed, representing a substantial volume of gas that has to be injected into the high-pressure combustion space and effectively distributed in the short time available. The very high difference in the density between the fuel gas and air, the resulting high flame speed, and the different turbulence characteristics all influence the mixing and combustion processes.

Since LPG can be maintained in the liquid state quite easily at ambient temperatures through the application of moderate pressures, it has been favored as a fuel for engines of the dual-fuel type, especially for transport applications. Its mode of introduction into the engine cylinder has followed largely conventional liquid fuels in spark ignition applications, while using suitable carburetion methods or injection into the cylinder without the need for additional compression. However, if it were to be injected directly into the later stages of compression, then the fact that it is liquid requires little effort, but there is a need to ensure its full vaporization and proper mixing with air before ignition. This often requires the liquid fuel to be warmed and fully vaporized before inlet valve closure. The liquid fuel, as with the case of LNG, is warmed up and vaporized using the engine coolant as the source for heating. Pressure regulators are installed while reducing the fuel vapor produced to moderately low pressures before being fed to the carburetor.

REFERENCES AND RECOMMENDED READING

Addy, J.M., Binng, A., Norton, P., Peterson, E., Campbell, K., and Bevillaqua, O., *Demonstration of Caterpillar C10 Dual Fuel Natural Gas Engines in Commuter Buses*, SAE Paper 2000-01-1386, 2000.

Baranescu, R., *Fumigation of Alcohol in a Diesel Engine*, SAE Paper 1080, 1980.

Barbour, T.R., Crouse, M.E., and Lestz, S.S., *Gaseous Fuel Utilization in a Light Duty Diesel Engine*, SAE Paper 860070, 1986.

Bechtold, R.L., *Alternative Fuels Guidebook*, SAE Publishing, Warrendale, PA, 1997.

Beck, N., Johnson, W., George, A., Peterson, P., vander Lee, B., and Klopp, G., *Electronic Fuel Injection for Dual Fuel Diesel Methane*, SAE Paper 891652, 1989.

Beppu, O., Fukuda, T., Komoda, T., Miyake, S., and Tanaka, T., Service Experience of Mitsui Gas Injection Diesel Engines, in *Proceedings of CIMAC Congress,* Copenhagen, 1998, pp. 187–202.

Blizzard, D., Schaub, F.S., and Smith, J., Development of the Cooper-Bessmer Clean Burn Gas-Diesel (Dual Fuel) Engine, in *ASME-ICE Division*, 1991, vol. 15, pp. 89–97.

Blyth, N., Development of the Fairbanks Morse Enviro-Design Opposed Piston Dual Fuel Engine, paper presented at ASME-ICE Division, 1994, Paper 100375.

Carlucci, A.P., Ficarella, A., and Laforgia, D., Control of the Combustion Behavior in a Diesel Engine Using Early Injection and Gas Addition, *Applied Thermal Engineering*, 26, 2279–2286, 2006.

Challen, B., and Barnescu, R., *Diesel Engine Handbook*, 2nd ed., SAE, Warrendale, PA, 1999.

Checkel, M., Newman, P., van der Lee, B., and Pollak, I., *Performance and Emissions of a Converted RABA 2356 Bus Engine in Diesel and Dual Fuel Diesel/Natural Gas Operation*, SAE Paper 931823, 1993.

Daisho, Y., Takahashi, Y.I., Iwashiro, Y., Nakayama, S., and Saito, T., *Controlling Combustion and Exhaust Emissions in a Direct-Injection Diesel Engine Dual Fuelled with Natural Gas*, SAE Paper 952436, 1995.

Danyluk, P.R., Development of a High Output Dual Fuel Engine, *ASME Journal of Engineering for Gas Turbines and Power*, 115, 728–733, 1993.

Douville, B., Ouellette, P., Touchette, A., and Ursu, B., *Performance and Emissions of a Two-Stroke Engine Fueled Using High Pressure Direct Injection of Natural Gas*, SAE Paper 981160, 1998.

Dumitrescu, S., Hill, P.G., Li, G.G., and Ouellette, P., *Effects of Injection Changes on Efficiency and Emissions of a Diesel Engine Fuelled by Direct Injection of Natural Gas*, SAE Paper 2000-01-1805, 2000.

Einang, P.H., Engja, H.E., and Vestergren, R., Medium Speed 4 Stroke Diesel Engine Using High Pressure Gas Injection Technology, in *Proceedings of the 8th International Congress Combustion Engines (CIMAC)*, Tienjing, PRC, 1989, pp. 916–932.

Green, R.K., and Glasson, N.D., High-Pressure Hydrogen Injection for Internal Combustion Engines, *International Journal of Hydrogen Energy*, 17(11), 895–901, 1992.

Grosshans, G., Development of a 1200 kW/Cyl. Low Pressure Dual Fuel Engine for LNG Carriers, in *Proceedings of CIMAC 1998*, 1998, pp. 1417–1428.

Harrington, J., Munashi, S., Nedelcu, C., Ouellette, P., Thompson, J., and Whitfield, S., *Direct Injection of Natural Gas in a Heavy-Duty Engine*, SAE Paper 2002-01-1630, 2002.

Hodgins, K.B., Gunawan, H., and Hill, P.G., *Intensifier-Injector for Natural Gas Fuelling Diesel Engines*, SAE Paper 921553, 1992.

Karim, G.A., Combustion in Gas Fueled Compression Ignition Engines of the Dual Fuel Type, *ASME Journal of Engineering for Gas Turbines and Power*, 125, 827–836, 2003.

Lee, J.T., Kim, Y.Y., Lee, C.W., and Caton, J.A., An Investigation of a Cause of Backfire and Its Control Due to Crevice Volumes in a Hydrogen Fueled Engine, *ASME Journal of Engineering for Gas Turbines and Power*, 123, 204–210, 2001.

Liu, Z., An Examination of the Combustion Characteristics in Compression Ignition Engines Fueled with Gaseous Fuels, PhD thesis, University of Calgary, Canada, 1995.

Liu, Z., and Karim, G.A., An Examination of the Ignition Delay Period in Gas Fuelled Diesel Engines, *ASME Journal of Gas Turbines and Power*, 120, 225–231, 1998.

Lom, E.J., and Ly, K.H., *High Injection of Natural Gas in a Two Stroke Diesel Engine*, SAE Paper 902230, 1990, p. 93.

Lowe, W., and Williamson, P.B., Combustion and Automatic Mixture Strength Control in Medium Speed Gaseous Fuel Engines, in *Proceedings of the 8th Congress de Machines a Combustion (CIMAC)*, 1996, pp. A14–A56.

Lowi, A., *Supplementary Fueling of Four Stroke Cycle Automotive Diesel Engines by Propane Fumigation*, SAE Paper 41398, 1984.

Meyer, D., High Tech Fuel Management and Fuel Control Systems, presented at IGT Conference Proceedings on Gaseous Fuel for Transportation, Vancouver, BC, August 1986.

Oester, U., and Wallace, J.S., *Liquid Propane Injection for Diesel Engines*, SAE Paper 872095, 1987.

Ouellette, P., High Pressure Direct Injection (HPDI) of Natural Gas in Diesel Engines, in *Proceedings of the 7th International Conference and Exhibition on Natural Gas Vehicles*, Yokohama, Japan, 2000, pp. 235–242.

Parikh, P.P., Bhave, A.G., and Shash, I.K., Performance Evaluation of a Diesel Engine Dual Fuelled on Process Gas and Diesel, in *Proceedings of the National Conference on ICE and Combustion*, National Small Industries Corp. Ltd., New Delhi, 1987, pp. Af179–Af186.

Park, T., Traver, M.L., Atkinson, R., Clark, N., and Atkinson, C.M., *Operation of a Compression Ignition Engine with a HEUI Injection System on Natural Gas with Diesel Pilot Injection*, SAE Paper 1999-01-3522, 1999.

Quigg, D., Pellegrin, V., and Rey, R., *Operational Experience of Compressed Natural Gas in Heavy Duty Transit Buses*, SAE Paper 931786, 1993.

Rain, R.R., and McFeatures, J.S., *New Zealand Experience with Natural Gas Fuelling of Heavy Transport Engines*, SAE Paper 892136, 1989.

Schiffgens, H.J., Brandt, D., Dier, L., and Glauber, R., Development of the New Man B&W 32/40 Dual Fuel Engine, in *ASME-ICE Division Technical Conference*, 1996, vol. 27-3, pp. 33–45.

Schiffgens, H.J., Brandt, D., Rieck, K., and Heider, G., Low-NOx Gas Engines from MAN B&W, in *CIMAC Congress*, Copenhagen, 1998, pp. 1399–1414.

Singh, S., Krishnan, S.R., Srinivasan, K.K., Midkiff, K.C., and Bell, S.R., Effect of Pilot Injection Timing, Pilot Quantity and Intake Charge Conditions on Performance and Emissions for an Advanced Low-Pilot-Ignited Natural Gas Engine, *International Journal of Engine Research*, 5(4), 329–348, 2004.

Steiger, A., Large Bore Sulzer Dual Fuel Engines, Their Development, Construction and Fields of Application, *Sulzer Technical Review*, 3, 1–8, 1970.

Stumpp, G., and Ricco, M., *Common Rail—An Attractive Fuel Injection System for Passenger Car DI Diesel Engines*, SAE Paper 960870, 1996.

Turner, S.H., and Weaver, C.S., *Dual-Fuel Natural/Diesel Engines: Technology, Performance and Emissions*, No. GRI-94/0094, Topical Report Gas Research Institute, Chicago, November 1994.

Umierski, M., and Stommel, P., *Fuel Efficient Natural Gas Engine with Common-Rail Micro-Pilot Injection*, SAE Paper 2000-01-3080, 2000.

Varde, K.S., and Frame, G.A., Hydrogen Aspiration in a Direct Injection Type Diesel Engine—Its Effects on Smoke and Other Engine Performance Parameters, *International Journal of Hydrogen Energy*, 8, 549–555, 1983.

Wong, Y., and Karim, G.A., *An Analytical Examination of the Effects of Exhaust Gas Recirculation on the Compression Ignition Process of Engines Fuelled with Gaseous Fuels*, SAE Paper 961396, 1998, pp. 45–53.

Xiao, F., and Karim, G.A., *Combustion in a Diesel Engine with Low Concentrations of Added Hydrogen*, SAE Paper 2011-01-0676, 2011, pp. 1–17.

Xiao, F., Soharabi, A., and Karim, G.A., Effect of Small Amounts of Fugitive Methane in the Air on Diesel Engine Performance and Its Combustion Characteristics, *International Journal of Green Energy*, 5, 334–345, 2008.

Zaidi, K., Andrews, G., and Greenhough, J., *Diesel Fumigation Partial Mixing for Reducing Ignition Delay and Amplitude of Pressure Fluctuations*, SAE Paper 980535, 1998.

13 Dual-Fuel Engine Combustion

13.1 COMBUSTION PROCESSES

The combustion processes in dual-fuel engines tend to be more complex than those of the common gas-fueled spark ignition or diesel engines. Some of the combustion features that can represent design and operational challenges are the following:

- The fuel gas tends to be less completely oxidized at light load, which tends to increase the specific energy consumption and exhaust emissions of those unburned hydrocarbons and carbon monoxide.
- Occasional combustion roughness is associated with a too rapid energy release, leading to an excessively rapid rate of pressure rise. Occasionally, significant preignition reactions take place within the gaseous fuel–air mixture ahead of the ignition of the pilot.
- There is an onset of knocking at high loads with certain fuels and operating conditions, attributed to a form of uncontrolled autoignition and subsequent very rapid combustion of part of the charge.

Generally, the in-cylinder mixing processes in the compression ignition engine are critical for the proper conduct of the combustion process. With the small-quantity pilot, ignition often starts after the end of injection. This would represent a significant opportunity for mixing much of the pilot fuel vapor with the premixed gas fuel and air. A somewhat earlier pilot fuel injection may help in initiating combustion of the lean mixtures since a longer time becomes available for mixing the pilot fuel vapor with the gaseous fuel and subsequent ignition. This is provided that the injection is not too early to adversely affect the preignition processes due to the lower prevailing temperatures and pressures and the greater opportunity for heat losses and dispersion of the pilot fuel vapor down to extremely low concentrations. Usually, for optimal combustion and engine performance, ignition should be such that the resulting peak pressure location needs to be around 10–15°C after the TDC position.

The combustion process in a typical engine during premixed dual-fuel operation depends on both the spray and ignition characteristics of the diesel pilot and on the type of gaseous fuel used and its overall concentration in the cylinder charge. The combustion energy release characteristics in a typical dual-fuel engine reflect throughout the relatively complex nature of the physical and chemical interactions that take place between the combustion processes of the two fuel systems. During the compression stroke, the premixed gaseous fuel–air charge becomes increasingly subjected with time to higher temperatures and pressures as the top dead center position is approached. Some preignition reaction activity may progress within the cylinder charge during compression, especially with certain reactive fuels and some exhaust gas recirculation (EGR) application. With operation on higher hydrocarbon fuels, some natural gases containing higher hydrocarbon components, liquefied petroleum gas (LPG), or even sometimes methane, partial oxidation reactions can proceed during the compression process, even in the absence of pilot injection, particularly in regions where high mixture and surface temperatures may be present. As can be demonstrated through computational modeling, while employing sufficiently detailed chemical kinetics for the oxidization reactions of the fuel vapor and air, the mean temperature of the mixture charge can increase perceptibly as the top dead center position is approached due to increasing chemical activity and the associated exothermic energy releases. This is especially prominent with mixtures that are not very lean. This increase in temperature may compensate to an extent sometimes greater than the lowering of the charge temperature due to changes in the thermodynamic properties of the cylinder content, as well as heat transfer effects. Such a reaction activity is made to proceed prominently and intensely in the absence of the pilot under homogeneous charge compression ignition (HCCI) operation.

The preignition reactions produce radicals and some partial oxidation products, such as those of aldehydes and carbon monoxide. The concentrations of these products can build up significantly during the latter stages of the compression stroke to influence the subsequent ignition and combustion processes of the high hydrocarbon diesel fuel pilot. They will also contribute directly to the combustion of some of the gaseous charge entrained into the pilot jet and in its immediate vicinity, and indirectly to that in the rest of the charge. After ignition of the liquid fuel pilot, the subsequent energy release would reflect that. However, turbulent flame propagation from the pilot ignition regions will not necessarily proceed throughout the charge within the short time available until the concentration of the gaseous fuel is beyond a limiting concentration that would vary with the fuel employed and operating conditions, notably those of temperature. The value of the limiting lean mixture concentration in the engine for flame propagation with methane operation appears to be richer than the corresponding value of the recognized lean flammability limit under the prevailing temperature and pressure conditions. The mixture limits in the engine tend to improve almost linearly with the increase of the size of the pilot.

An overall examination of the combustion process in a premixed dual-fuel engine can be made through the apparent heat release variation when derived from the corresponding cylinder pressure variation with time. For convenience, the energy release rate may be considered to be made up of essentially three overlapping major components. The first, as shown in Figure 13.1, may be that contributed by the combustion of the pilot. The second is due to the combustion of the gaseous fuel component that is considered to be in the immediate vicinity and influence of the ignition and combustion centers of the pilot. The third is due to any preignition reaction activity and subsequent turbulent flame propagation (and sometimes autoignition) within the overall lean mixture of mainly the gaseous fuel in air. With very lean gaseous fuel–air mixtures, the bulk of the energy release is expected to come from the ignition and subsequent rapid combustion of the relatively small pilot zone (the

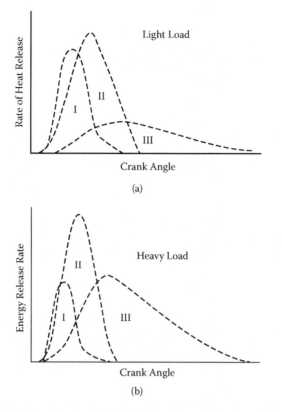

FIGURE 13.1 Schematic representation of the components of the heat release development at (a) light load and (b) heavy load.

first component). It also comes from the combustion of part of the gaseous fuel–air mixture entrained into the burning pilot jet spray, and from the immediate surroundings of such a zone where higher temperatures and relatively richer mixture regions are present. Under these conditions, only relatively little contribution to the energy release may come from the bulk of the gaseous fuel–air charge away from the influence of the pilot zone. Within the very lean mixtures, no consistent flame propagation may take place from the ignition centers and pilot-influenced burning regions. An increase in the size of the pilot zone, whether through increasing the mass of fuel injected or changes in its distribution in very lean gas–air mixtures, will tend to be more than proportional to the total energy released and its associated rates. Greater amounts of fuel gas–air mixtures, then, will be oxidized due to the larger amount of mixtures entrained within the pilot combustion zone, and as a result of the widening of burning regions in the vicinity. Greater energy releases and rates will also be evident due to some partial flame propagation and increased preignition reaction activity of the rest of the charge. Increasing the concentration of the gaseous fuel further while keeping the size of the injected pilot constant will eventually permit, after pilot ignition, flame propagation throughout the rest of the charge, resulting in a sudden increased contribution to the total energy release (Figure 13.2). A continued increase

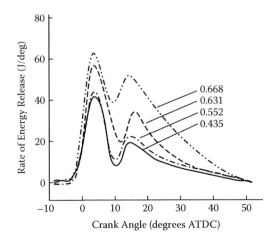

FIGURE 13.2 Calculated energy release rates variation for an engine operating on methane with a constant pilot size showing the typical changes in the release rates when the equivalence ratio is increased through further gas admission. (Adapted from Karim, G.A., and Khan, M.O., *Journal of Mechanical Engineering Science*, 10, 13–23, 1968.)

in the concentration of the gaseous fuel in air will result in a greater overlap between the second and third energy release segments and will lead to their amalgamation, further increasing much of the energy release immediately following the commencement of the autoignition of the pilot. Such intense rapid energy releases may then be associated with the onset of knock.

A consequent feature of the combustion process in the premixed dual-fuel engines operating on lean gaseous fuel–air mixtures is that part of the gaseous fuel and some of the species produced in the combustion process cannot get fully oxidized in time and survive to the exhaust stage. These species, through residual and exhaust gas recirculation, can play important chemical and thermal roles in the ignition and combustion processes of subsequent cycles, and may influence the associated extent of cyclic variations. The thermal effect of these residual gases is reflected in changing the mixture composition, its temperature at the commencement of compression, and influencing the temperature of the cylinder surfaces and associated heat transfer. The kinetic effect of these residuals may result in increasing the activity of preignition reactions during the latter stages of compression.

The oxidation reactions of methane, the major component of natural gas, are seriously affected by the presence of higher hydrocarbon vapors. Hence, the combustion of the droplets and associated vapor of the diesel fuel within a homogeneous mixture of methane in air, as takes place in dual-fuel engine applications, will be very much influenced by the resulting stratified concentrations of the species surrounding each droplet. Accordingly, the heterogeneous nature of the contents and combustion processes in a dual-fuel engine is reminiscent of the rich-lean combustion mixture arrangement devised in some combustion devices, mainly to minimize the production of oxides of nitrogen and particulates. In regions where the pilot vapor is intermixed with the premixed gaseous fuel–air charge, mainly combustion in stoichiometric proportions will take place. This would highlight the need to optimize the processes of pilot injection and of the different regions of the charge to maintain complete combustion, with the undesirable emissions kept at sufficiently low levels. A slightly earlier injection advance may be employed with methane in comparison to operation with propane, since methane is relatively slower burning while less prone to autoignition and knock.

13.2 THE IGNITION DELAY

In compression ignition engines of the diesel type, the injected liquid fuel cannot ignite instantaneously, despite the very high temperatures, pressures, and availability of excess air. A certain time period needs to elapse from the commencement of fuel spray injection before ignition can take place. This time lag to ignition, which is described as the ignition delay period, results from the fact that certain physically and chemically based requirements need to be satisfied before any autoignition of any resulting fuel–air mixture would ignite. These requirements are due to the time needed for injection, atomization, and evaporation of some of the injected liquid fuel, and for the mixing of any resulting fuel vapor with air before chemical gas phase reactions can actively begin somewhere within the cylinder. In turn, these take their own time to reach the ignition stage and begin releasing sufficient energy. It is well recognized that it is essential to keep the length of this delay period as short as possible, so as to effectively control the progress of the combustion process and associated energy release rate through the controlled fuel injection rates. Excessively delayed ignition will lead to reduced power output, low efficiency, increased emissions, and undesirably high rates of pressure rise, which arise mainly from the sudden autoignition of the relatively large amount of fuel that continues to be injected throughout the delay period.

The instant of ignition in engines can be established via a number of approaches that vary in complexity. A common approach is via examination of the changes in the cylinder pressure or its rate of pressure ~ time development records. The instant at which the pressure development shows a departure from the earlier normal trend during compression is taken as the beginning of ignition. Another convenient approach is to derive the effective polytropic index from the pressure ~ time record and observe the rapid change in its value due to the energy release during ignition. This approach can offer a more precise indication of the ignition point than reliance on the observed changes in pressure or its time derivatives. An equally effective approach is to determine the point of ignition from the corresponding rate of heat release when the onset of ignition is associated with the instant when the rate becomes positive. There are other approaches that require relatively elaborate procedures not commonly available in engine installations, except under special research settings. For example, the commencement of ignition may be noted through the associated changes in composition of a number of key reacting species. The ignition point can also be established via the first detectable emission of light within the cylinder of the firing engine when it is possible to suitably observe it.

The effective mean temperature of the charge during the ignition delay is very important and influences the processes going on, both physical and chemical. The mean value of this temperature is influenced by the intake temperature of the charge, heat transfer, water jacket temperature, residual gas effects, amount of fuel-injected preignition processes, and any associated energy releases due to preignition reactions and the extent of penetration by the pilot fuel jet. This is further complicated by the fact that in the cylinder of the internal combustion engine, some stratification in the temperature is produced during compression. This nonuniform temperature distribution would commonly lead to having hotter central regions within the charge than at the outer mixture regions. This resulting temperature difference would lead to preferentially faster reactions within the central region of the mixture, which will influence the overall progress of the ignition delay and subsequent combustion processes. Moreover, in common liquid fuels, which are usually made up of complex mixtures of different fuel components, it is the lighter fractions that are first to evaporate, and most likely would lead to the first formation of a fuel–air mixture and control the overall features of the ignition processes. Hence, this could point to the need for having fuel components that would ensure their easy volatilization, so as to lead to early chemical oxidation reactions in a number of centers that lead to early autoignition. However, a highly volatile fuel for the pilot may lead to its rapid dispersion into the surrounding mixture, to result in altogether undermining the processes leading to ignition.

In the homogeneous premixed gas-fueled dual-fuel engine, the introduction of the gaseous fuel with the intake air produces variations in the physical and transport properties of the mixture, such as the specific heat ratio and, to a lesser extent, the heat transfer parameters. Also, changes in the intake partial pressure of oxygen due to air displacement by the gaseous fuel, changes in the preignition reaction activity and its associated energy release, and the effects of residual gas can bring substantial changes to the preignition processes of the pilot and the gaseous fuel. Accordingly, in the dual-fuel engine the ignition delay displays trends significantly different from those observed in the corresponding diesel engine operation. The delay tends to increase initially with the increased gaseous fuel admission up to a detectable maximum value, and then drops to a minimum well before reaching the total stoichiometric ratio based on the combined gaseous and liquid fuels and the available air. For example, as seen in Figure 13.3, the admission of propane, which is normally regarded as a relatively more reactive fuel in air in comparison to methane, in fact tends to produce delay values greater than those observed with methane.

For any constant-compression-ratio, unthrottled engine operating at a fixed intake temperature in the absence of pilot injection, the peak temperature at the end of compression varies not only with the type of gaseous fuel, but also with its concentration in air. For example, with propane fumigation for the case shown in Figure 12.8, a drop in temperature of around 100 K can be observed with the stoichiometric mixture. The heat loss during compression influences the change in temperature also. On this basis, the contribution of the energy release of the preignition reaction activity to the development of the average temperature of the mixture is also strongly dependent on the gaseous fuel used. These reflect the net value of the reaction activity brought about by the lowering of the mean temperature due to thermodynamic and heat transfer effects compared to the lesser effects of the changes brought about by the increased fuel gas admission. Hence, the increased admission of either methane

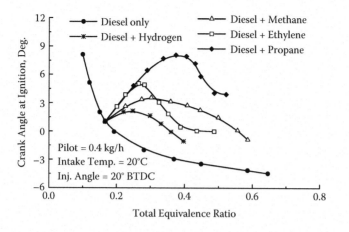

FIGURE 13.3 Variations of the point of ignition with total equivalence ratio of a dual-fuel engine operating on hydrogen, methane, ethylene, and propane using a fixed pilot quantity. (Adapted from Burn, K.S., The Effect of Cold Intake Temperatures on the Combustion of Gaseous Fuels in a Dual-Fuel Engine, MSc thesis, University of Calgary, Canada, 1977.)

or propane under some mixture conditions may not necessarily enhance the pre-ignition reactivity of the gaseous fuel–air mixture in a constant-compression-ratio engine. Such a trend of behavior shows that even in the absence of considering the contribution of the pilot injection to the length of the delay, the increased admission of a relatively reactive fuel such as propane can delay the ignition process, increasing the associated length of the delay and affecting dual-fuel performance in general, especially at low intake mixture temperatures.

The employment of exhaust gas recirculation (EGR), or the presence of diluent gases such as carbon dioxide or nitrogen in the fuel gas or in the air, into an engine charge can also produce different changes to the state of the mixture during compression. For example, the relative mean value of the mixture temperature during compression decreases virtually linearly with increasing the admitted concentrations of carbon dioxide, while the admission of nitrogen, in comparison, affects only a little the temperature development from that with only air. These trends result primarily from the marked changes in the effective specific heats. Also, it would be expected that the increase in the length of the delay with the increased admission or carbon dioxides can be considered to be largely a consequence of this lowering in charge temperature, as well as the effect of the displacement of some of the available oxygen by the diluent gas. Consequently, in dual-fuel engines, the small quantity of the pilot liquid fuel is injected into a mixture of gaseous fuel and air at mean temperature and pressure that may be different from the corresponding values for plain diesel operation. This is despite having the same compression ratio, engine speed, and initial temperature and pressure. On this basis, the processes of atomization, vaporization, and distribution of the small quantity of pilot fuel, as well as the pre-ignition reaction processes, would be affected by any changes in the flow, thermal, and transport characteristics of the charge. For optimum engine performance, there is a need to appropriately optimize the injection characteristics of the pilot for the homogeneously mixed gaseous fuel–air dual-fuel operation, instead of maintaining operation in accordance with that for pure diesel operation. This is becoming increasingly possible through the widespread reliance on in-line fuel common rail injection in diesel engines converted to dual-fuel operation.

The preignition reaction activity is widely different for the higher hydrocarbon components of the pilot fuel vapor than that of the premixed air of the gaseous fuel component, such as methane or hydrogen. For example, n-heptane, a fuel that may represent approximately the overall behavior of the light fractions of diesel fuel, begins to undergo complex multistage ignition reactions at much lower temperature levels than for the gaseous fuel, such as methane. The presence of any liquid pilot fuel vapor in various regions of the premixed gaseous fuel–air charge in a dual-fuel engine may bring about significant local changes to the overall rates of preignition reactions and the associated energy release rates. Figure 13.4 shows an example of the very rapid reduction in the ignition delay of methane and subsequent combustion time with increasing the mole fraction of n-heptane, β, in stoichiometric mixtures in air under constant volume conditions, with initial temperature and pressure similar to those that exist at the time of pilot fuel injection. Also, the presence of some relatively minute quantities of heptane vapor with the methane does substantially speed up the rates of ignition processes from those of methane on its own. The small concentrations of the heptane vapor begin to react with the oxygen well ahead of the methane, producing both exothermic energy releases and key transient products that serve to provide some of the radicals subsequently needed for the methane oxidation. It can be seen that the length of the ignition delay in the premixed dual-fuel engine is very much affected by the presence of a gaseous fuel with the air during the ignition of the pilot liquid fuel. Moreover, a relatively small lowering of the temperature can increase the delay substantially.

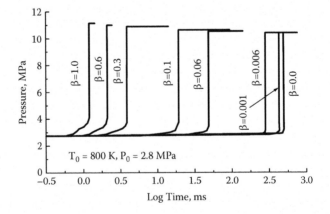

FIGURE 13.4 Calculated variation of the pressure development with the log of time due to autoignition reactions for a stoichiometric of various n-heptane-methane mixtures in air, under constant volume conditions with initial temperature and pressure of 800 K and 2.8 MPa respectively, β is the molar fraction of n-heptane in the fuel mixture. (Adapted from Khalil, E.B., Modeling the Chemical Kinetics of Combustion of Higher Hydrocarbon Fuels in Air, PhD thesis, University of Calgary, Canada, 1998.)

Since the extent of increase in the observed length of the ignition delay tends to be very sensitive to small changes in the effective value of the intake temperature of the mixture, very low intake temperature mixtures, especially those with small pilot quantities and at light engine loads, tend to produce increased erratic engine running that could even lead to ignition failure. This is especially noticeable with liquefied natural gas (LNG)-fueled engines.

Advancing the injection timing of the pilot can generally lead, up to a point, to an earlier ignition. However, the improvements brought about by such an advance of the pilot injection are rather limited and can lead to increased erratic running. In homogeneously premixed gaseous fuel–air dual-fuel engines, such as those burning methane, a relatively longer ignition delay does not necessarily produce a harsher knock or combustion noise in comparison to the corresponding diesel engine operation. This is because the autoignition energy release of the pilot is relatively small, and the preignition reactions of the lean gaseous charge are less active than those of the vapor-air mixtures in the diesel engine.

The energy released by the pilot is supplemented by the energy released from the local surrounding fuel and air, which would help in increasing the overall temperature level, as well as the mixing of the partial and complete combustion products of the usually richer mixture. On this basis, the high rate of pressure rise from pilot fuel ignition of relatively large size tends to contribute to the promotion of autoignition processes within some parts of the cylinder charge, which would increase the likelihood of the onset of knock, rather than primarily through turbulent flame propagation. Throughout, the effects of any energy release due to preignition reactions during compression will contribute to the ignition processes of the pilot, thermal load, efficiency, power output, and emissions.

For improved dual-fuel engine performance, the size of the pilot and its timing need to be adjusted optimally, depending on the associated operating condition and the fuel gas employed.

13.3   FLAME PROPAGATION LIMITS

The performance of a dual-fuel compression ignition engine improves with the increased admission of the gaseous fuel or the relative pilot size. This improvement appears to be dependent on the total equivalence ratio, based on both the pilot diesel and gaseous fuels. It is evident that there is a limiting leanness in the equivalence ratio, below which the exhaust emissions of mainly carbon monoxide and unconsumed fuel gas become hardly affected by changes in the pilot quantity. This has been suggested to be consistent with the existence of a limiting equivalence ratio or an operational mixture limit for successful flame propagation within the limited time available from the pilot ignition centers into the surrounding fuel gas–air mixture. The continued leaning of mixtures produces lower combustion temperatures, which leave a substantial amount of the carbon monoxide produced as a precursor for complete fuel oxidation reactions remaining unconverted within the time available for combustion. For sufficiently lean mixtures, the resulting combustion temperature may become so low that only little oxidation reactions of the fuel gas proceed within and very close to the pilot regions. The bulk of the fuel gas is left unconverted, while producing only small amounts of carbon monoxide, with much of the fuel gas remaining unconverted.

The volumetric concentration of the gaseous fuel in the intake air that identifies a conceived boundary for the commencement of improved engine operation may be taken to represent the operational limit for a dual-fuel engine under those operating conditions. This limit would represent the minimum concentration of the gaseous fuel in air for which the flame propagation appears just to begin to spread virtually throughout the entire cylinder charge within the time available. Such a limit for a specific engine can be determined experimentally. The reduction in the value of the limit with increasing the pilot quantity is due to a number of factors, which include correspondingly improved pilot injection characteristics, a larger size of pilot mixture envelope with a greater entrainment of the gaseous fuel, a higher number of ignition centers requiring shorter flame travel distances, a greater energy release on ignition, with higher rates of heat transfer to the unburned gaseous fuel–air mixture, and an increased positive contribution of the hot residual gases. The value of the limit is lowered as the intake charge temperature increases, aided mainly by the corresponding increase in the mean charge temperature during pilot ignition. Nevertheless, the limits can be viewed to correspond to conditions when the rate of energy release by combustion cannot keep pace with the demands for transporting sufficient energy to adjoining layers of the mixture and overcoming external thermal losses to the surrounding cylinder walls. Moreover, the values of the engine operational limits also depend on the associated mechanical and motoring losses. Normally, the operational lean mixture limits in the engine correspond to less lean values than the corresponding recognized flammability limit values of the fuel gas for the same operating conditions at around TDC position.

It is accepted that values of the lean flammability limits of different common fuel mixtures in air tend to correspond to mixtures having approximately the same calculated flame temperature. Of course, the effective flame front temperature in the engine is a complex function of most engine operating variables, such as those of the intake temperature and pressure, compression ratio, pilot size and its injection quantity and timing, heat transfer, and the physical properties of the mixture. For simplicity, the effective flame temperature of the cylinder lean limit charge in the case of dual-fuel compression ignition engines may be considered to be approximately proportional to the temperature when evaluated in terms of assuming it to be around the temperature at TDC, increased fractionally due to the combustion of the pilot. Some of the experimental results with methane operation showed that this fraction is around 40% of the heat release of the pilot (Figure 13.5). Such a procedure, however approximate it may be, can be helpful in estimating, for example, for an engine and operating conditions, how small a pilot can be employed for a lean mixture operation, or the minimum amount of preheating needed to obtain good combustion and acceptable levels of emissions.

Figure 13.6 shows that for a relatively small pilot quantity of around 3.5% of the full-load diesel quantity, while using the injection system of the original diesel engine, the bulk of methane fumigated into the engine remains unconverted and appears in the exhaust gas. However, when the concentration of the fumigated fuel gas in the intake reaches a certain value, the methane appears to get consumed rather

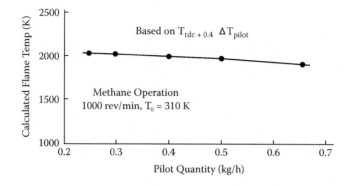

FIGURE 13.5 Variation of the calculated corresponding flame temperature mixture conditions with changes in the pilot quantity for methane operation in air. (From Bade Shrestha, O.M. and Karim, G.A., *ASME Journal of Energy Resources Technology*, 128, 223–228, 2006.)

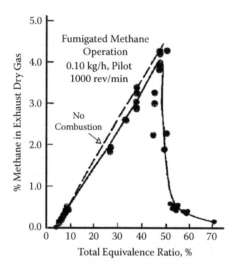

FIGURE 13.6 Variation of the exhaust concentration of methane with changes in gas equivalence ratio for methane operation with a relatively small constant pilot quantity. (Adapted from Jones, W., and Burn, K.S., The Effect of Cold Intake Temperatures on the Combustion of Gaseous Fuels in a Dual Fuel Engine, MSc thesis, University of Calgary, Canada, 1977.)

abruptly, indicating that its lean mixture with air has reached a concentration that supports combustion originating from the pilot ignition region within the time available. This apparent limiting concentration becomes of lower value as the size of the pilot is increased or a suitably dedicated pilot fuel injector for dual-fuel engine operation is employed.

An example of the changes in the specific energy consumption with the total equivalence ratio for three different pilot sizes when operating on fumigated methane is shown in Figure 13.7. The very substantial improvement in the fuel gas utilization with increasing the size of the pilot is clearly evident. It is also shown that

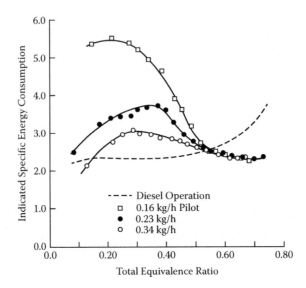

FIGURE 13.7 Variations in the specific energy consumption of a dual-fuel engine operating on methane for three different values of pilot quantity. (Adapted from Azzouz, D., Some Studies of the Combustion Processes in Dual Fuel Engines, MSc thesis, London University, UK, 1965.)

beyond a certain value of equivalence ratio, the increases in the size of the pilot make only a little improvement to the specific energy consumption, which indicates reductions in the size of the pilot are then acceptable. Moreover, at sufficiently high load, the dual-fuel engine energy utilization clearly appears superior to the corresponding values under diesel operation.

13.4 HOMOGENEOUS CHARGE COMPRESSION IGNITION (HCCI) ENGINE COMBUSTION

The homogeneous charge compression ignition (HCCI) engine has been suggested to have a number of key potential positive features that include high work production efficiency and low emissions of NOx and particulates. However, these engines still face a number of challenges, particularly for mobile applications. These include the need for effective control of the combustion process over a wide range of operating conditions, improving the operation of mixture limits at very low loads, developing wider turn-down ratios, and improving the start from cold and warm-up operation.

It is generally accepted that the combustion process in these engines is dominated by the autoignition processes and controlled mainly by the chemical-kinetic reaction rates, with a much lesser need to consider flame propagation. With the self-accelerating nature of autoignition reaction rates, the resulting heat release tends to become too high for acceptable operational conditions that take place in association with increased heat loss, noise, and vibrations.

There has been much attention focused on the problems associated with achieving effective control of the autoignition process in HCCI engines, and hence their performance. Some of these measures include controlling the mixture temperature during intake and compression, which include the employment of internal retention or external recirculation of the exhaust gas, EGR, some variation in the compression ratio, control of the development of the reacting species during compression and combustion, and fuel mixture stratification while using a wide variety of fuels and additives. It remains difficult to realize optimum control of the combustion process over wide ranges of engine loads and speeds, especially under transient operating conditions. Since the duration of combustion in HCCI engines tends to be very short in comparison to those of the spark ignition or diesel modes, HCCI engines' performance deteriorates significantly when the combustion process is not adequately optimized. High NOx emissions with knock, or misfire, with lower thermal efficiency and higher HC and CO emissions could be produced. It is also difficult to realize optimum combustion control when the engine load or speed is changed rapidly. Hence, it remains a challenge to develop a transient optimum control system and increase the power output of HCCI engines. So far, HCCI engine operation tends to be confined to part load and idling conditions. Various approaches to increase HCCI engine power have been reported. However, the improvements they produce tend to be insufficient to satisfy their practical requirements, including overcoming the significant limitation imposed by the onset of knock. A possible approach to improve HCCI engine power production is to change over to diesel operation under high load conditions. However, dealing satisfactorily with associated excessive particulates and NOx emissions would represent a challenge, as with traditional diesel engines. Should the HCCI engine be switched over to operation in the SI mode, then combustion becomes limited to operating at much lower compression ratios, bringing with it serious reductions to the thermal efficiency and the avoidance of knock. However, the changing of HCCI operation to a premixed dual-fuel mode while retaining the high compression ratios would appear to represent a more favorable alternative approach.

The deterioration in HCCI combustion under cold-start conditions is a consequence of the excessively long ignition delays and combustion durations, with high HC and CO emissions produced due to incomplete combustion. This would indicate that stable HCCI combustion most likely is only suitable for a very narrow range at part load, while the extension to the load limit range remains a challenge for a workable HCCI engine. On this basis, efforts to successfully develop high-performance commercial HCCI engines are continuing, with expenditure of time and resources to develop these engines to the practical mass production stage. It is quite likely that with the potential limitations facing the HCCI combustion engine, it will not be able to displace the traditional spark ignition, diesel, or dual-fuel engines. It is also unlikely that it can replace the dual-fuel engine in its diversity of operation on a wide range of different fuels.

13.5 COMBUSTION WITH DIRECT IN-CYLINDER HIGH-PRESSURE GAS INJECTION

There are engines dedicated to dual-fuel operation that employ high-pressure gas injection directly into the cylinder just after pilot fuel injection. These engines, which have been favored for transport applications, can use the same injector body housing with different passages for the two fuels, which are injected sequentially in multiple jets of symmetrical patterns. Such an approach has to be able to control many variables affecting the combustion process, perhaps well beyond those normally required in the premixed dual-fuel engine type. To keep the number of active control variables down, some of these may be kept fixed. Such an operational approach tends to reduce the potential to fully achieve some of the desirable features of dual-fuel operation, while the full range of performance in the normal diesel mode may not be achievable. For example, in order to ensure reliable ignition and combustion of the fuel gas over the whole load and speed ranges of the engine, the diesel pilot quantity injected per stroke may be kept constant while contributing a sizable fraction of the energy input. The load is then varied mainly by changing the quantity of the fuel gas injected. The time interval between the injections of the two fuels may be kept fixed, but changed with engine speed. The injection timing is made variable depending on speed and load.

The very high gas pressure needed relative to the prevailing pressure inside the cylinder after pilot ignition would have the fuel gas discharged under choked conditions, with the injection rate directly related to the gas injection pressure. A constant injection pressure is often employed for the fuel gas. Hence, increasing the amount of gas injected will be associated with a longer injection completion period and a longer combustion time that may extend at high loads well down the expansion stroke. Strict controls of the injection processes are also needed since a high injection pressure will produce jets that would overpenetrate the whole chamber and impinge on the cylinder and piston surfaces, leading to quenching, increased heat transfer, and lubricant losses. Also, it would result in much of the combustion taking place closer to the periphery of the cylinder.

It is evident that there is a critical need to properly orient the injection jets of the two fuels and control their respective times and discharge velocities. This needs to be carried out while ensuring the proper mixing of the several individual gas jets, with the burning centers originating from the pilot jets on one hand, and the provision of oxygen and the proper dispersion of the products on the other. This is quite a challenging task, particularly under transient operating conditions, when capital and operational costs are to be kept competitive with those of other types of prime movers.

REFERENCES AND RECOMMENDED READING

Aisho, Y., Yaeo, T., Koseki, T., Saito, T., and Kihara, R., *Combustion and Exhaust Emissions in a Direct Injection Diesel Engine Dual-Fueled with Natural Gas*, SAE Paper 950465, 1995.

Al-Himyary, T.J., Karim, G.A., and Dale, J.D., *An Examination of the Combustion Processes of Methane Fueled Engine when Employing Plasma Jet Ignition*, SAE Paper 891639, 1989.

Austen, A.E., and Lyn, W.T., Relation between Fuel Injection and Heat Release in a Direct Injection Engine and the Nature of the Combustion Processes, in *Proceedings of the Institute of Mechanical Engineers (AD)*, 1961, no. 1, pp. 47–62.

Azzouz, D., Some Studies of the Combustion Processes in Dual Fuel Engines, MSc thesis, London University, UK, 1965.

Bade Shrestha, O.M., and Karim, G.A., The Operational Mixture Limits in Engines Fuelled with Alternative Gaseous Fuels, *ASME Journal of Energy Resources Technology*, 128, 223–228, 2006.

Bade Shrestha, S.O., Wierzba, I., and Karim, G.A., An Approach for Predicting the Flammability Limits of Fuel-Diluent Mixtures in Air, *Journal of the Institute of Energy*, 122–130, 1998.

Badr, O., Karim, G.A., and Liu, B., An Examination of the Flame Spread Limits in a Dual Fuel Engine, *Applied Thermal Engineering*, 19, 1071–1080, 1999.

Burn, K.S., The Effect of Cold Intake Temperatures on the Combustion of Gaseous Fuels in a Dual-Fuel Engine, MSc thesis, University of Calgary, Canada, 1977.

Carlucci, A.P., Ficarella, A., and Laforgia, D., Control of the Combustion Behavior in a Diesel Engine Using Early Injection and Gas Addition, *Applied Thermal Engineering*, 26, 2279–2286, 2006.

Challen, B., and Barnescu, R., *Diesel Engine Handbook*, 2nd ed., SAE, Warrendale, PA, 1999.

Chen, K., and Karim, G.A., *An Examination of the Effects of Charge Inhomogeneity on the Compression Ignition of Fuel–Air Mixtures*, SAE Paper 982614, 1998.

Chrisman, B., Callaham, T., and Chiu, J., *Investigation of Macro Pilot Combustion in Stationary Gas Engine*, ASME Paper 98-ICE-106, 1998.

Daisho, Y., Takahashi, Y.I., Iwashiro, Y., Nakayama, S., and Saito, T., *Controlling Combustion and Exhaust Emissions in a Direct-Injection Diesel Engine Dual Fuelled with Natural Gas*, SAE Paper 952436, 1995.

Ebert, K., Beck, N.J., Barkhimer, R.L., and Wong, H., *Strategies to Improve Combustion and Emission Characteristics of Dual Fuel Pilot Ignited Natural Gas Engines*, SAE Paper 971712, 1997.

Elliot, O., and Davis, R.E., Dual Fuel Combustion in Diesel Engines, *Industrial and Engineering Chemistry*, 43, 2854–2863, 1951.

Gee, D., and Karim, G.A., Heat Release in a Compression-Ignition Engine, *The Engineer*, 222, 473–479, 1966.

Ishida, M., Chen, Z.L., Luo, G.F., and Ueki, H., *The Effect of Pilot Injection on Combustion in a Turbocharged D.I. Diesel Engine*, SAE Paper 841692, 1994.

Jones, W., and Burn, K.S., The Effect of Cold Intake Temperature on the Combustion of Gaseous Fuels in a Dual-Fuel Engine, MsC thesis, University of Calgary, Canada, 1977.

Jones, W., Liu, Z., and Karim, G.A., *Exhaust Emissions from Dual Fuel Engines at Light Load*, SAE Paper 932822, 1993.

Jones, W., Raine, R.R., and Karim, G.A., *An Examination of the Ignition Delay Period in Dual Fuel Engines*, SAE Paper 892140, 1989.

Karim, G.A., A Review of Combustion Processes in the Dual Fuel Engine—The Gas Diesel Engine, *Progress in Energy and Combustion Science*, 6, 277–285, 1980.

Karim, G.A., Combustion in Gas Fueled Compression Ignition Engines of the Dual Fuel Type, *ASME Journal of Engineering for Gas Turbines and Power*, 125, 827–836, 2003.

Karim, G.A., and Khan, M.O., Examination of Effective Rates of Combustion Heat Release in a Dual Fuel Engine, *Journal of Mechanical Engineering Science*, 10, 13–23, 1968.

Karim, G.A., and Ward, S., The Examination of the Combustion Processes in a Compression-Ignition Engine by Changing the Partial Pressure of Oxygen in the Intake Charge, *SAE Transactions*, 77, 3008–3016, 1968.

Karim, G.A., Klat, S.R., and Moore, N.P.W., Knock in Dual Fuel Engines, in *Proceedings of the Institute of Mechanical Engineers*, 1967, vol. 181, pp. 453–468.

Khalil, E., and Karim, G.A., A Kinetic Investigation of the Role of Changes in the Composition of Natural Gas in Engine Applications, *ASME Journal of Engineering for Gas Turbines and Power*, 124, 404–411, 2002.

Khalil, E.B., Modeling the Chemical Kinetics of Combustion of Higher Hydrocarbon Fuels in Air, PhD thesis, University of Calgary, Canada, 1998.

Kubish, J., and Brehob, D.D., *Analysis of Knock in a Dual Fuel Engine*, SAE Paper 922367, 1992.

Kusaka, J., Daisho, Y., Shimonagata, T., Kihara, R., and Saito, T., Combustion and Exhaust Characteristics of a Diesel Engine Dual-Fuelled with Natural Gas, in *Proceedings of the 7th International Conference and Exhibition on Natural Gas Vehicles*, Yokohama, Japan, October 17–19, 2000, pp. 23–31.

Larson, C.R., Bushe, W.K., Hill, P.G., and Munshi, S.R., Relative Injection Timing Effects in a Diesel Engine Fuelled with Pilot-Ignited, Directly Injected Natural Gas, in *Canadian Section of the Combustion Institute (CICS) Spring Technical Meeting*, 2003, p. 86.

Liu, Z., and Karim, G.A., *The Ignition Delay Period in Dual Fuel Engines*, SAE Paper 950466, 1995.

Lowe, W., and Williamson, P.B., Combustion and Automatic Mixture Strength Control in Medium Speed Gaseous Fuel Engines, in *Proceedings of the 8th Congress de Machines a Combustion (CIMAC)*, 1996, pp. A14–A56.

Nielson, O.B., Qvale, B., and Sorenson, S., *Ignition Delay in the Dual Fuel Engine*, SAE Paper 870589, 1987.

Tomita, E., Fukatani, N., Kawahara, N., and Maruyama, K., Combustion in a Supercharged Biomass Gas Engine with Micro Pilot Ignition Effects of Injection Pressure and Amount of Diesel Fuel, *Journal of Kones Powertrain and Transport*, 14(2), 513–520, 2007.

Xiao, F., and Karim, G.A., *Combustion in a Diesel Engine with Low Concentrations of Added Hydrogen*, SAE Paper 2011-01-0676, 2011, pp. 1–17.

14 Dual-Fuel Engine Performance

14.1 PERFORMANCE UNDER LIGHT LOAD CONDITIONS

The concentration of the fuel and air within and around the combustion zone enveloping a jet diffusion flame in a premixed fuel–air surrounding, as shown schematically in Figure 14.1, extends significantly both radially and axially as the concentration of the gaseous fuel in the surrounding air is increased. This would go on until a concentration is reached that would permit the propagating flame to sweep through.

Similarly, the ignition centers formed off the injected pilot liquid fuel will be affected by the presence of the gaseous fuel and any increase in its concentrations. Both the reaction rates and the associated energy release rates will be increasingly modified by the presence and any increase in the concentration of the gaseous fuel in the liquid fuel surroundings. Figure 14.2 shows the changes in the energy release rate with increasing the concentration of methane for a dual-fuel engine operating on methane with a constant pilot quantity. A consequence of this is a feature of dual-fuel engines where relatively poor performance under light load conditions is displayed when very lean mixtures of the gaseous fuel in air are employed, especially with a relatively small pilot. The extent of this relative deterioration in engine performance depends on numerous affecting factors, but it is largely dependent on the pilot quantity employed, the gaseous fuel used and its concentration in air, the engine employed, and the prevailing local operating conditions. Under these circumstances, with the introduction of the gaseous fuel, a significant proportion of the fuel will remain unchanged to appear in the exhaust gases, which indicates a failure of the propagating flame to

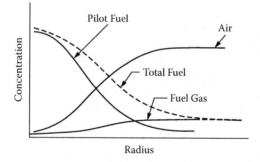

FIGURE 14.1 Schematic representation of the distribution of fuel concentrations radially along a fuel jet with and without the presence of gaseous fuel in its surroundings.

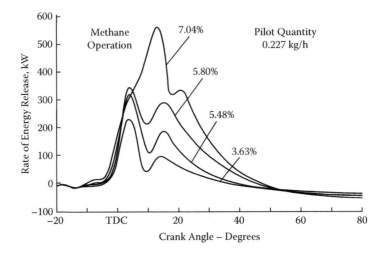

FIGURE 14.2   An example showing changes in the gross heat release rate of an engine fueled with methane as the intake concentration of the gas is increased for constant pilot operation. (From Karim, G.A., and Khan, M.O., *Journal of Mechanical Engineering Science*, 10, 13–23, 1968.)

cover the entire charge within the time available. This goes on despite the availability of much excess air and a significant consistent energy release arising from pilot ignition. Evidently, the flames that could have originated from the pilot ignition regions cannot propagate fully and far enough into the surrounding lean mixtures, leaving some of the gaseous fuel unconverted or only partially oxidized.

Normally, associated with these low gaseous fuel oxidation rates within lean premixed gaseous fuel–air mixtures is a significant increase in partial oxidation products, notably those of carbon monoxide, to an extent well beyond the low values that are normally observed with the corresponding diesel operation. However, there is hardly any indication of the production and survival of particulates to the exhaust stage. Figure 14.3 shows an example of the variation in the concentration of carbon monoxide in the exhaust gas of a methane-fueled engine operating on a constant pilot quantity and speed with the increased admission of methane. Figure 14.4 shows an example of the variation in the exhaust concentration of the unconverted methane with changes in the extent of conversion with three different size constant pilots. Through a continuing increase in the intake gaseous fuel concentrations, a value will be reached when the flame manages to spread within the time available through the whole mixture, which can be then considered to be associated with an effective operational flame spread limit. Beyond this concentration, the fuel conversion approaches completion with an associated improvement in engine performance. Thus, as a remedial measure, a common practice when converting diesel engines to premixed dual-fuel operation has been to retain diesel operation for idling and very low load operations, or to operate with sufficiently large pilots. Otherwise, the specific energy consumption, when based on the pilot diesel fuel and the gaseous fuel supplied, together with the associated power output and emissions, will be markedly

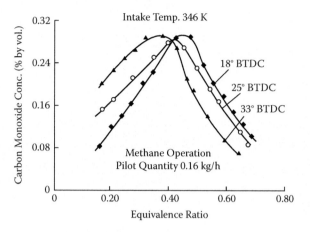

FIGURE 14.3 An example of the variation of the concentration of carbon monoxide with total equivalence ratio of a dual-fuel engine operating with constant pilot on fumigated methane for three different pilot injection timing. (Adapted from Azzouz, D., Some Studies of the Combustion Processes in Dual Fuel Engines, MSc thesis, London University, UK, 1965.)

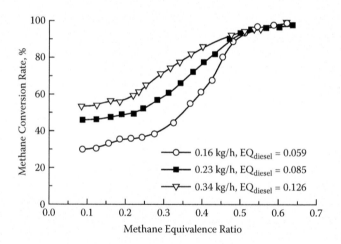

FIGURE 14.4 Typical variation of the extent of methane conversion with changes in methane equivalence ratio for three different pilot sizes, fumigated methane operation, and constant speed. (Adapted from Xiao, F., Experimental and Numerical Investigation of Diesel Combustion Processes in Homogeneously Premixed Lean Methane or Hydrogen–Air Mixtures, PhD thesis, University of Calgary, Canada, 2011.)

inferior to those values obtained with the corresponding straight diesel operation. This is particularly evident for the relatively slow burning methane and at low intake mixture temperature operation.

When very lean mixtures are employed, the associated relatively slow flame propagation and quenching are the main factors affecting dual-fuel engine performance. The mixing of some higher hydrocarbon vapor originating from the pilot with some of the lean gaseous fuel–air mixture tends to enhance flame propagation locally due to both local fuel enrichment and the presence of more reactive higher hydrocarbons. Accordingly, dual-fuel engines, when destined mainly for low load operation, become associated with high installed and operational costs per unit of power produced.

14.2 SOME REMEDIAL MEASURES FOR IMPROVING LIGHT LOAD OPERATION

Light load dual-fuel operation for premixed gaseous fuel–air charges at light load, when using relatively small pilot quantities, may be improved through the following possible measures:

- Using relatively large pilot sizes with optimum injection characteristics, such as by employing low injection nozzle opening pressure and advancing the injection timing somewhat without undermining the corresponding diesel operation. Increasing the effective cetane number of the liquid pilot will also enhance light load operation and permit operation with gaseous fuel mixtures that may contain high concentrations of the diluents' nitrogen or carbon dioxide, such as with biogases.

- Attempts to produce an effectively richer mixture are implemented through partial restriction of the air component of the charge via throttling for the same mass of fuel gas. This simple procedure, which requires some modifications to the diesel engine controls when applied at light load, can improve the combustion process, reduce the emissions of unburned fuel gas and carbon monoxide, and improve the effective specific energy consumption. However, the throttling procedure needs to be implemented with care so as not to undermine pilot ignition processes and its emissions. Also, the associated increase in pumping losses with the excessive employment of throttling can reduce the extent of improvement to engine efficiency. In turbocharged dual-fuel engine applications at light load, the excess air may be made to controllably bypass the engine altogether or sometimes redirected to the intake manifold, helping to slightly warm the lean mixture.

- In multicylinder engine applications with fuel gas injection, the procedure known as skip firing may be applied to provide improved overall performance. With this approach, some cylinders may be made to operate continuously or alternately, wholly in the diesel mode. The remaining cylinders, while operating in the dual-fuel mode, employ less lean gas fuel–air mixtures than had the same mass of gas been distributed uniformly to all cylinders. Procedures such as these may not be readily available in all engine instillations since they require careful control of the operating variables, so as to achieve the desired improvements to engine performance. With the recent advances made in the control and injection of liquid and gaseous fuels, the application of this procedure has been increasingly applied.

- Slight preheating of the lean intake gas–air mixture may be made, such as through heat exchange with the exhaust gas or increasing the water jacket temperature. This can result in a higher mixture temperature at the end of compression and during the pilot injection period, leading to reductions in the ignition delay and the fraction of the gaseous fuel surviving the combustion process and the associated emission of carbon monoxide. This procedure is also consistent with the tendency of the lean flammability limit values of all common fuels and their mixtures in air to widen approximately linearly with the average mixture temperature. In turbocharged engine applications, the feeding back of some of the excess air that has been warmed through rapid compression into the compressor intake at light load may also help when executed properly.
- The selective addition at light load of small amounts of a suitable auxiliary fuel to the main gaseous fuel supply, such as hydrogen, higher hydrocarbons, or some gasoline vapor, may improve dual-fuel engine operations at light load. Such an approach should also be carefully executed and may even be avoided beyond the light load range since apart from adding complexity to controls, it adversely affects costs and the specific energy consumption. It can also bring about an earlier onset of knock.
- Some stratification of the gaseous fuel component in relation to the air, if it can be implemented effectively in practice, such as through proper control of direct in-cylinder gas injection, may improve light load operation. This is achieved through arranging for a richer mixture in the vicinity of the pilot fuel so that following combustion of these richer regions, flames may be able to propagate through much of the rest of the leaner mixtures. Resorting to some uncooled exhaust gas recirculation may also help. However, with lean mixture operation, the exhaust temperature level is still rather low, limiting the preheating of the mixture that can be achieved.
- An effective method for improved operation on overall very lean mixtures through stratified charge combustion is through having a prechamber to the engine cylinder. The prechamber is supplied directly with fuel so as to produce a richer mixture than that formed in the main combustion chamber. On pilot injection into the richer mixture, the resulting high-velocity burning jet will emerge into the main chamber to ignite and burn more completely the leaner mixture.

Resorting to a judicial combination of these measures, bearing in mind the question of cost and increased complexity, can be effective in improving light load performance of dual-fuel engines while retaining the capacity to operate effectively as a diesel, when needed. It is also evident that dual-fuel engine operation is more suited to the high load range, where less lean mixtures are employed. Otherwise, the engine may be started in the diesel mode and switched over to dual-fuel operation once the engine reaches normal operating conditions, such as with respect to temperature and load.

Some other additional potential measures for improving dual-fuel performance in general and at light load may include the employment of variable valve timing, variable geometry turbochargers, turbocharger bypass, and controlled adjustment of heating some cylinder surfaces.

The conversion of diesel engines to dual-fuel operation results in overall lean mixture operation. This will be conducive at high loads to achieving high efficiency with reduced emissions, especially those of the oxides of nitrogen and particulates. It may even be possible, particularly outside transport applications, to dispense with some of the exhaust gas treatment measures, such as through the employment of catalytic converters. The engine will tend to run cooler, with less tendency to knocking, and can contribute to improved durability. However, there are some associated potential drawbacks with such an operation. These include having the combustion and flame propagation processes proceeding at relatively slower rates with increased cyclic variation than a diesel-only operation. A compromise may be to operate on sufficiently lean mixtures to reduce oxides of nitrogen emission while avoiding operation on excessively too lean mixtures, so as to reduce the impact of some of these undesirable operational features.

14.3 THE DISPENSING OF FUGITIVE GAS EMISSIONS THROUGH THE OPERATION OF DIESEL ENGINES ON ULTRA-LEAN METHANE–AIR MIXTURES

The fuel injection characteristics in diesel engines that are critically important to engine performance are controlled by numerous operating and design variables. These would include the injection pressure, number of orifices employed, spray injection angles, fuel properties, and fluid dynamics of the mixture within the cylinder. Much time, effort, and resources go into attempts to diagnose performance shortfalls and optimize the liquid fuel injection system. It would be highly desirable to develop a simple and relatively fast approach that can give an indication of whether any change in the variables of the system would have a beneficial effect on engine combustion or not, so as to lead to improved performance and reduced emissions. Such a simple approach may exploit the known features of dual-fuel engine behavior under light load conditions. This may be obtained through operating the engine on ultra-lean methane–air mixtures at concentrations well below those corresponding to the effective flame spread limit values under the prevailing combustion chamber conditions. The very slow rates of oxidation reactions of the very lean mixtures of premixed methane in air are greatly speeded up by mixing with vapors emanating from the light fractions in diesel fuels, even when present at very low concentrations. The boundaries of jet diffusion flames, when burning in homogeneously mixed atmospheres of sufficiently lean methane–air, cannot extend beyond those with pure air until sufficiently high methane concentrations in the surrounding air are employed. On this basis, it can be suggested that in a firing diesel engine operating on air, very small amounts of methane are introduced to function as a marker, so as to produce ultra-lean homogeneous methane–air mixtures. The measured concentrations of the unconverted methane appearing in the exhaust gases can be considered an indication of the extent and size of regions where diesel fuel combustion could not extend to. The small fraction of methane converted may be indicative to have been directly present within the combustion regions of the diesel fuel. On this basis, any positive change in engine design and operating parameters would be reflected in a relatively higher extent of relative methane conversion.

Experience derived from investigating the ignition of homogeneous mixtures of methane–air in engines using high-energy sparks or plasma jets indicates that it is possible to have successful local ignition of a very lean mixture when a sufficiently high energy ignition and temperature source is employed. However, once the initiated flame departs well away from the ignition source, it becomes increasingly difficult for the flame to continue propagating independently throughout the rest of the mixture. However, having a cluster or multiple high ignition sources would result in successful ignition of ultra-lean mixtures, with the resulting flames initiated from these centers not needing to propagate far before encountering and being supported by other flame fronts. This can take place in the case of pilot ignition of very lean mixtures, in comparison to ignition from a single energy source, such as an electric spark. Moreover, the energy released at ignition helps increase the overall mixture temperature and pressure within the closed volume of the charge and produce sufficient high-temperature products from the ignition zones that manage to disperse into the rest of the mixture, aiding the flame in its propagation. The employment of a large pilot in a direct injection engine or a smaller pilot injected into the prechamber of an indirect injection engine, where the energy released by the pilot is supplemented by the energy released from the local surrounding fuel and air, would help in increasing the overall temperature level, as well as in the mixing of the partial and complete combustion products of the usually richer mixture. Such a mode of combustion can be exploited to oxidize by burning any small and intermittent releases of methane or other gases from a variety of sources. The gas may be fed into the intake of a fully operational diesel engine at relatively high load, ensuring that the bulk of the methane becomes oxidized in a manner similar to that of a dual-fuel engine operating on an excessively large pilot with ultra-lean fuel gas–air mixtures. Such an approach becomes an effective method of rendering fugitive gas releases environmentally harmless through converting the methane, a very potent greenhouse gas, to its products of combustion, which have a much weaker greenhouse gas effect, while usefully exploiting the heating value of the fugitive gases.

14.4   RICH MIXTURE OPERATION IN DUAL-FUEL ENGINES

The provision of excess air is an essential condition for acceptable operation of any conventional diesel engine. Moreover, the operation of spark ignition (SI) engines on rich fuel mixtures, even those of gaseous fuels, is wasteful of fuel and seldom followed nowadays. There is always a need to maintain acceptable fuel economy and reduce the associated undesirable exhaust emissions, which would require the use of lean or stoichiometrically controlled fuel–air mixtures. Accordingly, there is a relative lack of information about the performance and operational limits of dual-fuel engines when operated on rich fuel mixtures or when some oxygen is added to extend such operations. Such information can help in better understanding the complexities of the combustion processes. It would also help to provide a basis for contrasting rich mixture operation with the operation on low-heating-value fuel mixtures containing excessive amounts of carbon dioxide and nitrogen.

Unlike conventional diesel engine operation, the gas-fueled dual-fuel engine tends to have a very extended operational mixture range up to the corresponding rich mixture limits. In comparison, there is a negligible extent of exhaust smoke and particulate emissions with gas fuel operation, especially with methane. However, in fuel-fumigated, high-compression-ratio engines, the excessive presence of fuel displacing air will bring about very substantial changes in the thermodynamic and transport properties of the resulting mixture. This would produce in the constant-compression-ratio engine a marked reduction of the mean temperature at the end of compression. Such a reduction, unlike in the air-only charged diesel engine, will produce a marked reduction in the effective compression ratio from the geometrical value associated with diesel operation. This effect will lower the compression temperature and pressure levels and bring about a marked undermining of the ignition and flame propagation processes. Figure 14.5 shows an example of the drop in effective compression ratio of a diesel engine of 14.2:1 when operating on excessively rich mixtures. In principle, there may be some less common occasions to expose a dual-fuel engine, albeit for a short time, to operation on fuel-rich mixtures that are well beyond the normal lean mixture operational range. These may include, for example, the need to obtain a greater power output as used to take place in the quality-controlled spark ignition engine. It may also include situations where uncontrolled fuel admission arising through the uncontrolled release of some gaseous fuel in the immediate vicinity of the engine intake leads momentarily to rich mixture induction. There are also applications of dual-fuel engines where a specific exhaust gas composition may be needed for industrial type processes, such as to produce highly reducing atmospheres or for the production of hydrogen/synthesis gas through the partial oxidation of methane, while operating on sufficiently rich mixtures, including operation on oxygenated air.

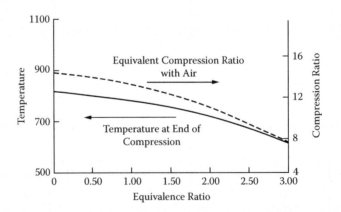

FIGURE 14.5 A typical example showing the potential reduction in the calculated temperature at the end of compression with increasing the admission of methane. The corresponding potential reduction in the resulting effective compression ratio is also shown. (Adapted from Karim, G.A., *British Chemical Engineering*, 8, 392–396, 1963.)

The high-compression-ratio dual-fuel engine, unlike its diesel version, can operate on rich mixtures in a manner unlike that of a corresponding SI version. A continued increase in the relative gas admission at constant engine speed may lead to the onset of knock under some operating conditions. When sufficiently rich mixtures are employed, knock-free operation may be restored beyond a rich mixture knock limit. Excessively rich mixtures can eventually lead to erratic running, and ultimately ignition failure. This failure appears to arise primarily from a failure of the pilot to ignite properly, rather than a failure of flame propagation through the rich mixture (Figure 14.6). An example of the corresponding changes to the ignition delay with the increased fuel concentrations under conditions outside the knocking operational range is shown in Figure 14.7. It displays a gradual increase in the delay to a maximum value, and then drops to a minimum before approaching the stoichiometric mixture region. Continuing with the increase in fuel gas admission increases the length of the delay at an increasing rate until very late, and irregular ignition is encountered, which would eventually lead to ignition failure.

Variations in the indicated power output over the rich equivalence ratio range indicate that the maximum power is obtained at equivalence ratios just beyond the stoichiometric value. As shown in Figure 14.6, the employment of larger pilots produces slightly higher power output, but with the onset of irregular running, is then encountered with less rich mixtures. Earlier pilot injection can extend the operational rich mixture range, while delayed injection, to beyond the TDC position, leads to ignition failure. Similarly, the rich operational mixture limit becomes reduced as the engine speed is increased since there is less time to complete the processes needed for pilot ignition and subsequent flame propagation.

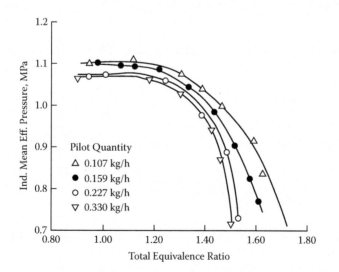

FIGURE 14.6 Schematic representation of the variation in the indicated power output with changes in total equivalence ratio of a dual-fuel engine operating on rich gas fuel–air mixtures with the increase in pilot size.

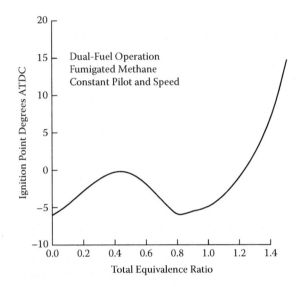

FIGURE 14.7 Schematic of the typical variation in the ignition delay of a dual-fuel engine fueled with fumigated methane at constant pilot and speed over the whole total equivalence ratio range.

The rich ignition limit becomes extended into richer mixture regions as the intake mixture temperature is increased.. The rate of increase is much greater than that observed for the lean operational mixture limit. This is consistent with the trend normally observed of the variation in the flammability limits of fuel–air mixtures with increasing initial mixture temperature.

Moreover, the introduction of some hydrogen, such as through exhaust gas recirculation (EGR), or additional oxygen for the rich methane–air mixture can extend the operational mixture range markedly. On the other hand, the increased presence of cold diluents in the intake mixture, notably carbon dioxide, as in biogases, markedly reduces the rich operational mixture range of the engine.

14.5 SIMULTANEOUS PRODUCTION OF SYNTHESIS GAS AND POWER IN A DUAL-FUEL ENGINE

One of the avenues for the production of hydrogen from natural gas, apart from its reforming with steam, is through its partial oxidation while using oxygen or oxygenated air. There is a good potential to carry out such oxidation processes without the benefit of catalysts in a suitably converted diesel engine operating in the dual-fuel mode. Figure 14.8 shows a schematic representation for a possible arrangement of such an operation, where an air separation unit is driven by the engine to produce the oxygen additive to methane needed to partially oxidize very rich feed mixtures. This way, the engine is capable of continuously producing hot exhaust gases made up of mainly synthesis gas, with the simultaneous production of power, together with good prospects for active EGR and waste heat recovery. An important practical

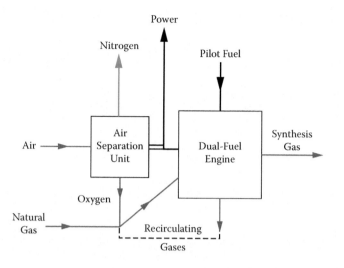

FIGURE 14.8   Possible arrangement for a dual-fuel engine installation for the simultaneous production of synthesis gas and power.

consideration apart from operational and capital costs is ensuring the safety of the operation with the employment of oxygen while having the hot exhaust containing much hydrogen.

The dual-fuel engine, in comparison to the spark ignition engine, may be considered to be more suited for synthesis gas production since higher reaction temperatures are involved with a much more powerful ignition source in the form of pilot ignition. The high-compression-ratio engine employed is more rugged and can better withstand the rapid rates of pressure rise that may be generated by the use of oxygen, while maintaining high efficiencies for the production of associated power. The combustion of sufficiently rich mixtures of methane and air in a dual-fuel engine with pilot ignition produces increasingly with equivalence ratio high concentrations of hydrogen and carbon monoxide, in addition to the unconverted methane. The concentration of the carbon dioxide is less than that of H_2O, with the H_2/CO volumetric ratio increasing with the equivalence ratio and temperature. The concentration of the unconsumed methane appears to be affected little by changes in the pilot size when employing mixtures producing relatively fast burning rates. It is only when excessively rich mixtures are employed, producing late ignition and combustion, that the unconverted methane concentration increases. Only with large pilots (e.g., 30% of full-load diesel) do the exhaust gases show significant concentrations of particulates, which could have originated mainly from the poor and extended combustion period of the large diesel pilot in atmospheres increasingly deficient in oxygen.

The production of synthesis gas, as shown in Figure 14.9, while using oxygenated air in a fixed pilot operation at ambient intake conditions, shows a very high relative yield that approaches 90% of the dry exhaust gas by volume. The associated power produced displays high net efficiency values when based on the net energy released.

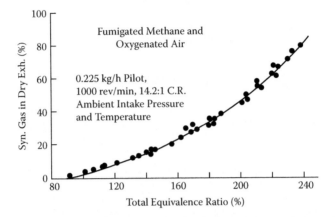

FIGURE 14.9 Variation of the relative amount of synthesis gas produced by volume in the dry exhaust with total equivalence ratio of a dual-fuel engine operating on methane and variable concentration of oxygenated air operating at ambient intake conditions while employing a constant pilot quantity. (Adapted from Karim, G.A., *British Chemical Engineering*, 8, 392–396, 1963.)

Such an approach may also point to the possibilities of different novel approaches that aim at modifying the combustion process through measures such as having thermally and chemically activated EGR or through seeding the cylinder charge with activated partial oxidation products. These may be produced, for example, from an independently controlled source, such as a heated reactor that may use the same fuel or yet another fuel, or from a cylinder or group of cylinders in a multicylinder engine.

REFERENCES AND RECOMMENDED READING

Azzouz, D., Some Studies of the Combustion Processes in Dual Fuel Engines, MSc thesis, London University, UK, 1965.

Callahan, T., Survey of Gas Engine Performance and Future Trends, presented at Proceedings of the ASME-ICE Division Conference, Salzburg, Austria, 2003, Paper ICES2003-628.

Challen, B., and Barnescu, R., *Diesel Engine Handbook*, 2nd ed., SAE, Warrendale, PA, 1999.

Danyluk, P.R., Development of a High Output Dual Fuel Engine, *ASME Journal of Engineering for Gas Turbines and Power*, 115, 728–733, 1993.

Gunea, C., Razavi, M.R., and Karim, G.A., *The Effects of Pilot Fuel Quality on Dual Fuel Engine Ignition Delay*, SAE Paper 982453, 1998.

Karim, G.A., Production of Synthesis Gas, *British Chemical Engineering*, 8, 392–396, 1963.

Karim, G.A., *An Examination of Some Measures for Improving the Performance of Gas Fuelled Diesel Engines at Light Load*, SAE Paper 912366, 1991.

Karim, G.A., Combustion in Gas Fueled Compression Ignition Engines of the Dual Fuel Type, *ASME Journal of Engineering for Gas Turbines and Power*, 125, 827–836, 2003.

Karim, G.A., and Khan, M.O., Examination of Effective Rates of Combustion Heat Release in a Dual Fuel Engine, *Journal of Mechanical Engineering Science*, 10, 13–23, 1968.

Karim, G.A., and Moore, N.P.W., *Examination of Rich Mixture Operation of a Dual Fuel Engine*, SAE Paper No. 901500, 1990.

Karim, G.A. and Moore, N.P.W., *The Production of Hydrogen by the Partial Oxidation of Methane in a Dual Fuel Engine*, SAE Paper 901501, 1990.

Kusaka, J., Daisho, Y., Shimonagata, T., Kihara, R., and Saito, T., Combustion and Exhaust Characteristics of a Diesel Engine Dual-Fuelled with Natural Gas, in *Proceedings of the 7th International Conference and Exhibition on Natural Gas Vehicles*, Yokohama, Japan, October 17–19, 2000, pp. 23–31.

Lin, Z., and Su, W., *A Study on the Determination of the Amount of Pilot Injection and Lean and Rich Boundaries of the Premixed CNG-Air Mixtures for a CNG/Diesel Dual Fuel Engine*, SAE Paper 2003-01-0765, 2003.

Ryan, T.W., Lestz, S.S., and Meyer, E., *Extension of the Lean Misfire Limit and Reduction of Exhaust Emission of a S.I. Engine by Modification of Ignition and Intake Systems*, SAE Paper 740105, 1974.

Saito, H., Sakurai, T., Sakaoji, T., Hirashima, T., and Karnno, K., *Study on Lean Burn Gas Engine Using Pilot Oil as the Ignition Source*, SAE Paper 2001-01-0143, 2001.

Turner, S.H., and Weaver, C.S., *Dual-Fuel Natural/Diesel Engines: Technology, Performance and Emissions*, No. GRI-94/0094, Topical Report Gas Research Institute, November 1994.

Xiao, F., Experimental and Numerical Investigation of Diesel Combustion Processes in Homogeneously Premixed Lean Methane or Hydrogen-Air Mixtures, PhD thesis, University of Calgary, Canada, 2011.

Xiao, F., and Karim, G.A., *Combustion in a Diesel Engine with Low Concentrations of Added Hydrogen*, SAE Paper 2011-01-0676, 2011, pp. 1–17.

Xiao, F., Sohrabi, A., and Karim, G.A., *Reducing the Environmental Impact of Fugitive Gas Emissions through Combustion in Diesel Engines*, SAE Paper 2007-01-2048, 2007.

Xiao, F., Sohrabi, A., and Karim, G.A., Effect of Small Amounts of Fugitive Methane in the Air on Diesel Engine Performance and Its Combustion Characteristics, *International Journal of Green Energy*, 5, 334–345, 2008.

Zhou, G., and Karim, G.A., The Uncatalyzed Partial Oxidation of Methane for the Production of Hydrogen with Recirculation, 5, 115, 307–313, 1993.

15 Knock in Dual-Fuel Engines

15.1 KNOCK IN GAS-FUELED SPARK IGNITION ENGINES

The uncontrolled combustion phenomenon commonly encountered in spark ignition (SI) gas engines, known as knock, is associated with sudden exceedingly high rates of energy release, excessive heat transfer to the walls, and rapid rates of pressure rise. These must be avoided to ensure acceptable engine operation. Serious operational limits are imposed by engine knock to the increase in power output, efficiency, type of fuel that can be used, and any further reductions in emissions. Much effort is usually expended in the design, operation, and control of spark ignition engines to reduce the likelihood of encountering knock.

The phenomenon of knock results from the uncontrolled and rapid combustion energy release well ahead of the turbulent propagating flame due to autoignition of some of the unburned fuel–air mixture. This autoignition is the outcome of a complex thermal and chemical interaction that takes place between the turbulent flame propagation processes and the preignition oxidation reactions of the "end gas" region of the mixture that is yet to be consumed by the flame. Figure 15.1 shows a schematic representation of the onset of knock in a typical spark-ignited engine operating on premixed fuel gas and air.

The onset of knocking severely restricts the type of fuel that can be employed while undermining engine operation, its useful life, and effective lubrication. Often, design and operating parameters are selected conservatively to avoid knock even without achieving optimum performance.

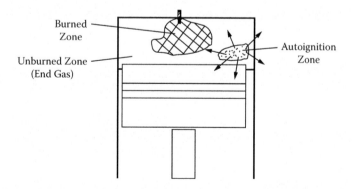

FIGURE 15.1 Schematic representation of the onset of knock in an SI engine. (From Karim, G.A., *Fuels, Energy, and the Environment*, CRC Press, Boca Raton, FL, 2012.)

The onset of knock and its intensity can be detected through measures such as monitoring the characteristic resulting engine vibrations, the distinctive noise emitted, the rapid change in the rate of cylinder pressure rise displaying the appearance of high-frequency pressure pulsations, the distinctive noise emitted, a rapid drop in power output and efficiency, excessive heat loss to the walls, or the changes in the exhaust temperature and the concentrations of the exhaust gases. The intensity of the knock is related to the intensity of the resulting rate of pressure rise. This in turn will be a function of the net energy released by autoignition, which mainly controls the intensity of the temporal changes in cylinder pressure. These may be through monitoring measures such as the output of strain gauges mounted on the engine cylinder stud. In general, automatic detection and control measures for knock, when fitted, have been somewhat effective, yet they tend to have their limitations.

Knock may be avoided by measures that include sufficiently reducing the equivalence ratio, ignition timing advance, compression ratio, intake mixture temperature, and boost pressure in turbocharged engines. Figure 15.2 shows a typical representation of the knock mixture region relative to the normal operational mixture boundaries with changes in the compression ratio for a spark-ignited engine operating on methane at ambient intake temperature and pressure conditions. It is seen that knocking is encountered with methane only when very high compression ratios are employed. Figure 15.3 shows the changes with intake mixture temperature in both the lean and rich mixture knock limits for a spark-ignited engine when operated on fumigated propane for a range of compression ratios. The knock resistance of propane, compared to methane operation, is lower. The knock producing mixture region widens significantly with the increase in compression ratio and intake mixture temperature, which restricts the knock-free operating range and reduces the knock-free power. When a sufficiently

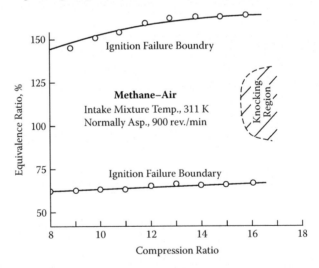

FIGURE 15.2 The variation of the ignition failure mixture boundaries and the knocking mixture region of a spark ignition engine operating on methane–air with changes in compression ratio for an intake mixture temperature of 311 K. (From Karim, G.A., and Klat, S.R., *Journal of the Institute of Fuel*, 39, 109–119, 1966.)

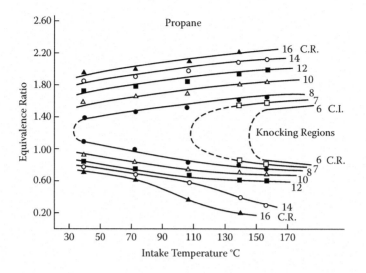

FIGURE 15.3 Variation of the lean and rich mixture operational limits of a propane-fueled spark ignition engine with changes in intake temperature for a range of equivalence ratios. (From Karim, G.A., and Klat, S.R., *Journal of the Institute of Fuel*, 39, 109–119, 1966.)

high compression ratio or high intake temperature is employed, homogeneous charge compression ignition (HCCI), which requires no external ignition source, will be encountered, as shown schematically in Figure 15.4 for methane operation.

The chemical nature of the fuel used is of paramount importance in avoiding the onset of knock. The blending of some gaseous fuels may undermine their knock resistance qualities. For example, the presence of small amounts of a higher hydrocarbon fuel vapor such as n-butane or n-hexane with methane significantly lowers its

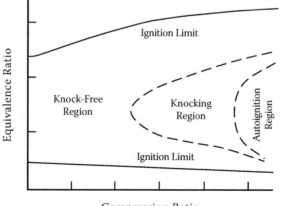

FIGURE 15.4 The schematic variation of the ignition, knock, and autoignition limits with compression ratio for methane–air operation in a spark ignition. (Adapted from Karim, G.A., and Klat, S.R., *Journal of the Institute of Fuel*, 39, 109–111, 1966.)

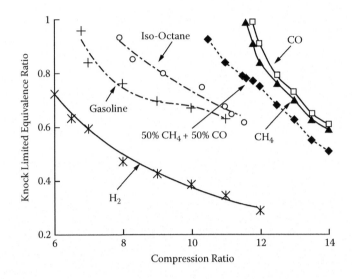

FIGURE 15.5 Variation of the knock-limited equivalence ratio with compression ratio for a number of fuels in a variable-compression-ratio test engine at fully open throttle. (From Li, H., and Karim, G.A., *Journal of Energy and Power*, 220, 459–473, 2005.)

excellent knock resistance. Figure 15.5 shows, for a range of common gaseous fuels in a variable-compression-ratio cooperative fuel research (CFR) engine, the reduction in the knock-free limit mixture equivalence ratio with increasing the compression ratio. The superior knock resistance of methane and dry carbon monoxide is clearly evident. However, the blending of methane with carbon monoxide produces knock-resisting qualities that are lower than those of methane or dry carbon monoxide on their own.

15.2 THE KNOCK RATING OF FUELS

A number of approaches have been available for the rating of gaseous fuels, such as through the use of the octane number and the methane number. These, especially for gaseous fuel applications, have their limitations, and there is still room for developing improved procedures for the knock rating of gaseous fuels and their mixtures, especially for dual-fuel operation.

The octane number (ON), which uses iso-octane (2,2,4-trimethylpentane) and n-heptane as the reference fuels, is employed primarily for the knock rating of liquid fuels. Another method proposed for the rating of gaseous fuels is the methane number (MN), which is based correspondingly on methane and hydrogen as reference fuels. A 100 MN is given to pure methane and 0 MN for pure hydrogen. A methane number is assigned to a gaseous fuel on the basis of the percentage by volume of methane in a blend of hydrogen and methane that matches the knock intensity characteristics when using the test fuel under a specified set of operating conditions in a test engine. Similarly, a butane number (BN) has been proposed based on a blend of methane and n-butane, with pure methane assigned a value of 0, while the easier to autoignite n-butane is given a value of 100 (Table 15.1).

TABLE 15.1
Knock Rating of C1 to C4 Normal Paraffins

Gas	Motor ON	MN	BN
CH_4	122	100	0.0
C_2H_6	101	44	7.5
C_3H_8	97	34	10
C_4H_{10}	89	10	100

Source: Ryan, T.W., Callahan, T.J., and King, S.R., *ASME Journal of Engineering for Gas Turbines and Power*, 115, 922–930, 1985.

There are some fundamental deficiencies for these empirically devised numbers, especially when applied to fuel mixtures other than those of natural gas and those containing diluents.

15.3 DIESEL ENGINE KNOCK

Depending on the intensity of the resulting rates of pressure rise following ignition in diesel engines, the characteristic intense knocking sound is sometimes heard. It is described somewhat loosely as diesel knock. The intensity of the resulting noise and rough running produces shock loading to working parts and high peak values of maximum cycle pressure. These are particularly evident following long ignition delays, such as at low intake temperature conditions or while operating at very high loads (Figure 15.6).

The commonly assigned knock to diesel engines is fundamentally quite different from that known for spark ignition engines. Diesel knock is associated with the beginning phase of combustion, while in spark ignition and often in dual-fuel

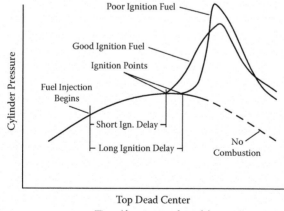

FIGURE 15.6 Schematic representation of the pressure development records for two diesel fuels with different ignition characteristics and resulting ignition delay. (From Karim, G.A., *Fuels, Energy, and the Environment*, CRC Press, Boca Raton, FL, 2012.)

engines it is associated with the latter parts of the charge to burn. The intensity of diesel knock is affected by factors that would include engine design, such as direct vs. indirect injection, compression ratio, engine size, its speed, wall temperature, sound insulation provided, and the state and condition of the engine.

Some of the remedial actions to reduce the intensity of diesel knock include reduction of the ignition delay and the amount of fuel that has accumulated in the chamber during the delay time at the instant of ignition.

15.4   DUAL-FUEL ENGINE KNOCK

An important limitation associated with the performance of the premixed type of dual-fuel engines is their tendency to encounter a type of knocking phenomenon mainly with some gaseous fuels or under highly rated power output conditions. Its avoidance, while retaining in full many of the positive features of the efficient high compression ratios of diesel engines, can be a challenge to would-be diesel engine convertors to dual-fuel operation.

The energy release in dual-fuel engines represents the combined contribution of the combustion of the pilot fuel regions and the resulting rapid turbulent flame front propagation within much of the rest of the charge originating from these regions. Such a mode of combustion is primarily responsible for the ability of dual-fuel engines to benefit from the employment of pilots of sufficiently large size and burn fuel–air mixtures that are much leaner than those normally possible in gas-fueled spark ignition engines. These features help to contribute to the high efficiencies associated with dual-fuel operation.

In a typical premixed type of dual-fuel engine, the pilot liquid fuel is injected into the premixed charge commonly in the form of individual small jets through the multiple holes of the injector body. After a short delay, ignition takes place and burning commences very quickly and virtually simultaneously at a number of points within the charge. Turbulent flames then begin to spread from these ignition centers to the local adjacent regions of the fuel vapor–gaseous fuel–air mixtures. With relatively small size pilots, a flame starting from an ignition point would have to propagate through the adjacent mixture before it meets flames from similar neighboring regions. Accordingly, despite the high compression ratios employed in dual-fuel engine applications, there is less likelihood for an autoignition type of knock to take place in the manner of spark ignition engines. However, at certain operating conditions with some gaseous fuels, such as when using sufficiently large pilots or less lean gaseous fuel–air mixtures, very fast rates of pressure rise can be encountered following the commencement of ignition. This is due to the rapid energy release rates arising mainly from the autoignition of the diesel fuel augmented by the energy releases associated with the multiple fast flame propagation fronts formed within the adjacent mixtures. When these rates of pressure rise are sufficiently high, then this may lead to a knocking behavior that is somewhat akin to that encountered in diesel engines.

A prime requirement of any alternative gaseous fuel for satisfactory operation in the premixed dual engines is that its mixture with air should not have sufficient time to autoignite spontaneously prior to, during, or following the rapid energy releases of the burning pilot and the mixture in its immediate neighborhood. Failing to do this can lead to the onset of intense knocking, which manifests itself with excessively rapid rates of pressure rise. These would induce vibrations in the combustion chamber components, which then convey the oscillations to the external walls. It will result in subsequent overheating of the cylinder and piston surfaces, leading to significant loss in power and efficiency with increased cyclic variations and heat loss. Persistent knock is highly objectionable and needs to be avoided. As a result of the effects of cyclic variations, knock may appear in only a few cycles among many, but soon knocking may increase in frequency and intensify to cover all cycles. The vibration level changes with changes in the state and intensity of combustion. These vibrations may be detected via measures that include the appearance of intense and high-frequency cylinder pressure oscillations following pilot ignition and the distinctive accompanying knocking sound. An effective detection measure may be through monitoring the output of an acceleration sensor that can be mounted on engine parts, such as a stud bolt of each cylinder head.

Figure 15.7 shows a typical example of the pressure development record of a dual-fuel engine operating on methane, together with a corresponding record under knocking conditions. Figure 15.8 shows a typical example of the rapidly increasing pressure development leading to the onset of knock with different pilot quantities when operating a dual-fuel engine on methane.

The type of knocking encountered in gas-fueled dual-fuel engines may be due to both diesel and end gas type knocks. The presence of much premixed albeit lean fuel–air mixture that can be considered an end gas in relation to the autoigniting pilot fuel mixture regions may undergo sufficient autoignition reactions aided by the rise in temperature following pilot ignition, to lead to an end gas type of knock.

Depending on the size of the pilot and its mode of injection, knock can be perceived to involve autoignition of that portion of the charge in the neighborhood of

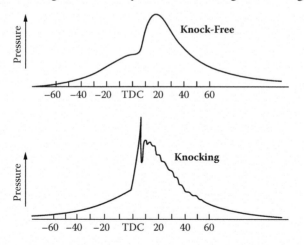

FIGURE 15.7 Pressure time development records for a dual-fuel engine when operating on methane under knock-free and knocking conditions. (From Karim, G.A., Klat, S.R., and Moore, N.P.W., Knock in Dual Fuel Engines, in *Proceedings of the Institute of Mechanical Engineers*, 181, 453–466, 1967.)

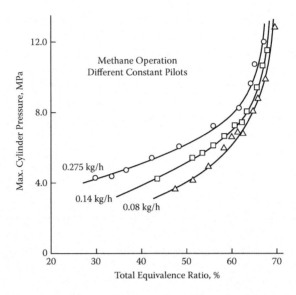

FIGURE 15.8 Variation in the values of maximum pressure with the total equivalence ratio as it is approaching the knocking condition for three different pilot quantities with methane as the fuel. (From Karim, G.A., Klat, S.R., and Moore, N.P.W., Knock in Dual-Fuel Engines, in *Proceedings of the Institute of Mechanical Engineers*, 181, 453–466, 1967.)

these ignition centers leading to very high rates of pressure rise and a consequent very rapid burning of the remaining parts of the charge. With small pilots, the energy release during the initial stages of ignition and the resulting turbulent flame propagation can lead, under certain conditions, to an autoignition of the charge well away from these ignition centers in the end gas regions ahead of these flames, in a manner resembling the occurrence of knock in spark ignition engines.

As far as knock is concerned, it is mainly the temperature level of the mixture that is of prime importance. The effects of changes in the values of the fuel–air ratio tend, in comparison, to be relatively less severe and regulate primarily the resulting combustion temperature. Figure 15.9 shows the changes in the heat release rate diagram of a dual-fuel engine fueled with propane as the concentration of the fuel gas in the intake increases to produce knocking while maintaining the size of the pilot fixed.

The knock-limited power output for different fuel and pilot settings appears to deteriorate logarithmically with the inverse of the intake mixture absolute temperature for fuel-fumigated dual-fuel engines, over a wide range of pilot fuel sizes (Figure 15.10).

Methane is relatively more tolerant to increases in intake mixture temperature than other, more reactive gaseous fuels. Operation on propane and butane also shows the log knock-limited power output to be linearly related to the inverse of the absolute intake mixture temperature. Increases in the size of the pilot tend to produce an earlier onset of knock with a corresponding reduction in the knock-free output. The knock-limited power output in hydrogen-fueled dual-fuel engines is especially more sensitive to changes in intake mixture temperature. The associated very fast flame propagation following pilot ignition in hydrogen operation can lead to a knocking behavior.

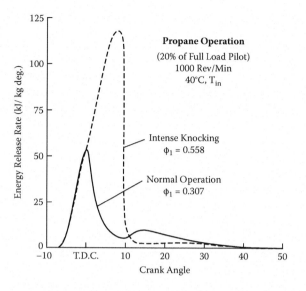

FIGURE 15.9 Variation in the energy release with crank angle for a knocking condition and for normal operation for a dual-fuel engine operating on propane using a constant pilot. (From Karim, G.A., and Rogers, A., *Journal of the Institute of Fuels*, 40, 513–522, 1967.)

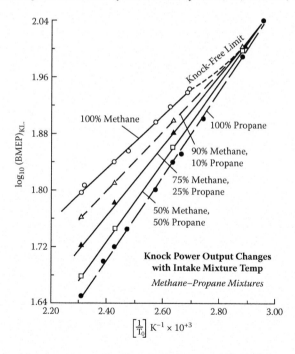

FIGURE 15.10 The logarithmic variation of the knock limited brake power output with the inverse of the absolute intake mixture temperature for different methane–propane mixtures under constant pilot and speed condition in a normally aspirated engine. (Adapted from Karim, G.A., *Journal of the Institute of Fuel*, 37, 530–636, 1964.)

15.5   SOME MEASURES TO REDUCE THE INCIDENCE OF KNOCK

With the increased availability of different gaseous fuels and their mixtures with wide variations in composition, manufacturers of dual-fuel engines, especially those of the highly turbocharged type, often tend to operate their engines at ratings well below their potential optimum. This is mainly to avoid the incidence of knock and overheating of working parts while ensuring acceptable levels of emissions and trouble-free operation.

The incidence of knock may be avoided or its intensity, if it does occur, reduced through a number of possible design and operational measures. These include the lowering of the intake mixture and water jacket temperatures and somewhat delaying the commencement of pilot injection. Some other possible measures, such as lowering the compression ratio, are not commonly resorted to since they could undermine pilot ignition and the operation of the engine as a diesel. However, a prime means for avoiding the onset of knock in a natural gas-fueled dual-fuel engine is through careful and proper controls, such as the composition of the gaseous fuel mixtures and the size and injection characteristics of the pilot. This can be accomplished through measures such as reducing and avoiding having droplets of fuel condensates, which are usually mostly the more reactive components of higher hydrocarbons present in the original gas. The employment of cooled exhaust gas recirculation (EGR) would also be an effective practical approach. Of course, in principle, an effective procedure for reducing the incidence of knock, if it can be achieved, is through the optimum distribution of the gaseous fuel-to-air ratio by a proper stratification of the gaseous fuel component within the cylinder, where less reactive, leaner gaseous fuel–air mixtures are within regions of the ignition centers of the pilot where richer mixtures are located. This would require careful control of the gaseous fuel distribution such that it will not hinder pilot fuel ignition or encourage the autoignition of regions adjacent to the pilot, while reducing the tendency of the charge away from these centers to autoignite despite the high rates of increase in cylinder pressure and temperature. This may be accomplished via timed gaseous fuel injection during the latter parts of the compression stroke or, to a lesser degree, of effectiveness via its timed injection just outside the cylinder.

Normally, the knocking operational regions with methane and some natural gases are beyond common operational conditions unless highly supercharged large-bore engines and insufficiently optimized pilot quantity and injection characteristics are employed. However, for some other gaseous fuel mixtures containing the more reactive higher hydrocarbons, such as in propane-fueled engines, the occurrence of knock in an unsuitably modified diesel engine may be commonly encountered.

Enhancing the cetane number of the pilot fuel and resorting to cooled exhaust gas recirculation, for example, can have a beneficial effect in reducing the incidence of knock by delaying the onset of knock a little. Turbochargers are normally used in conjunction with intercoolers to help keep the incoming mixture temperature relatively low, aiding in improving engine performance and reducing the tendency to knock.

Of course avoiding the fumigation of the fuel gas or its injection in the early part of the compression stroke would help in delaying the progress of the gaseous fuel–air mixtures toward early autoignition. Moreover, the high-pressure injection of the fuel gas after pilot ignition will drastically reduce the tendency to encounter knock in unpremixed dual-fuel engines, but adds much complexity to engine design and control.

REFERENCES AND RECOMMENDED READING

Ali, A.I., and Karim, G.A., Combustion, Knock and Emission Characteristics of a Natural Gas Fuelled Spark Ignition Engine with Particular Reference to Low Intake Temperature Conditions, in *Proceedings of the Institute of Mechanical Engineers*, London, 1975, vol. 189, pp. 139–147.

Attar, A., and Karim, G.A., Knock Rating of Gaseous Fuels, *ASME Journal of Gas Turbines and Power*, 125, 500–504, 2003.

Azzouz, D., Some Studies of the Combustion Processes in Dual Fuel Engines, MSc thesis, London University, UK, 1965.

Challen, B., and Barnescu, R., *Diesel Engine Handbook*, 2nd ed., SAE, Warrendale, PA, 1999.

Karim, G.A., An Analytical Approach to the Uncontrolled Combustion Phenomena in Dual Fuel Engines, *Journal of the Institute of Fuel*, 37, 530–536, 1964.

Karim, G.A., Combustion in Gas Fueled Compression Ignition Engines of the Dual Fuel Type, *ASME Journal of Engineering for Gas Turbines and Power*, 125, 827–836, 2003.

Karim, G.A., *Fuels, Energy and the Environment*, CRC Press, Boca Raton, FL, 2012.

Karim, G.A., and Klat, S.R., Knock and Autoignition Characteristics of Some Gaseous Fuels and Their Mixtures, *Journal of the Institute of Fuel*, 39, 109–119, 1966.

Karim, G.A., and Lui, Z., *A Prediction Model of Knock in Dual Fuel Engines*, SAE Paper 921550, 1992.

Karim, G.A., and Rogers, A., Comparative Studies of Propane and Butane as Dual Fuel Engine Fuels, *Journal of the Institute of Fuels*, 40, 513–522, 1967.

Karim, G.A., and Zhoada, Y., *An Analytical Model for Knock in Dual Fuel Engines of the Compression Ignition Type*, SAE Paper 880151, 1988.

Karim, G.A., Klat, S.R., and Moore, N.P.W., Knock in Dual Fuel Engines, in *Proceedings of the Institute of Mechanical Engineers*, 1967, vol. 181, p. 453.

Klimstra, J., Heranaez, A.B., Gerard, A., Karti, B., Quinto, V., Roberts, G.R., and Schollmeyer, H., Classification Methods for Knock Resistance of Gaseous Fuels, in *ASME-ICE Division*, 1999, vol. 33-1, Paper 99-ICE-214.

Kubish, J., and Brehob, D.D., *Analysis of Knock in a Dual Fuel Engine*, SAE Paper 922367, 1992.

Kubish, J., King, S.R., and Liss, W.E., *Effect of Gas Composition on the Octane Number of Gaseous Fuels*, SAE Paper 922359, 1992.

Li, H., and Karim, G.A., Experimental Investigation of the Knock and Combustion Characteristics of CH4, H2, CO and Some of Their Mixtures, *Journal of Energy and Power*, 220, 459–473, 2005.

Li, S.C., and Williams, F.A., *A Reduced Reaction Mechanism for Predicting Knock in Dual Fuel Engines*, SAE Paper 2000-01-0957, 2000.

Ryan, T.W., Callahan, T.J., and King, S.R., Engine Knock Rating of Natural Gases–Methane Number, *ASME Journal of Engineering for Gas Turbines and Power*, 115, 922–930, 1985.

16 Exhaust Emissions

16.1 EXHAUST PRODUCTS OF ENGINE COMBUSTION

Ideally, it would be expected that the presence of excess air with a hydrocarbon fuel at high temperatures and pressures, as in dual-fuel operation, would see, following combustion, the carbon in the fuel appearing as CO_2 and the hydrogen appearing as H_2O, together with unutilized oxygen and unchanged nitrogen. However, when much excess air is supplied, then combustion will take place at correspondingly lower temperatures and slower rates, which may lead to the incomplete production of combustion products and even ignition failure. In regions where there is an excess of fuel, such as those associated with the pilot fuel and its immediate surroundings, incomplete combustion products will be produced, with some of the fuel remaining unconverted. Much of the hydrogen in the fuel will be oxidized to H_2O, and with some appearing as molecular H_2. The carbon in the converted fuel, because of the insufficient amount of oxygen locally available, will produce soot, CO, and CO_2, with hardly any unutilized oxygen remaining. In practice, depending on the operational conditions and the fuels employed, a departure from this idealized composition will be encountered for a variety of reasons, producing a range of other products, albeit often in small concentrations. Some of these survive to the final product stage, while others are transiently unstable and would not last to the exhaust stage. In reality, they can still play a critical part in the conduct of combustion processes. Some of the main factors that bring about such a behavior are the following:

- Insufficient mixing of the fuel and air and of the time available to fully complete combustion; some of the flame propagation does not extend throughout the whole mixture, such as through a rapid drop in temperatures.
- Excessive heat loss quenching combustion reactions, such as by contact with cold surfaces, mixing with colder air, or rapid expansion.
- Dissociation effects at high temperatures, and with some of the nitrogen and other elements present in the fuel becoming sufficiently reactive.

Accordingly, the products of combustion contain not only CO_2, H_2O, O_2, and N_2, but also CO, H_2, unburned fuel, oxides of nitrogen and sulfur, and some other partial oxidation products and particulates. Figure 16.1 shows a schematic representation of typical variations in the components of the exhaust gas of a dual-fuel engine operating on methane with a constant pilot size with changes in the total equivalence ratio.

To be able to control the composition of the combustion products and reduce the undesirable corresponding species present remains a challenge. This is needed to ensure optimum efficiency and to emit the minimum of undesirable pollutants.

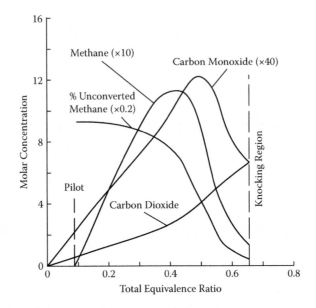

FIGURE 16.1 Typical variation in the concentrations of the exhaust gas with total equivalence ratio for a dual-fuel engine fumigated with methane using a constant pilot value. (Adapted from Azzouz, D., Some Studies of the Combustion Processes in Dual Fuel Engines, MSc thesis, London University, UK, 1965.)

With the continuing emphasis on the reduction of greenhouse gas emissions, such as those of CO_2, even the complete combustion of hydrocarbon fuels will yield products that are considered to be potentially harmful in the long term to the environment, and their releases need curtailing. This remains, of course, a challenge at present and no doubt will also in the future.

In dual-fuel engines, especially when operating with relatively large pilots, a varying transient spectrum of mixture ratios normally exists within the combustion zone due to the heterogeneity of the combustion process. Accordingly, an increasing variety of measures are incorporated into the design and operation of engines with the aim of reducing exhaust emissions before they leave the cylinder. These try to suitably improve the conduct and control of the combustion process by measures such as attending to the fuel quality with ultra-low sulfur content while optimizing the processes of fuel delivery, spray systems, and mixing. Also, turbocharging is employed with appropriate modifications to the combustion chamber and controlled exhaust gas recirculation. Much effort has also been focused on improving the exhaust gas after-treatment. These include measures such as the provision of an oxidation catalytic converter, particulate matter traps and oxidizer, selective catalytic reduction of NOx, and employing optimum control of exhaust gas recirculation. However, controlling the emissions of NOx in diesel engines has been a more challenging task than in spark ignition (SI) engines. This is because diesel combustion is of the diffusion type, where localized high temperatures near stoichiometric combustion operations continue to dominate, while there is always much excess supply of oxygen. The common SI automotive three-way catalytic converter under these conditions tends to be relatively less effective.

16.2   DUAL-FUEL ENGINE EMISSIONS

The nature and extent of exhaust gas emissions in dual-fuel engine applications are controlled by changes in numerous factors that are more varied than for spark ignition or diesel engines. Some of the main factors follow:

- The type and composition of the gaseous fuel employed, its properties, mode of introduction, temperature and pressure, air temperature, and the nature of the mixing processes
- Size and properties of the pilot fuel employed, injection timing and pressure, nozzle geometry, and jet velocity
- Engine type and size, cylinder geometry, engine speed and its range, compression ratio, water jacket temperature, turbocharging, exhaust gas recirculation (EGR), and any skip firing employed

The main constituents of the exhaust gas of dual-fuel engines are normally made up of unburned hydrocarbons, mainly in the form of unconverted gaseous fuel, carbon monoxide, carbon dioxide, oxides of nitrogen, and particulate emissions. As was indicated earlier, turbulent flame propagation from the ignition regions of the pilot may not proceed in time throughout the fuel–air charge until the concentration of the gaseous fuel in air becomes sufficiently less lean beyond a limiting value that varies with the type of fuel, engine, and operating conditions.

When operating with a charge of sufficiently lean fuel–air mixture, a significant amount of the gaseous fuel and the products of any preignition and partial combustion processes may survive to the exhaust stage. As shown in lean mixture operations, the concentration of the unconverted fuel gas, methane, appearing in the exhaust gas initially increases almost proportionally with the extent of the relative gas admission. This is indicative of a minimal combustion through flame propagation taking place within the time available beyond the immediate combustion regions dominated by the constant size pilot fuel. With the continued increase in the concentration of the gaseous fuel, a limiting value is reached when the exhaust concentration of methane drops rapidly. This indicates an increased progression of the combustion zone and some flame propagation into the bulk of the gas fuel–air charge consuming a greater proportion of the fuel. On continuing with increasing the relative concentration of the fuel gas in the intake charge, eventually much of the fuel gas admitted gets consumed. The unconverted methane concentration in the exhaust gas then drops, aided by the burning pilot, to well below the corresponding values usually encountered in its gas-fueled spark ignition counterpart. Such a limiting methane concentration is associated with the combustion proceeding through virtually all the lean mixture regions that are well below those corresponding to the stoichiometric value. However, estimates of this mixture limit on the basis of the flammability limit values at the corresponding local temperature and pressure conditions will underestimate the unconverted amount of methane exhaust concentration.

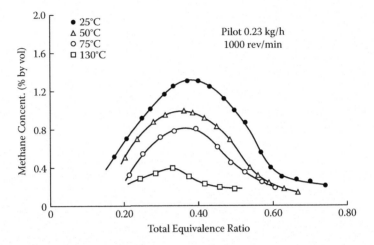

FIGURE 16.2 An example of the variation of the exhaust concentration of methane with total equivalence ratio for different intake mixture temperatures and a constant value of pilot for an engine operating on fumigated methane. (Adapted from Azzouz, D., Some Studies of the Combustion Processes in Dual Fuel Engines, MSc thesis, London University, UK, 1965.)

The releases of the unconsumed methane tend to be of much less significance as a contributor to the production of photochemical smog than the releases of other hydrocarbons. However, methane emissions from gas-fueled engines do not respond too well to the catalytic oxidation methods normally applied in gasoline spark ignition engines, requiring special measures. Moreover, the presence of methane in the air is known to have a much more potent influence as a greenhouse gas than the corresponding presence of carbon dioxide, and requires increasingly effective control. Figure 16.2 shows for any operating equivalence ratio the significant drop in the methane releases with increasing the intake mixture temperature and the associated bulk temperature at the end of compression.

As part of the sequence of the oxidation reactions of methane in air, carbon monoxide, rather than carbon dioxide, is produced in the early stages of the oxidation sequence. Much of this carbon monoxide becomes oxidized further on with time to carbon dioxide.

For very lean mixtures where combustion temperatures are rather low and the reaction rates are relatively slow, some of this carbon monoxide may not get oxidized sufficiently rapidly in the time available due to the combined effects of the failure of the combustion process to encompass and oxidize the whole charge, the relative slowness of the reaction processes, and quenching effects. With the very lean mixtures employed at very light loads, and especially when using small pilot quantities, not only does much methane remain unconverted, but also some of the carbon monoxide produced transiently may remain unoxidized to the final exhaust phase, as shown in Figure 16.3. The percentage of the total carbon present in both the pilot fuel and the fuel gas that appears as carbon monoxide in the exhaust gas remains, as shown in Figure 16.4, relatively small. The extent of carbon monoxide in the exhaust gases that originates from the gaseous fuel component in regions mainly

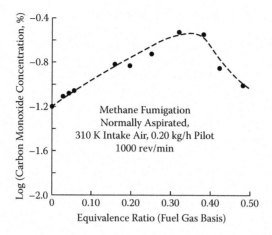

FIGURE 16.3 Logarithmic variation in the concentration of carbon monoxide with gas equivalent ratio during light load operation of a dual-fuel engine operating on fumigated methane with a constant pilot quantity. (Adapted from Liu, Z., An Examination of the Combustion Characteristics in Compression Ignition Engines Fueled with Gaseous Fuels, PhD thesis, University of Calgary, Canada, 1995.)

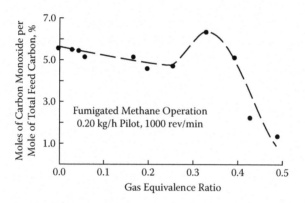

FIGURE 16.4 Variation of the percentage of carbon in the charge appearing as carbon monoxide with changes in a gas-based equivalence ratio for a methane fumigated engine with constant pilot quantity, using the data of Figure 18.4.

within and adjoining the burning pilot will depend on the size and concentrations in these regions. For the same total equivalence ratio, the use of a larger pilot, which is associated with leaner gas fuel–air mixtures, tends to produce higher carbon monoxide concentrations. An increase in the intake mixture temperature enhances the oxidation of carbon monoxide for the same pilot mass and overall equivalence ratio. Changes in the injection characteristics of the pilot can modify the extent of methane conversion and the exhaust concentrations of carbon monoxide. The contribution of the preignition reactions of the bulk of the charge to carbon monoxide exhaust emissions tends to be variable, but remains generally small.

16.3 OXIDES OF NITROGEN EMISSIONS

The production of oxides of nitrogen in engine combustion depends primarily on the peak value and distribution of the temperature within the combustion zone and its effective volume, the availability of oxygen, and whether there is sufficient time for the oxygen-nitrogen reactions to proceed to significant levels of completion. In dual-fuel engines, much less NOx is produced than in the corresponding diesel or SI engine operation. Production is associated primarily with the pilot diffusion combustion zone, where very high local temperatures are produced and longer reaction times are possible (Figure 16.5). Some further NOx production, but to a lesser extent, will be from heated mixture regions in the vicinity of the pilot combustion zone. These will tend to involve somewhat leaner mixtures and lower temperatures than those associated with the high-temperature pilot combustion zones. With the prevailing very lean overall mixture operation, relatively little NOx production is expected to be produced from regions within the remaining gaseous fuel–air mixture charge. Increasing the size of the pilot, as shown in Figure 16.5, will have an important contribution to the increased production of NOx, as the size of the hot combustion zone will be similarly increased. Also, increasing the effective equivalence ratio of the premixed fuel gas–air charge will produce higher temperatures with an increase in NOx production. Thus, the extent of the production of oxides of nitrogen will be markedly influenced by both the quantity of the pilot employed and the resulting overall equivalence ratio (Figure 16.6). Accordingly, premixed dual-fuel engines may employ relatively large pilots at light load when very lean gaseous fuel–air mixtures are employed, but as the engine load is increased, increasingly smaller pilots may be employed with less lean fuel–air mixtures. This tends to reduce the size of

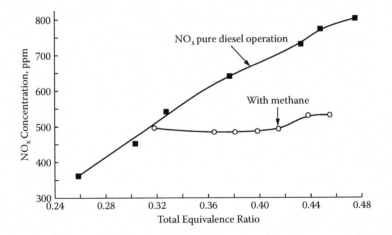

FIGURE 16.5 The concentration of oxides of nitrogen of a dual-fuel engine operating on fumigated methane employing a constant pilot quantity as the concentration of methane is increased, the corresponding concentrations of NOx emissions when operating as a diesel is shown. (Adapted from Xiao, F., Experimental and Numerical Investigation of Diesel Combustion Processes in Homogeneously Premixed Lean Methane or Hydrogen–Air Mixtures, PhD thesis, University of Calgary, Canada, 2011.)

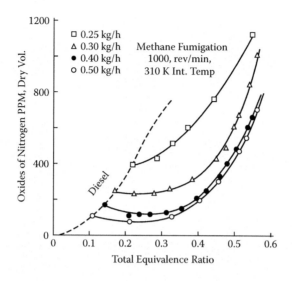

FIGURE 16.6 Changes in the concentration of oxides of nitrogen with the total equivalence ratio for a dual-fuel engine operating on fumigated methane for a number of fixed pilot quantities. (Adapted from Burn, K.S., The Effect of Cold Intake Temperatures on the Combustion of Gaseous Fuels in a Dual-Fuel Engine, MSc thesis, University of Calgary, Canada, 1977.)

the high-temperature zones while increasing only by a relatively small amount the overall combustion temperature of the whole charge and reducing the availability of free oxygen. The levels of oxides of nitrogen emissions can be maintained well below those encountered at similar loads for the corresponding diesel operation. With the use of small pilots and the overall lean mixture operation employed in dual-fuel engines, the level of oxides of nitrogen produced may also be lower than in comparable spark ignition engine operation, so long as the onset of knock is avoided. Slow-speed engines tend to exhibit relatively higher emissions of oxides of nitrogen. Also, as shown in Figure 16.7, earlier pilot injection also leads to increased NOx emissions. This is due to the longer combustion time period available for the formation reactions, combined with having excess air and high temperatures, particularly at high loads and turbocharged operation.

The emission of NOx is reduced through procedures that would include EGR, reduced pilot sizes with delayed injection, and employing after-treatment of the exhaust gas emissions.

16.4 PARTICULATE EMISSIONS

Diesel engine combustion has been noted for its production of a significant amount of particulates and smoke, especially at high loads and under cold-start conditions, which require strict control. Various regulatory measures have been enacted, especially in recent years, to control such emissions. The major source of particulate emissions in diesel engines is the high-temperature pyrolysis of the heavy hydrocarbons

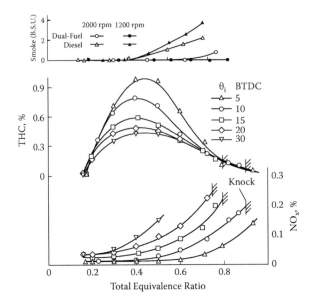

FIGURE 16.7 Variations in the concentrations of oxides of nitrogen unburned hydrocarbons and smoke density with total equivalence ratio for a dual-fuel engine operating on natural gas and constant pilot for different pilot injection timings and speed. (Adapted from Ishyama, T., Shioji, M., Mitani, S., Shibata, H., and Ikegami, M., *Improvement of Performance and Exhaust Emissions in a Converted Dual Fuel Natural Gas Engine,* SAE paper 2000-01-1866, 2000.)

in the diesel fuel, which is burning primarily in a diffusion combustion mode involving a very wide localized variation in the fuel–air ratio.

In the dual-fuel engine for the same overall equivalence ratio as that of a corresponding diesel engine at high loads, a major part of the fuel mixture is lean and will burn in a premixed mode, with primarily light hydrocarbons employed such as methane. This leads to a very substantial reduction in the amount of particulates emitted in the dual-fuel engine mode. Even then, the bulk of any smoke that may be detected will originate mainly from the small amount of pilot liquid fuel employed and some lubricants burning. A very important feature of dual-fuel operation is that the extent of smoke and particulate emissions will be much lower than that encountered for the same load with the corresponding diesel operation. This is obtained over the whole power output range, even under high levels of turbocharging and when very low intake mixture temperatures are involved, such as when operating with the boil-off gas in liquefied natural gas (LNG) operation (Figure 16.8).

This feature obviously contributes to considering the conversion of diesel engines to dual-fuel engine operation distinctly environmentally attractive over the whole load range and immensely simplifies the treatment of the exhaust gas. Also, to reduce particulate emissions, it is necessary to minimize the associated amount of diesel fuel injected. Particulate emissions need to be considered not only for steady engine operation, but also, more importantly, for transient operation.

The measurement and determining the nature of the small amount of particulate matter emissions from dual-fuel engine operation is more challenging than in the

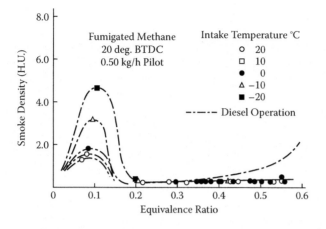

FIGURE 16.8 Variations of the smoke density of a methane-fueled engine with equivalence ratio while operating on a constant pilot quantity and different intake mixture temperatures. The corresponding concentrations with diesel operation are also shown. (Adapted from Burn, K.S., The Effect of Cold Intake Temperatures on the Combustion of Gaseous Fuels in a Dual-Fuel Engine, MSc thesis, University of Calgary, Canada, 1977.)

case of diesel engines. Mustafi (2008) evaluated a variety of methods for particulate measurement in a methane-fueled dual-fuel engine and showed that about 77% reduction in particulate mass emissions was achieved with dual fueling, compared to the corresponding values for diesel-only operation. A chain-like elongated structure is observed for the dual-fuel engine carbon, while that for diesel engines was more disordered than that for the dual-fuel engine.

The total aldehyde emissions in dual-fuel engines are very small, with formaldehyde being the principal component of the total aldehyde emissions. Their emissions tend to be higher with the fumigation of the gaseous fuel than its timed injection. Natural gas, with its high concentration of methane, tends to produce more aldehydes than propane.

16.5 OXIDES OF SULFUR EMISSIONS

The concentrations of sulfur oxides together with oxides of nitrogen in the air, apart from being toxic and greenhouse gases, are a main source of acid rain and produce a variety of negative effects on health and the environment. They tend to also accelerate the corrosion of metals and may damage stone, masonry, paint, some fibers, leather, and electrical components. Moreover, sulfur compounds react with the catalysts in exhaust gas catalytic converters of current design, greatly reducing their effectiveness and durability.

Since most living matter contains some sulfur compounds, it is to be expected that sulfur compounds would be encountered naturally, to varying extents, in most fossil fuels that have originated from bio-sources. Also, natural gas and biogases often contain some sulfur compounds, usually in the form of hydrogen sulfide, that need removing or are kept below certain very low levels; otherwise, it would appear in the products of combustion.

It has been somewhat technically challenging and costly to ensure the full removal of sulfur from fuels. This is complicated by the higher sulfur contents of some crude oils becoming increasingly available and the prevalence of hydrogen sulfide in natural gas and biogases. In recent years, the sulfur contents of diesel fuels have been reduced in most countries to very low concentrations through their removal by mainly chemical processing. Also, virtually all the hydrogen sulfide is removed out of gaseous fuels before delivery for their combustion. Accordingly, dual-fuel engine exhaust gas, when operating on fuels having had their sulfur removed, will display virtually no emissions of sulfur compounds. An exception is those compounds that could originate from the very small concentrations of the mercaptans commonly used as odorants in gaseous fuels, or any sulfur that could have originated from the small amount of diesel fuel used as a pilot.

16.6   EMISSION CONTROL MEASURES

There are several strategies that can be implemented to reduce all forms of emissions out of dual-fuel engines besides exhaust gas after-treatment. These include measures such as further refinement and optimization of the fuel injection system and its control, better lubricating oil controls, using higher-cetane-number pilot fuels, and resorting to better quality, well-filtered, and processed fuel gases.

The heavy hydrocarbon pilot fuel produces relatively high production of NOx since its burning takes place mainly around its stoichiometric mixture envelope. However, because this envelope in dual-fuel engines is made up of a mixture with fuel gas, it would involve a bigger mass, but would tend to lower the peak temperature a little, which contributes toward reduction of NOx emissions. With higher injection pressure, injection rate shaping, and variable injection timing, NOx emissions and particulates can be reduced.

A combination of control measures has been implemented with varying degrees of effectiveness to reduce the undesirable emissions. These measures have included increasingly complex controls such as including effective particulate traps, application of very high pressure liquid fuel injection through common rail systems, bringing about improved fuel atomization and distribution, optimal variation of the geometry of turbochargers, variable valve timing, optimized combustion chamber geometry, electronic unit injection and shaping its rate, occasional water injection, and exhaust gas recirculation.

Generally, primary controls of operating variables of the engine are adjusted to bring about reductions in emissions. However, the extent of this reduction is limited and may lead to some negative effects, such as undermining the power output and increasing the specific energy consumption. Accordingly, since emission constraints are becoming increasingly severe and restrictive, secondary treatment methods are being increasingly applied. These increase the complexity of control and operating, capital, and maintenance costs. They rely on approaches such as the catalytic treatment of the exhaust gases. This way, the overall performance of the engine would be penalized less, but at the expense of increased complexity and cost.

In dual-fuel engines the relative size of the pilot employed, as well as its injection characteristics, has a controlling influence, particularly on the emission of fuel gas at any equivalence ratio. Relatively large pilots can be used at very light loads when very lean mixtures of the fuel gas and air are employed. But the size of the pilot needs to be reduced substantially in size, with increasingly less lean fuel gas–air mixtures admitted at high loads. When direct injection of the gaseous fuel is employed in multicylinder installations or two-stroke engine applications, though increasing the complexity of control and associated costs, it can lead to a reduction in the exhaust emissions of the unconverted gaseous fuel and carbon monoxide.

Other sources for HC exhaust emissions in engines include the scavenging phase and the crevice volumes. Unlike conventional diesel engines, dual-fuel engines have the fuel gas mixed with the air as it is admitted into the cylinder. A fraction of the gaseous fuel is forced into crevices during compression and may escape through the exhaust valve during the valve overlap period. To reduce HC emissions, valve overlap is reduced through changes to valve timing. However, this adds to increased complexity and contributes to some overheating, while reducing the effective volumetric efficiency, which affects power output. Of course, timed gas injection, whether into the manifold or directly into the engine, is an added positive feature for better combustion control. It would enable taking advantage, by employing skip firing, when less than the full complement of cylinders are employed for combustion while the remaining cylinders are motored. This way, the firing cylinders will be operating at much richer mixtures and smaller pilots than the overall total mixture reckoned on the basis of the fuel and air introduced. This approach, which requires great care in its proper implementation in multiple cylinder engines, would increase the motoring and frictional losses due to the none-firing cylinders. However, such an approach can result in improving the fuel utilization efficiency and reducing emissions at light loads.

Suitable catalytic converters for engine applications have been developed for simultaneously removing the three major pollutants of the products of combustion: unburned hydrocarbons, carbon monoxide, and oxides of nitrogen. These catalytic devices can then reduce the oxides of nitrogen to harmless molecular nitrogen while at the same time oxidizing the hydrocarbons and carbon monoxide to carbon dioxide and water vapor. Commonly, the active catalysts in this arrangement, at least for gasoline-fueled spark ignition engines, are those of mixtures of platinum and rhodium, and sometimes palladium, applied to a largely inert wash coat of aluminum oxide supported on a ceramic or metallic substrate. Catalyst performance would depend on many factors beyond the role of exhaust gas composition, such as temperature, space velocity, equivalence ratio and its modulation, and other factors, such as aging, poisoning, and thermal.

It is to be noted that if there is too much oxygen in the exhaust gas, then the reduction of the oxides of nitrogen will be slow and incomplete. When the exhaust gas does not contain sufficient oxygen, then the carbon monoxide and the hydrocarbons will not be fully oxidized. On this basis, a compromise has been reached in spark ignition engines by operating on stoichiometric mixtures. Such an operational procedure is not adopted in dual-fuel engine applications. In such a setup, methane would tend to react quite slowly and would require higher levels of gas temperature to effect suitable conversion. Dual-fuel engines, which normally operate on overall lean gas fuel–air mixtures and often involve gaseous fuels that contain very high proportions of methane, produce somewhat lower exhaust gas temperatures than corresponding gas-fueled spark ignition or diesel engines. The operating and design features of the dual-fuel engine make it have different emission composition characteristics, which require different approaches for catalytically controlling its exhaust emissions. The associated exhaust gas products are significantly less responsive to the conventional catalytic treatment of SI engine operation. Among the common catalytic materials, palladium (Pd)-based catalysts tend to be somewhat more effective in methane oxidation applications. However, the effectiveness of the palladium tends to drop rapidly as leaner and cooler mixtures are used. Dual-bed catalysts tend to show better CH_4, CO, and NOx conversion and light-off characteristics than single catalysts.

Ammonia gas at sufficiently high temperatures can strip oxygen out of NOx, in the presence of suitable catalysts. Urea, which can generate ammonia, may be stored and carried to be applied in transport applications as a fluid. It is controllably squirted into the exhaust system to carry out the reduction reactions. The fluid needs to be replenished periodically, and the system needs to be actively maintained with diesel and dual-fuel engines.

Exhaust gases containing relatively high concentrations of oxygen and oxides of nitrogen may be removed through the addition of the reducing agent ammonia when the exhaust gases are sufficiently hot in the presence of a catalyst according to the following overall reaction equation:

$$a\,NO_x + b\,NH_3 \rightarrow (a + b)/2.N_2 + 3b/2\,H_2O + d\,O_2$$

At high b/a molar ratios, a high removal of oxides of nitrogen is obtained, but with the potential of an undesirable increase in the emission of any unconverted ammonia. Obviously, this is adding to the complexity of controls and increases both operating and capital costs. The exhaust gas reactor needs to be equipped with means for preheating the control system during start-up. In engines using urea injection in the exhaust system, the amount of urea employed is usually small and amounts to around less than a few percent of that of the quantity of the liquid diesel fuel employed. However, ammonia is itself toxic, corrosive, and adds to operational costs.

The injection of diesel fuel sufficiently early in the compression stroke can lead to a reduction in NOx emissions, but would lead to substantial increases in the emissions of total hydrocarbons (THC) and CO, and fuel consumption, with the operating region narrowed to only partial load.

In addition to the catalytic treatment of the exhaust gases, the extent of NOx emissions can be reduced through proper changes in a number of operational controls that would bring about reductions in the peak level of combustion temperature, the available time for completing the reactions, and excess oxygen, and the elimination of fuel nitrogen. An additional control measure is an increase in the exhaust and inert gas recirculation while reducing its temperature through cooling, or ensuring the filling of the engine cylinder with cool fresh charge and expelling all combustion products from previous cycles. This is an important measure in the design and operation of engines to ensure high power output and efficiency while avoiding the incidence of knock. However, the role played by the increased retention of hot products of combustion or their recirculation tends to be more complex than merely lowering the overall temperature levels of the charge. Thermal and kinetic contributions are involved.

In general, faster rates of combustion of the cylinder contents are desirable since the bulk of the mixture is burned and the energy released early in the expansion stroke, resulting in high efficiencies and power. But this will lead to high gas temperatures and high production of oxides of nitrogen. This indicates that there is a need for a trade between aiming for very high efficiencies and low oxides of nitrogen emissions and for a strict control of engine operation.

The amount of particulates in the exhaust gas of a diesel engine is generally determined using a dilution tunnel, which ensures homogeneous mixing as well as cooling the diluted exhaust sample below the maximum filtering temperature. Elaborate methods have been devised recently to deal with diesel engine particulate emissions. These (particulate matter (PM)) are usually found in the order of less than 1 μm and consist of mainly unburned carbon particles, soot, which make up the largest portion of the total PM, and soluble organic fractions (SOFs), which consist of unburned hydrocarbons that have condensed into liquid droplets or have condensed on the soot particles.

Monolithic particulate filters are being fitted to engine exhaust systems. They are usually made up of small-diameter parallel channels that are closed at one end or the other (Figure 16.9). The channel walls are produced from a porous ceramic material, to which a suitable catalyst is applied at the inlet end. During the flow of the exhaust gases through the channel walls, the particulates are held back and oxidized by the catalyst and burned off, leaving the exhaust gas soot-free. In the outlet channels, an oxidizing catalytic coating is applied, which oxidizes any hydrocarbons and carbon monoxide to carbon dioxide and water vapor. The filter is loaded again after the soot has been burned off. However, if the particulates contain no combustible components, then these are left behind to contribute to clogging of the filter.

In the quest to reduce the extent of exhaust emissions in dual-fuel engines, some applications have been investigated where a reversing flow reactor is applied to methane-fueled dual-fuel engines (Figure 16.10). In such an application the feed of exhaust gas is switched periodically between the ends of the reactor, thus raising

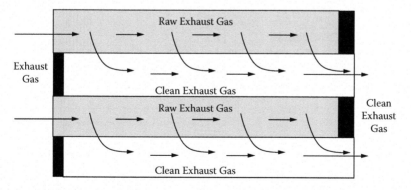

FIGURE 16.9 Schematic representation of a catalytic reactor for oxidizing particulates in the exhaust gas of a diesel engine. (From Karim, G.A., *Fuels, Energy, and the Environment*, CRC Press, Boca Raton, FL, 2012.)

the mean reactor bed temperature and enhancing its overall reactivity and methane conversion. However, it appears that such applications, due to limiting issues such as their complexity of operation, durability problems, and increased capital and operational costs, have not seen wide-scale adoption so far (Figures 16.10 and 16.11).

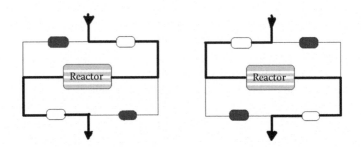

FIGURE 16.10   Schematic representation of the periodically switching exhaust gas catalytic reactor. (From Karim, G.A., *Fuels, Energy, and the Environment*, CRC Press, Boca Raton, FL, 2012.)

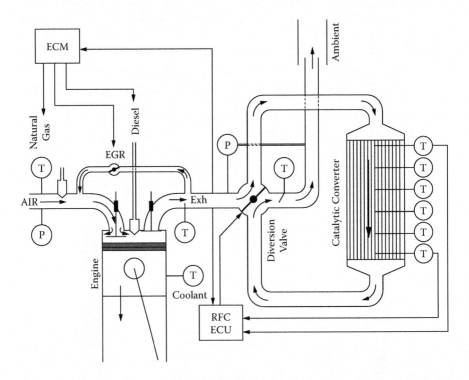

FIGURE 16.11   The schematic layout of the periodically switching reactor. (From Zheng, M., Mirosh, E.A., Ulan, D.A., Klopp, W.E., Pardell, M.E., Newman, P.E., and Nishimura, A., A Novel Reverse Flow Catalytic Converter Operated on an Isuzu-6HH1 Diesel Dual Fuel Engine, presented at ASME-ICE Division Fall Technical Conference, Ann Arbor, MI, October 1999.)

16.7 EMISSIONS OF GREENHOUSE GASES

Some of the increases in the concentrations of greenhouse gases in the earth's atmosphere have been attributed to the combustion of carbon-containing fuels releasing carbon dioxide, as well as the discharges of fugitive greenhouse gas emissions, such as those of methane. These additions to the atmosphere may contribute to gradual warming. There is an urgent need to control the emissions and reduce the intensity of these effects.

The carbon dioxide produced on combustion of common fuels varies very widely, depending on the fuel and its mode of processing. Take the combustion of hydrogen as an example. Although it does not appear to directly produce any carbon dioxide, the fact that the fuel was produced most likely through processes involving the gasification of coal, or the reforming of natural gas, nevertheless will be associated indirectly with relatively large greenhouse gas emissions at its source of production. The burning of liquid hydrogen is especially associated indirectly with serious emissions of greenhouse gas emissions at the source and stages of its manufacture since the liquefaction process is highly energy-intensive. Similarly, electric vehicles, by virtue of using electric power produced most likely by fossil fuel combustion, are associated with greenhouse emissions, albeit indirectly. Table 16.1 shows an estimate of greenhouse emissions of carbon dioxide from the combustion of fuels.

The amount of carbon dioxide produced depends, of course, on the amount of carbon fuel oxidized. However, methane is a much stronger greenhouse gas than carbon dioxide. Its releases into the atmosphere can come about from a number of sources that would include installations such as furnaces and engines that operate mainly on natural gas with methane its main constituent. However, the continued reduction in natural gas flaring and diverting much of the gas to the production of useful power, such as in engines, represents an indirect positive step toward globally reducing greenhouse emissions.

Some of the main measures that may be employed to reduce greenhouse gas emissions would include further improving the conduct of the combustion processes and the utilization of the energy produced. Also, whenever possible, lower

TABLE 16.1
Estimated Greenhouse Emissions of CO_2 from Fuel Combustion

Combustion Process	Average CO_2 Emissions (g/kJ)
Burn gasoline	68
Burn natural gas	54
Burn coal	125
Burn hydrogen (from reformed natural gas)	72
Burn hydrogen (from coal gasification)	207
Burn liquid hydrogen (from natural gas)	117
Burn liquid hydrogen (from coal)	316

Source: Adapted from Kukkonen, C.A., and Shelef, M., *Hydrogen as an Alternative Automobile Fuel*, SAE Paper 940766, 1994.

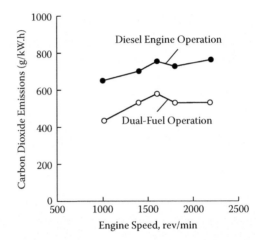

FIGURE 16.12 Variations in the extent of carbon dioxide emissions for a dual-fuel engine operating on natural gas and fixed pilot with changes in engine speed. The corresponding values for diesel operation are also shown. (From Checkel, M., Newman, P., van der Lee, B., and Pollak, I., *Performance and Emissions of a Converted RABA 2356 Bus Engine in Diesel and Dual Fuel Diesel/Natural Gas Operation*, SAE Paper 931823, 1993.)

carbon-to-hydrogen ratio fuels need to be used with devising effective methods for the economic disposal of the greenhouse gas methane before being emitted into the open atmosphere. This, however, continues to be not so easy and adds much to costs.

The continued reduction or elimination of exhaust gas pollutants, especially those of oxides of nitrogen, which are also strong greenhouse gases, will also be a contributory factor.

The dual-fuel engine scores well in comparison to other prime movers in the reduction of emissions of greenhouse gases through its high efficiency and the consumption of gaseous fuels that have relatively low carbon-to-hydrogen ratios, and with some, may have been produced from renewable sources. These fuels, and especially natural gas, also require a minimum amount of energy for their processing. Hence, they tend to be responsible overall for relatively lesser amounts of greenhouse gas releases. This is also aided by the need to use only small quantities of diesel fuel, which serve merely as the ignition agent. Figure 16.12 shows an example of the substantial reduction in carbon dioxide emissions with dual-fuel operation relative to that when the engine is operated in the diesel mode. This superiority in dual-fuel performance extends over the whole speed range.

An effective approach to reduce the emissions of greenhouse gases in dual-fuel engine operation is through the oxidation of any unconverted methane, such as through its catalytic oxidation, and minimizing the absolute amount of fuel consumed per unit of power produced. This would require maximizing the efficiency without increasing the amount of emissions produced, which requires a compromise between these two rather conflicting requirements.

Some of the general measures that have been employed to improve engine efficiency while reducing the associated emissions are the following:

- Resorting to operation on sufficiently fuel lean mixtures and employing direct gas injection per cylinder, rather than fumigation, to ensure better control of the fuel introduction and its distribution among the different cylinders, with less fuel passing unreacted directly to the exhaust stage
- Using optimized exhaust gas recirculation maintained warm at light load and cooled at high loads, for emissions and knock controls
- Employment of properly stratified mixture combustion where relatively rich mixtures are arranged near the ignition source, while the remaining mixture is fuel lean, such that the overall mixture remains fuel lean
- Employment of optimized variable valve timings and controlled exhaust gas turbocharging, including through variable geometry and exhaust gas bypass under selective conditions
- Retaining the high compression ratio of the engine subject throughout to the avoidance of the undesirable onset of knock or excessive NOx emissions
- Using sufficiently high quality processed fuels for both the gas and the liquid pilot and improving engine controls, especially under transient conditions
- Using cogeneration and hybrid operation with mixed staged sources for power production and storage and considering the use of homogeneous charge compression ignition (HCCI) operation, once it has been developed satisfactorily

REFERENCES AND RECOMMENDED READING

Aisho, Y., Yaeo, T., Koseki, T., Saito, T., and Kihara, R., *Combustion and Exhaust Emissions in a Direct Injection Diesel Engine Dual-Fueled with Natural Gas*, SAE Paper 950465, 1995.

Azzouz, D., Some Studies of the Combustion Processes in Dual Fuel Engines, MSc thesis, London University, UK, 1965.

Beck, N., Johnson, W., George, A., Peterson, P., van der Lee, B., and Klopp, G., *Electronic Fuel Injection for Dual Fuel Diesel Methane*, SAE Paper 891652, 1989.

Beppu, O., Fukuda, T., Komoda, T., Miyake, S., and Tanaka, T., Service Experience of Mitsui Gas Injection Diesel Engines, in *Proceedings of the CIMAC Congress*, 1998, pp. 187–202.

Bergman, H., and Busenthur, B., Facts Concerning the Utilization of Gaseous Fuels in Heavy Duty Vehicles, in *Proceedings of the Conference on Gaseous Fuels for Transportation*, August 1986, pp. 813–849.

Bittner, R.W., and Aboujaoude, F., Catalytic Control of NOx, CO, and NMHC Emissions from Stationary Diesel and Dual Fuel Engines, *ASME Journal of Engineering for Gas Turbines and Power*, 114, 597–601, 1992.

Boisvert, J., Gettel, L.E., and Perry, G.C., Particulate Emissions of a Dual Fuel Caterpillar 3208 Engine, in *ASME-ICE Division*, 1988, pp. 1–7, Paper 88-ICE-18.

Burn, K.S., The Effect of Cold Intake Temperatures on the Combustion of Gaseous Fuels in a Dual-Fuel Engine, MSc thesis, University of Calgary, Canada, 1977.

Checkel, M., Newman, P., van der Lee, B., and Pollak, I., *Performance and Emissions of a Converted RABA 2356 Bus Engine in Diesel and Dual Fuel Diesel/Natural Gas Operation*, SAE Paper 931823, 1993.

Choi, S., Yoon, Y.K., Kim, S., Yeo, G., and Han, H., *Development of Urea—SCR System for Light Duty Diesel Passenger Car*, SAE Paper 2001-01-0519, 2001.

Daisho, Y., Takahashi, Y.I., Iwashiro, Y., Nakayama, S., and Saito, T., *Controlling Combustion and Exhaust Emissions in a Direct-Injection Diesel Engine Dual Fuelled with Natural Gas*, SAE Paper 952436, 1995.

Ding, X., and Hill, P.G., *Emissions and Fuel Economy of a Prechamber Diesel Engine with Natural Gas Dual Fuelling*, SAE Paper 860069, 1988.

Dumitrescu, S., Hill, P.G., Li, G.G., and Ouellette, P., *Effects of Injection Changes on Efficiency and Emissions of a Diesel Engine Fuelled by Direct Injection of Natural Gas*, SAE Paper 2000-01-1805, 2000.

Ebert, K., Beck, N.J., Barkhimer, R.L., and Wong, H., *Strategies to Improve Combustion and Emission Characteristics of Dual Fuel Pilot Ignited Natural Gas Engines*, SAE Paper 971712, 1997.

Heenan, J., and Gettel, L., *Dual Fueling Diesel/NGV Technology*, SAE Paper 881655, 1988.

Imitrescu, S., and Hill, P.G., *Effects of Injection Changes on Efficiency and Emissions of a Diesel Engine Fuelled by Direct Injection of Natural Gas*, SAE Paper 2000-01-1805, 2000.

In, C.B., Kim, S.H., Kim, C.D., and Cho, W.S., *Catalyst Technology Satisfying Low Emission of Natural Gas*, SAE Paper 970744, 1997.

International Panel on Climate Change (IPCC), *Climate Change 2001: The Scientific Basis*, Cambridge University Press, Cambridge, 2001.

Ishyama, T., Shioji, M., Mitani, S., Shibata, H., and Ikegami, M., *Improvement of Performance and Exhaust Emissions in a Converted Dual Fuel Natural Gas Engine*, SAE Paper 2000-01-1866, 2000.

Jones, W., Liu, Z., and Karim, G.A., *Exhaust Emissions from Dual Fuel Engines at Light Load*, SAE Paper 932822, 1993.

Karim, G.A., *An Examination of Some Measures for Improving the Performance of Gas Fuelled Diesel Engines at Light Load*, SAE Paper 912366, 1991.

Karim, G.A., Combustion in Gas Fueled Compression Ignition Engines of the Dual Fuel Type, *ASME Journal of Engineering for Gas Turbines and Power*, 125, 827–836, 2003.

Karim, G.A., *Fuels, Energy and the Environment*, CRC Press, Boca Raton, FL, 2012.

Karim, G.A., and Burn, K.S., *Combustion of Gaseous Fuels in a Dual Fuel Engine of the Compression Ignition Type with Particular Reference to Cold Intake Temperature Conditions*, SAE Paper 800263, 1980.

Karim, G.A., and Rogers, A., Comparative Studies of Propane and Butane as Dual Fuel Engine Fuels, *Journal of the Institute of Fuel*, 40, 513–522, 1967.

Karim, G.A., and Wierzba, I., Comparative Studies of Methane and Propane as Fuels for Spark Ignition and Compression Ignition Engines, *SAE Transactions*, 92, 3677–3688, 1983.

Klimstra, J., *Catalytic Converters for Natural Gas Fueled Engines—A Measurement and Control Problem*, SAE Paper 872165, 1987.

Kukkonen, C.A., and Shelef, M., *Hydrogen as an Alternative Automobile Fuel*, SAE Paper 940766, 1994.

Kusaka, J., Daisho, Y., Shimonagata, T., Kihara, R., and Saito, T., Combustion and Exhaust Characteristics of a Diesel Engine Dual-Fuelled with Natural Gas, in *Proceedings of the 7th International Conference and Exhibition on Natural Gas Vehicles*, Yokohama, Japan, October 17–19, 2000, pp. 23–31.

Lambe, S.M., and Watson, H.C., Low Pollution, Energy Efficient C.I. Hydrogen Engine, *International Journal of Hydrogen Energy*, 17(7), 513–525, 1992.

Liu, B., and Checkel, D., *Experimental and Modeling Study of Variable Cycle Time for a Reversing Flow Catalytic Converter of Natural Gas/Diesel Dual Fuel Engines*, SAE Paper 2000-01-0213, 2000.

Liu, Z., An Examination of the Combustion Characteristics in Compression Ignition Engines Fueled with Gaseous Fuels, PhD thesis, University of Calgary, Canada, 1995.

Liu, Z., and Karim, G.A., Examination of the Exhaust Emissions of Gas Fuelled Diesel Engines, in *Proceedings of the 18th ASME Fall Conference, ICE Division*, 1996, Part 3, pp. 9–22.

Maiboom, A., Tauzia, X., and Hétet, J.F., Experimental Study of Various Effects of Exhaust Gas Recirculation (EGR) on Combustion and Emissions of an Automotive Direct Injection Diesel Engine, *Energy*, 33, 22–34, 2008.

Masood, M., and Ishrat, M.M., Computer Simulation of Hydrogen–Diesel Dual Fuel Exhaust Gas Emissions with Experimental Verification, *Fuel*, 87, 1372–1378, 2008.

Maxwell, T., and Jones, J., *Alternative Fuels: Emissions, Economics and Performance*, SAE, Warrendale, PA, 1995.

Mayer, R., Meyers, D., Shahed, S.M., and Duggal, V.K., *Development of a Heavy Duty On-Highway Natural Gas Fueled Engine*, SAE Paper 922362, 1992.

Miller, W., Klein, J., Mueller, R., Doelling, W., and Zuerbig, J., *The Development of Urea—SCR Technology for US Heavy Duty Trailers*, SAE Paper 2000-01-0190, 2000.

Mustafi, N.N., Particulates Emissions of a Dual Fuel Engine, PhD thesis, University of Auckland, New Zealand, January 2008.

Needham, J.R., May, M.P., Doyle, D.M., Faulkner, S.A., and Ishiwata, H., *Injection Timing and Rate Control—A Solution for Low Emissions*, SAE Paper 900854, 1990.

Rente, T., Golovichev, V.I., and Denbratt, I., *Effect of Injection Parameters on Auto-Ignition and Soot Formation in Diesel Sprays*, SAE Paper 2001-01-3687, 2001.

Sakai, T., Choi, B.C., Ko, Y., and Kim, E., *Unburned Fuel and Formaldehyde Purification Characteristics of Catalytic Converters for Natural Gas Fueled Automotive Engine*, SAE Paper 920596, 1992.

Schiffgens, H.J., Brandt, D., Rieck, K., and Heider, G., Low-NOx Gas Engines from MAN B&W, in *CIMAC Congress*, Copenhagen, 1998, pp.1399–1414.

Shioji, M., Ishiyama, T., and Ikegami, M., Approaches to High Thermal-Efficiency in High Compression-Ratio Natural-Gas Engines, in *Proceedings of the 7th International Conference on Natural Gas Vehicles*, Yokohama, Japan, 2000, pp. 13–21.

Singh, S., Krishnan, S.R., Srinivasan, K.K., Midkiff, K.C., and Bell, S.R., Effect of Pilot Injection Timing, Pilot Quantity and Intake Charge Conditions on Performance and Emissions for an Advanced Low-Pilot-Ignited Natural Gas Engine, *International Journal of Engine Research*, 5(4), 329–348, 2004.

Tsolakis, A., Hernandez, J.J., Megaritis, A., and Crampton, M., Dual Fuel Diesel Engine Operation Using H2 Effect on Particulate Emissions, *Energy and Fuels*, 19, 418–425, 2005.

Turner, S.H., and Weaver, C.S., *Dual-Fuel Natural/Diesel Engines: Technology, Performance and Emissions*, No. GRI-94/0094, Topical Report Gas Research Institute, November 1994.

USEPA, *Inventory of U.S. Greenhouse Gas Emissions and Sinks (1990–2002)*, USEPA/30/R/04/003, 2004.

Xiao, F., Experimental and Numerical Investigation of Diesel Combustion Processes in Homogeneously Premixed Lean Methane or Hydrogen-Air Mixtures, PhD thesis, University of Calgary, Canada, 2011.

Xiao, F., and Karim, G.A., *Combustion in a Diesel Engine with Low Concentrations of Added Hydrogen*, SAE Paper 2011-01-0676, 2011, pp. 1–17.

Xiao, F., Lu, C., Hu, Y., and Yang, B., *Experimental Study on Diesel Nitrogen Oxide Reduction by Exhaust Gas Recirculation*, SAE Paper 2000-05-0335, 2000.

Xiao, F., Sohrabi, A., and Karim, G.A., *Reducing the Environmental Impact of Fugitive Gas Emissions through Combustion in Diesel Engines*, SAE Paper 2007-01-2048, 2007.

Xiao, F., Sohrabi, A., and Karim, G.A., Effect of Small Amounts of Fugitive Methane in the Air on Diesel Engine Performance and Its Combustion Characteristics, *International Journal of Green Energy*, 5, 334–345, 2008.

Yonetani, H., Hara, K., and Fukatani, I., *Hybrid Combustion Engine with Premixed Gasoline Homogeneous Charge and Ignition by Injected Diesel Fuel-Exhaust Emission Characteristics*, SAE Paper 940268, 1994.

Yoshida, K., Shoji, H., and Tanaka, H., *Study on Combustion and Exhaust Emission Characteristics of Lean Gasoline-Air Mixture Ignited by Diesel Fuel Direct Injection*, SAE Paper 982482, 1998.

Zheng, M., Mirosh, E.A., Ulan, D.A., Klopp, W.E., Pardell, M.E., Newman, P.E., and Nishimura, A., A Novel Reverse Flow Catalytic Converter Operated on an Isuzu-6HH1 Diesel Dual Fuel Engine, in *ASME-ICE Division Fall Technical Conference*, Ann Arbor, MI, October 1999.

17 Some Operational Features of Dual-Fuel Engines

17.1 DUAL-FUEL ENGINE PERFORMANCE

Results and corresponding conclusions reported in the technical literature relating to the employment of gaseous fuels in dual-fuel engines often describe tendencies that are specific to the engine setup and the gaseous fuel employed. Since gaseous fuels, including natural gases and liquefied petroleum gases (LPGs), can sometimes vary widely in composition, especially in terms of the concentrations of higher hydrocarbons and diluents, it is necessary whenever evaluating results of engine performance to keep the effects of such possible variations in mind and to account for their possible contribution. Accordingly, it is preferable when investigating some aspects of the combustion processes in dual-fuel engines to refer, whenever possible, to fuels of known composition or, even better, to pure fuels rather than, for example, a generic natural gas or biogas fuel.

A typical comparison of the values of the specific energy consumption with changes in the total equivalence ratio of a dual fuel operating on different gaseous fuels at constant pilot quantity with the corresponding values when operating as a diesel would show that the performance of the dual-fuel version is much inferior to that of the diesel at low loads. However, at high loads the performance of the dual-fuel engine improves and may become superior to that of the diesel. The employment of reduced pilot sizes increases the consumption differences at light load. However, at high load the dual-fuel consumption becomes less affected by the relative pilot size employed compared to the fuel gas. Engine performance can be correlated adequately in terms of overall equivalence ratio when based on both the pilot and fuel gas. Figure 17.1 shows schematically the typical variation in dual-fuel engine performance parameters when operating on fumigated methane with constant pilot and engine speed over the operational total equivalence ratio. The full load is often limited by the onset of knock.

The performance of dual-fuel engines in relation to that of typical spark ignition gas engines may be considered to have the following superior characteristics:

- It can accept a wider spectrum of gaseous fuels and has superior control and safety characteristics.

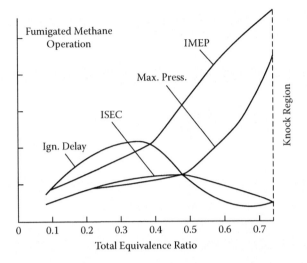

FIGURE 17.1   Schematic of typical variations in IMEP, maximum pressure, ignition delay, and indicated specific energy consumption, with changes in total equivalence ratio of a dual-fuel engine operating at constant speed on methane using a constant pilot. The extent of operational range is limited by the onset of knock at high load. (Not to scale.)

- It displays less cyclic variation even at light load, aided mainly by the deliberate and reliable source of the energetic pilot ignition.
- It requires fewer changes in the pilot fuel injection timing in comparison to the changes in the timings of electric spark ignition needed with changes in engine speed and load.
- Quality control is normally obtained by changes in the fuel concentration rather than throttling, as in common spark ignition engines.
- The engine can operate over a wider overall range of mixture strength and intensity of turbocharging while displaying superior emissions.
- Higher compression ratios and torque can be obtained, leading to improvements in power output and efficiency, with less tendency to encounter the incidence of knock. This is because of the overall lean mixtures normally employed and the multiple ignition centers provided by the injection of the pilot.

An example that compares the specific energy consumption over the whole load range of a spark ignition gas-fueled engine when exclusively operating on stoichiometric mixtures of natural gas with that when operated on a variable lean mixture range is shown in Figure 17.2. The superior performance of operation on variable lean mixtures is evident. Similarly, the corresponding specific energy consumption when operating as a diesel is compared with the corresponding values when the engine is operated as a dual-fuel engine. It is shown that the specific consumption under spark ignition operation is inferior throughout to the corresponding operation as a diesel or dual fuel. It can also be seen that dual-fuel engine performance may become even superior to the corresponding values of the diesel mode at high loads. It

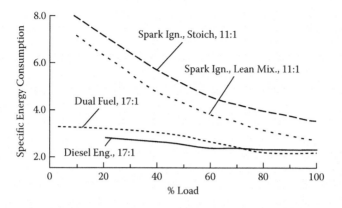

FIGURE 17.2 A comparison of variations of the specific energy consumption with load for spark ignition, diesel, and dual-fuel engines operating on natural gas. (From Beck, J., Karim, G.A., Pronin, E., and Mirosh, E., The Diesel Dual-Fuel Engine Practical Experience and Future Trends Update, Paper no. 96EL070, *Proceedings of the International Symposium of Automotive Technology and Automation,* ISATA, Florence, Italy, pp. 1–6, 1996.)

is to be noted that the compression ratio for the spark ignition operational mode had to be lowered substantially, primarily to avoid the incidence of knock.

An example of the variation of the brake specific energy consumption with load for a dual-fuel engine fueled with a natural gas is also shown in Figure 17.3. In this specific example, it can be seen that such performance is superior to diesel operation. The performance of the dual fuel can be substantially improved even further when skip firing and turbocharger bypass are implemented.

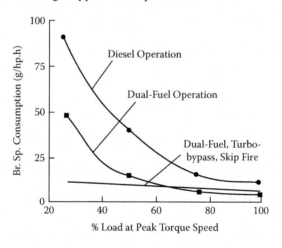

FIGURE 17.3 Comparison of the brake specific energy consumption with load for dual-fuel operation with that of diesel showing the superior performance of the dual-fuel engine, especially when fitted with skip firing and turbocharger bypass. (From Beck, J., Karim, G.A., Mirosh, E., and Pronin, E., *Bus Fuel Efficiency Local Emissions and Impact on Global Emissions*, presented at NGV 96 Conference and Exhibition, Kuala Lumpur, Malaysia, October 1996.)

17.2   EXHAUST GAS RECIRCULATION (EGR)

Some form of exhaust gas recirculation has been increasingly incorporated in designs and operation of engines in recent years. It is mainly employed as a form of aiding in the effective control of the generation of pollutants during combustion, especially those of NOx. Retention of products of combustion or exhaust gas reintroduction is normally implemented through measures that include direct controlled recirculating of some of the exhaust gas to the engine intake, either cooled or uncooled. Alternatively, it can be made through incorporating suitable adjustment to the valves' timing and their overlap period. The net effect is to increase the dilution of the fresh charge with combustion products that tend to act effectively as inert diluents. This will produce some changes in the composition of the cylinder contents and lead to an overall reduction in the values of the peak combustion temperature. In turn, this reduction leads to a lowering in the production of oxides of nitrogen, which is heavily dependent on values of the local temperature and the partial pressure of the available oxygen. Since the thermodynamic properties of the residual and exhaust gases are sufficiently different from those of the fresh charge mixture, they can also modify the temporal variations of the mean temperature and the associated heat transfer processes during and beyond the end of compression. Additionally, the effects of EGR can be viewed in terms of its contributions through the increased presence of diluents, higher initial compression temperatures that may lead to higher overall combustion temperatures, burning of some of the recirculated unburned fuel in the exhaust gas, and a reduction in the oxygen concentration of the charge. Depending on how the EGR is implemented, there will be a slight lag in introducing the exhaust gas from previous cycles. This is especially important under transient operating conditions, and it is much influenced by the design and mode of EGR application.

There are broadly three types of EGR. First is the hot EGR, where hot exhaust gases are recycled without external cooling before mixing with the intake charge, to result in an increase in its overall temperature. This may be achieved by insulating and shortening the EGR delivery line. The second is a fully cooled EGR, obtained when the exhaust gas to be recirculated is cooled essentially down toward ambient conditions. This is implemented, for example, through employing a specially installed water-cooled heat exchanger and introducing the gases directly into the intake pipe. In such approaches, much of the water vapor in the exhaust gas is condensed and removed so as not to enter the cylinder. The third approach, which may be described as partially cooled EGR, avoids some of the problems of the cooled or hot EGRs. The exhaust gas may be partially cooled to such an extent that its temperature remains above the dew point of the exhaust gas. In turbocharged engine applications, the exhaust gas may be extracted from the outlet of the exhaust turbocharger, cooled, and the excess water extracted. The EGR routing in turbocharged dual-fuel engines can also be made from somewhere before entry into the turbine to the inlet manifold, where there is a sufficiently positive pressure difference between the exhaust and inlet manifolds. This is aided by the use of a waste-gate valve and may permit a small reduction in the size of the turbine. The engine controls need to be adjusted to account for the intensity and extent of EGR employed. This is usually established mainly through experimentation.

The cooling of the recirculated gases before their introduction into the intake charge is commonly desirable, not only to reduce the production of oxides of nitrogen, but also to lower the tendency to knock at high loads. However, the increased proportion of exhaust gas into the engine intake charge, whether carried out through some form of gas retention or recirculation actions, displaces some of the fresh charge and vitiates it. Occasionally, the recirculated gases, even when cooled, may raise the temperature toward the end of compression, such as through the presence of some unconverted fuel from previous cycles. There can also be some changes to the chemical kinetic activity since a wide range of quenched and partial oxidation products may become mixed with the fresh fuel–air mixture. This form of mixing can lead to the production and survival during compression of a variety of actively reacting species that may influence the subsequent combustion processes of both the pilot and the main gaseous fuel–air mixture. Hence, the ignition delay, pressure development, tendency to knock, and level and nature of exhaust emissions can be affected.

In dual-fuel engines the employment of hot EGR can improve the part load efficiency and power output. It can lead to a reduction in the level of unburned hydrocarbons in the presence of the excess air charge. However, hot EGR can bring about an increase in thermal loading and, eventually at high loads, an overall reduction in the power output and efficiency, with an earlier encounter of the onset of knock. For practical purposes, in dual-fuel applications it is necessary to maintain a low EGR ratio at light load since the temperature of the exhaust gases is relatively low due to the very lean mixtures employed and the incomplete combustion of the gaseous fuel component. Hence, EGR may be implemented optimally with the different gaseous fuels mainly at high loads. The use of a high amount of EGR increases the presence of diluents in the charge, which leads to longer combustion periods and reduction in the efficiency and power output. Also, a high EGR ratio reduces the average combustion temperature and the heat loss during expansion. A reduction in the exhaust gas temperature is favored for maintaining the reliability and durability of equipment.

In summary, the effects of EGR on the combustion processes can be considered to various degrees due to a number of combined effects. These are thermal diluting the charge, and thereby changing the equivalence ratio and composition. There are also thermodynamic effects, such as changing the effective polytropic index during compression and other properties, as well as kinetic influences that may provide some partial oxidation products and active species while affecting the extent and mode of heat transfer.

17.3   HEAT TRANSFER IN DUAL-FUEL ENGINES

It is usually a challenge to experimentally conduct heat transfer measurements reliably in reciprocating internal combustion engines. This is mainly because the measurement of the extent of heat transfer within the engine can be made only over specific small locations, while the heat flux distribution over the cylinder surfaces is nonuniform and highly periodic. Moreover, the values and distributions of the key controlling factors, such as fluid flow, temperature, and composition, together with the action of deposits and lubricants, are uncertain. Largely, approximate quasi-empirical approaches have been devised that tend to give results that are indicative of only the average global trends. This represents a significant limitation to the knowledge of the processes involved. Such knowledge is needed not only for predictive modeling, but also when considering the operational and design problems associated with the control of knock, emissions, and the gas-liquid fuel injection processes in general, and in particular in engines of the precombustion chamber type.

As a result of the specific relationship between the cylinder volume and temperature during compression, typically about half the temperature rise during compression takes place within the last part of the piston travel, as the piston slows down considerably when it approaches its end of travel. The extent of heat transfer is then increased as a result of the high temperatures, larger relative surface area to volume, and increased forced convective turbulent flows. The diesel engine, even with its high compression ratio and when turbocharged, tends to have cooler surfaces than those of the stoichiometrically operated spark ignition gasoline engines. This is mainly since the diesel engine operates with much excess air and with combustion, and the hot regions are usually centered farther away from the chamber surfaces. The conventional dual-fuel engine has a different distribution of its mixtures. The overall lean premixed charge is distributed throughout the whole cylinder, with only a much smaller region associated with pilot combustion more centrally located. The dual-fuel engine is associated with much less injected liquid fuel penetration and surface deposition because of its small size. This would lead to having the extent of heat transfer in dual-fuel engines generally lower than in both spark ignition engines and diesel engines. A greater portion of the energy release takes place through flame propagation, with combustion possibly proceeding into some of the cylinder crevices. There will also be less cooling than in diesel engines via liquid fuel injection, leading to the injector body tending to run hotter. Radiation heat transfer would be lower and less persistent down the expansion stroke. These features would positively reflect the efficiency and increased durability of the engine.

There are a number of distinct features of the working and design of dual-fuel engines that reflect, to a large extent, the unknown extent of the heat transfer. For example, the fact that the engine runs on a mixture of a gaseous fuel and high compression ratio affects its temperature-time development during compression because of the changes in the thermodynamic properties, unlike the diesel engine, where compression takes place with air only. Also, depending on the gaseous fuel employed and the extent of any exhaust gas recirculation employed, this temperature will be further modified by the energy releases, if any, during the preignition processes. In some gaseous fuels, such as LPG, some chemical energy releases can take place somewhat earlier in the later stages of the compression stroke than with methane or hydrogen operation. Accordingly, in dual-fuel engines, the heat flux appears to increase with load at higher rates than those in the corresponding diesel engine. The temperature at the upper part of the cylinder liner and the heat flux in the cylinder head are lower, but with the heat flux at the lower part of the liner somewhat higher.

For dual-fuel engines of the prechamber type, the temperature of the tip increases, but the valve seat and piston crown temperatures for natural gas applications are lower than for diesel operation. This is due to the less heterogeneous combustion taking place, in contrast to that of the diesel engine, where the combustion is of the liquid fuel diffusion type burning. In the cases of high-pressure fuel gas injection, the temperature, composition, and mixture velocity distribution are markedly different from those in the case of the corresponding diesel engine or the premixed version of the dual-fuel engine. With very high pressure injection of the fuel gas, it is to be expected that the introduction of an expanding high-velocity and very cold gas jet will substantially modify the temperature distribution, and thus the heat transfer processes.

17.4   ENGINE OPERATION ON LIQUEFIED NATURAL GAS (LNG)

Control of the many variables in LNG engine applications is increasingly being made through the use of computers. The LNG fuel needs to be evaporated, requiring a suitable heat exchanger that often uses engine hot circulating water. This has to be accomplished without much reduction in the water jacket temperature. Otherwise, it can adversely affect engine performance. Arrangements are usually made to bypass this heat exchanger when needed. The hot exhaust gas, in comparison, tends to be a less suitable source for such fuel heating. The gas injection system needs to be capable of having to occasionally accept not fully vaporized methane in the form of an aerosol, especially near engine full-load conditions, where high fuel flow rates are encountered. Normally, this is less common, as heat extraction increases as full-load conditions are approached with a reduced tendency for thermal stressing of engine parts.

Figure 17.4 shows the variations with the total equivalence ratio at three fixed values of diesel pilot of the concentrations of the unreacted methane appearing in the dry exhaust gas of a normally aspirated direct injection dual-fuel engine fumigated with methane. Different constant intake mixture temperatures are shown, including those of sufficiently low values that may arise from operation on LNG. It can be seen that maximum unconverted methane concentrations are encountered at low loads, but are then reduced as less lean mixtures are employed. Cooling the intake mixture temperature and reducing the size of the pilot lead to increased emissions of unconverted methane. This can lead to ignition difficulties when sufficiently cold mixtures with small pilots are used. An example of the corresponding substantial changes in the ignition point over a range of total equivalence ratios of cooled intake methane air mixtures is shown in Figure 17.5, displaying excessively long delays as the size of the pilot is reduced. The corresponding uncooled intake operation is also shown.

This chilling effect of the intake charge will lead, especially at light loads, to a higher proportion of the fuel remaining unoxidized, with high concentrations partially converted to increase the concentrations of exhaust carbon monoxide. Of course, the emissions of oxides of nitrogen tend to be insignificant. In the case of using excessively large pilots, the presence of particulates in the exhaust gases tends to increase. However, the injection of the cold-boiled gas into the intake manifold or into the engine cylinder would enhance the volumetric efficiency of the engine through increasing the density of the charge.

Some engine modifications are needed with LNG operation. These can include the potential to employ higher compression ratios since there is less likelihood for knock. The injection timing of the pilot may benefit from a certain amount of advance to compensate somewhat for the increased cooling of the intake charge. Engine performance can also benefit from having the water jacket temperature

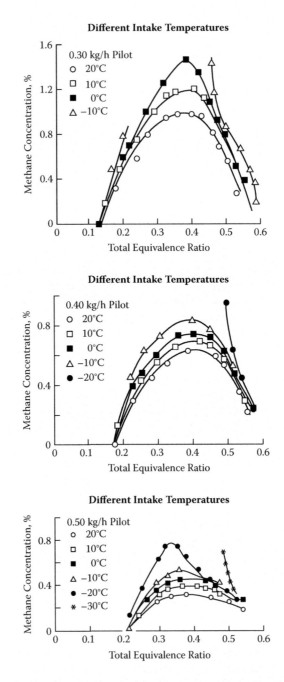

FIGURE 17.4 Variations of the concentration of unconsumed methane in the exhaust gas of a methane-fumigated engine with changes in the total equivalence ratio. Different intake mixture temperatures are employed that include the range of temperatures arising from mixing with LNG vapor. Three different cases employing increasingly larger pilots are shown. (From Karim, G.A., *Progress in Energy and Combustion Science*, 6, 277–285, 1980.)

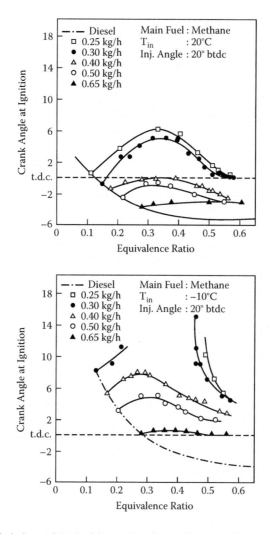

FIGURE 17.5 Variations of the ignition angle values with changes in the total equivalence ratio of a dual-fuel engine operating on fumigated methane at a range of constant pilot quantities, when the intake mixture is both uncooled and cooled. (Adapted from Burns, K.S., The Effect of Cold Intake Temperatures on the Combustion of Gaseous Fuels in a Dual-Fuel Engine, MSc thesis, University of Calgary, Canada, 1977.)

controlled suitably according to the load and the size of the pilot employed. The size of the radiator may be reduced, while greater attention is given to deal with potential increases in lubricating oil problems, increased corrosion of the exhaust system, and materials compatibility difficulties. Also, some water icing problems may need to be attended to, especially in operations in regions where high-humidity air is normally encountered.

17.5   CYCLIC VARIATIONS

Cyclic variation in engine performance parameters needs to be reduced to a minimum and is an important source of limiting engine power output, increasing fuel consumption and the levels of emissions, and undermining the optimization of engine performance. The extent of variation in the values of performance parameters, which tend to be interrelated, is commonly described by the corresponding coefficient of cyclic variations (COV). It is given as the standard deviation of the values of the observed parameter normalized by the mean of the observed values. A statistically adequate number of observations needs to be employed. Often, the coefficient of cyclic variations for the indicated work output (COV_{IMEP}) is employed to describe the stability of engine performance, while the variation in the values of peak pressure indicates combustion stability. In general, the extent of cyclic variations in diesel engines and also in dual-fuel engines tends to be much lower than in gas-fueled spark ignition engines, especially for those operating not stoichiometrically.

Numerous factors influence the mode and extent of engine cyclic variation. For example, these include any variations in intake mixture composition and temperature, the concentrations of residuals retained and any reactive species that may be formed during compression, values of the temperature in the vicinity of the ignition centers, the mode and intensity of charge motion, the extent of exhaust gas recirculation, and the heat loss. High cycle-to-cycle variations are usually associated with the occurrence of substantial combustion instabilities, such as when operating on excessively lean mixtures, throttled operation, employing excessive EGR, or cold running. In such situations, the injection characteristics and the pilot fuel quantity, their fluctuations, vaporization, mixing processes, and aerodynamic variations collectively exert significant influence on the extent of these variations. Larger pilots and less lean gaseous fuel–air mixtures produce much improved stability of operation. Fast-burning fuels such as hydrogen display less variation than the slower-burning methane or biogases containing diluents. Typically, cyclic variation in the output of dual-fuel engines at light load with very lean mixture operation needs to be kept well below a value of 5–10%.

It can be seen in Figure 17.6 that the power output at constant pilot and speed varies essentially linearly with the extent of fuel gas admission. Also, the noise at very light load tends to be mostly associated with the rapid combustion of the diesel pilot, but decreases rapidly with the increased admission of methane. The noise later becomes much less dependent on the amount of gas burned.

An example of the extent of variations of peak pressure, rate of pressure rise, and indicated mean effective pressure (IMEP) with changes in the equivalence ratio of the fuel gas feed at a constant pilot quantity with methane as the fuel is shown in Figure 17.7. The corresponding power output is also shown.

Random operational variations in the values of most of the key influencing variables, such as the following, contribute to different degrees to the cyclic variations that may be observed in dual-fuel engines, especially within the low load operating region:

- Very lean mixtures and small amounts of liquid fuel pilot injection are employed in dual-fuel operation. These produce long delays that lead to slow post-ignition energy releases, with flame propagation becoming more

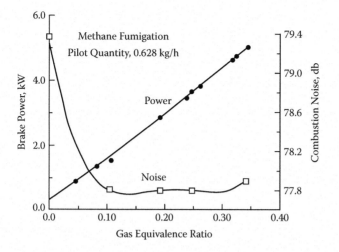

FIGURE 17.6 Variation of brake power output with gas equivalence ratio at constant pilot and speed. The corresponding variation in the combustion noise is also shown. (Adapted from Liu, C., An Experimental and Analytical Investigation into the Combustion Characteristics of HCCI and Dual Fuel Engines with Pilot Injection, PhD thesis, University of Calgary, Canada, 2008.)

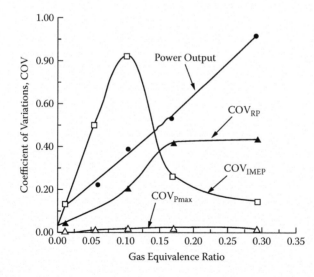

FIGURE 17.7 An example of the extent of variations in peak pressure, rate of pressure rise, and IMEP with equivalence ratio based on the gas admitted at a constant pilot with a methane fumigated engine. (Adapted from Liu, C., An Experimental and Analytical Investigation into the Combustion Characteristics of HCCI and Dual-Fuel Engines with Pilot Injection, PhD thesis, University of Calgary, Canada, 2008.)

prone to variations. Also, there will be contributions to the observed variation in engine performance from poor pilot fuel vaporization and mixing,
which lead to fluctuations in the nonuniformity in equivalence ratio distribution during combustion.

- There are significant nonuniformities in temperature, composition, and
  velocity during the compression stroke that change with time and produce
  dynamic changes from one cycle to another in the behavior of the values
  of the hottest region of the mixture and its location, such as relative to pilot
  fuel injection and cylinder surfaces.

- The liquid diesel fuel commonly used as pilot usually displays a very prominent multistage ignition behavior, rather than the single-stage continuous
  acceleration of reactions associated with methane. These multistage reactive characteristics can lead to amplifying the observed cyclic variation
  when small changes in influencing variables take place, and can prolong
  the apparent interruption of the energy release processes between the stages
  of reactions. The fraction and reactivity of residual gases also can vary
  from one cycle to another, which can have a strong effect on the course of
  the oxidation reactions.

- The cooling of compressed natural gas on its expansion through the pressure regulators can contribute to the variations. The extent of this cooling
  of course depends on the flow rate and how high the gas supply pressure
  is. Gas regulators may need some heating using engine coolant to prevent
  potential icing or hoar frost formation, which may lead to variations in the
  value of the fuel/air ratio of the delivered mixture.

- At the borderline for lean mixture ignition limits, any variation in heat
  transfer patterns will make a contribution aiding or impeding the reactions.
  Also, the development of a hot spot or part of the cylinder surface becoming
  slightly hotter than the other surfaces will be an influencing factor.

- When the fuel gas is introduced into the incoming air via a simple carburetion system, the intense pulsating flow into the cylinder is not necessarily
  consistently repeatable and in proper phase cyclically with engine valve closure or opening. This can be a contributory factor toward cyclic variations
  in the resulting equivalence ratio and its distribution, especially around the
  mixture ignition limits.

- Mixtures in the neighborhood of the lean limit values produce very low
  combustion temperatures. Hence, a very slight change in temperature in
  one region relative to another will make a big difference in the relative reaction rate. This is not the case with much less lean mixtures, such as those
  near full load or those leading to knock, where a small change in temperature changes the reaction rates only relatively little. Moreover, for very lean
  mixtures, a slight shift in equivalence ratio makes a very large change in the
  flame speed and the energy release rates. On the other hand, for mixtures
  around the stoichiometric value, the flame speed is of course very much
  higher, but relatively less sensitive to small variations in that region of the
  equivalence ratio.

REFERENCES AND RECOMMENDED READING

Beck, J., Karim, G.A., Mirosh, E., and Pronin, E., *Bus Fuel Efficiency Local Emissions and Impact on Global Emissions,* presented at NGV 96 Conference and Exhibition, Kuala Lumpur, Malaysia, October 1996.

Beck, J., Karim, G.A., Pronin, E., and Mirosh, E., The Diesel Dual-Fuel Engine-Practical Experience and Future Trends Update, paper no. 96EL070, *Proceedings of the International Symposium on Automotive Technology and Automation,* ISATA, Florence, Italy, pp. 1–6, 1996.

Beck, N., Johnson, W., George, A., Peterson, P., vander Lee, B., and Klopp, G., *Electronic Fuel Injection for Dual Fuel Diesel Methane,* SAE Paper 891652, 1989.

Bell, S., *Natural Gas as a Transportation Fuel,* SAE Paper 931829, 1993.

Beppu, O., Fukuda, T., Komoda, T., Miyake, S., and Tanaka, T., Service Experience of Mitsui Gas Injection Diesel Engines, in *Proceedings of CIMAC Congress,* 1998, pp. 187–202.

Blizzard, D., Schaub, F.S., and Smith, J., Development of the Cooper-Bessemer Cleanburn Gas-Diesel (Dual-Fuel) Engine, in *ASME-ICE Division,* 1991, vol. 15, pp. 89–97.

Callahan, T., Survey of Gas Engine Performance and Future Trends, presented at Proceedings of the ASME Conference, ICE Division, Salzburg, Austria, 2003, Paper ICES2003-628.

Ebert, K., Beck, N.J., Barkhimer, R.L., and Wong, H., *Strategies to Improve Combustion and Emission Characteristics of Dual Fuel Pilot Ignited Natural Gas Engines,* SAE Paper 971712, 1997.

Heenan, J., and Gettel, L., *Dual Fueling Diesel/NGV Technology,* SAE Paper 881655, 1988.

Ishyama, T., Shioji, M., Mitani, S., Shibata, H., and Ikegami, M., *Improvement of Performance and Exhaust Emissions in a Converted Dual Fuel Natural Gas Engine,* SAE Paper 2000-01-1866, 2000.

Jacobs, T., Assanis, D., and Filipi, Z., *The Impact of Exhaust Gas Recirculation on Performance and Emissions of a Heavy-Duty Diesel Engine,* SAE Paper 2003-01-1068, 2003.

Karim, G.A., A Review of Combustion Processes in the Dual Fuel Engine—The Gas Diesel Engine, *Progress in Energy and Combustion Science,* 6, 277–285, 1980.

Karim, G.A., Combustion in Gas Fueled Compression Ignition Engines of the Dual Fuel Type, *ASME Journal of Engineering for Gas Turbines and Power,* 125, 827–836, 2003.

Karim, G.A., and Burn, K.S., *Combustion of Gaseous Fuels in a Dual Fuel Engine of the Compression Ignition Type with Particular Reference to Cold Intake Temperature Conditions,* SAE Paper 800263, 1980.

Klimstra, J., Hattar, C., Nylund, I., and Sillanpää, H., The Technology and Benefits of Skip Firing for Large Reciprocating Engines, presented at ASME-ICE Division, Chicago, April 2005, Paper ICES2005-1073.

Li, H., and Karim, G.A., Modeling the Performance of a Turbo-Charged Spark Ignition Natural Gas Engine with Cooled Exhaust Gas Recirculation, *ASME Journal of Engineering for Gas Turbines and Power,* 130, 328041-10, 2008.

Liu, C., An Experimental and Analytical Investigation into the Combustion Characteristics of HCCI and Dual Fuel Engines with Pilot Injection, PhD thesis, University of Calgary, Canada, 2008.

Liu, Z., and Karim, G.A., *An Examination of the Role of Residual Gases in the Combustion Processes of Motored Engines Fuelled with Gaseous Fuels,* SAE Paper 961081, 1996.

Mayer, R., Meyers, D., Shahed, S.M., and Duggal, V.K., *Development of a Heavy Duty On-Highway Natural Gas Fueled Engine,* SAE Paper 922362, 1992.

Selim, M., Thermal Loading and Temperature Distribution of a Precombustion Chamber Diesel Engine Running on Gasoil/Natural Gas, in *ASME-ICE Division,* 1998, vol. 31-3, pp. 113–130, Paper 98-ICE-159.

Shudo, T., Improving Thermal Efficiency by Reducing Cooling Losses in Hydrogen Combustion Engines, *International Journal of Hydrogen Energy,* 32, 4285–4293, 2007.

Singh, S., Krishnan, S.R., Srinivasan, K.K., Midkiff, K.C., and Bell, S.R., Effect of Pilot Injection Timing, Pilot Quantity and Intake Charge Conditions on Performance and Emissions for an Advanced Low-Pilot-Ignited Natural Gas Engine, *International Journal of Engine Research*, 5(4), 329–348, 2004.

Steiger, A., Large Bore Sulzer Dual Fuel Engines, Their Development, Construction and Fields of Application, *Sulzer Technical Review*, 3, 1–8, 1970.

Turner, S.H., and Weaver, C.S., *Dual-Fuel Natural/Diesel Engines: Technology, Performance and Emissions*, No. GRI-94/0094, Topical Report Gas Research Institute, Chicago, November 1994.

Yoshida, K., Shoji, H., and Tanaka, H., *Study on Combustion and Exhaust Emission Characteristics of Lean Gasoline-Air Mixture Ignited by Diesel Fuel Direct Injection*, SAE Paper 982482, 1998.

18 Dual-Fuel Engine Management

18.1 ENGINE MANAGEMENT SYSTEMS

Much progress has been made in recent years in the development of electronically controlled liquid and gaseous fuel injection equipment and methods for dual-fuel engine applications. These permit, for example, independent variations in the pilot fuel quantity and injection timing, with various approaches for the controlled introduction of the gaseous fuel, including its direct injection into the cylinder under sufficiently high pressure. Engines in service that are not fitted with up-to-date control technologies tend not to achieve the full potential of dual-fuel operation in terms of most key performance parameters. These engines tend to display poor reliability and incur high maintenance costs.

Diesel engines converted to dual-fuel operation are normally required to retain the capacity to also perform satisfactorily as a diesel engine when needed. Any changes or additions to render the engine capable of operating as a dual-fuel engine should not undermine this capacity. The peak cylinder pressure, rates of pressure rise, rates of energy release, and torque ~ speed characteristics need to be kept at the same order as those for the corresponding diesel operation. Controls for engine starting, idling, overload, and shutdown are preserved, and the switchover to diesel operation must be sufficiently prompt, with proper diesel performance fully restored. Additionally, to keep costs down and avoid increased complexity, no major modifications to the engine need to be made. Nevertheless, to ensure that control units fitted to dual-fuel engines remain optimum is relatively challenging and costly, given the low volume of dual-fuel engines produced in comparison to the huge numbers of the mass-produced conventional liquid-fueled diesel engines. The requirements of relatively complex gaseous fuel infrastructure, the need to match the specific characteristics of the diesel engine to be converted, the constraints imposed by the increasingly stringent regulations to control emissions, and the potential variations in the quality of the gaseous fuels that may become locally available economically are additional challenges.

Figure 18.1 shows schematically a typical relatively simple installation of a dual-fuel engine operating on fumigated fuel gas while employing the fuel pump and injector of the original diesel version to provide pilot injection, as may be used in some transport applications.

Dual-fuel engine operation needs to ensure the precise metering of the gas and diesel fuel flows, the liquid-to-gas fuel ratio, and the total fuel energy supply while safeguarding against overfueling. The following controls are employed:

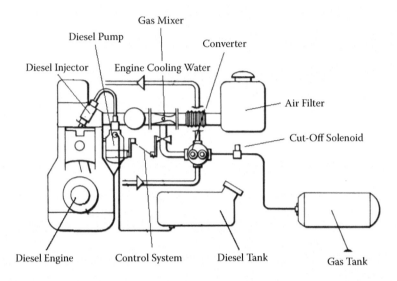

FIGURE 18.1   An example of dual-fuel engine installation with fumigated fuel gas. (From Bergman, H., and Busenthur, B., Facts Concerning the Utilization of Gaseous Fuels in Heavy Duty Vehicles, in *Proceedings of the Conference on Gaseous Fuels for Transportation*, August 1986, pp. 813–849.)

- The pilot with mechanical pump and governor
- The pilot with mechanical pump and electronic governor
- The electrohydraulic pilot with full electronic control, permitting variable injection timing
- The pilot high-pressure fuel gas injected directly into the cylinder, with full electronic control and permitting variable injection timing

Mechanical type control systems were employed for a long time in the control of dual-fuel engines, especially for stationary applications. These usually required only limited changes in speed and load, while typically employing a relatively substantial pilot, such as around 10% of the full-load diesel quantity. However, in comparison to transport applications, they tend to be much more demanding, such as in city buses, where there is a need for frequent transient operation over wide ranges of speed and load while adhering increasingly to restrictive emission controls. This has produced a need for more exacting controls while catering for more frequent returns of the engine to operation in the diesel mode.

As has been indicated earlier, in order to obtain the potential benefits of converting diesel engines to the dual-fuel mode, the size of the pilot fuel and its injection characteristics need to be varied suitably, depending on the engine type, the fuel gas, and other operating conditions. This would be achievable through electronic control systems, which generate the necessary nonlinear relationships in the different operating parameters needed for proper control. It can also ensure that the pressure-time development under dual-fuel operation is similar to that for operation in the diesel mode for the same engine and corresponding conditions. On this basis, electrical controls of dual-fuel engines have been increasingly preferred in recent years, in comparison to the older mechanical approaches. Moreover, control systems based on digital microprocessors are usually preferred over those of the analog type due to their relative flexibility and improved reliability, while keeping conversion costs down. In any case, because of the inherent nonlinearity of the desirable fuel control operational map, an electrically based control system is needed. There are many different versions of ever-evolving electronic controls, together with the development and wide application of more efficient injection/combustion systems of the common rail type of diesel engines. These are aiding in hastening the movement to the increased conversion of diesel engine to dual-fuel operation and becoming an effective and attractive type of a power producing device.

There is an assembly of electronic and electromechanical components that continuously monitor engine operation and vary engine calibration to meet requirements. A processor is the heart of the electronic engine control (EEC), where information, including a simulation of the operational and driving conditions that could be encountered by the engine operator, is programmed. Sufficiently detailed engine performance maps are stored in the computer memory, and the processor's task is to select the required engine operating conditions.

Various types of servos provide information signals to the processor. These include thermistors, variable resistance sensors, and thermocouples for variations in temperature, such as that of the engine coolant, exhaust gas, or inlet mixture. Potentiometers and variable resistances indicate position, such as that of the throttle valve. Different switches send voltage signals to indicate off and on positions of a specified condition, while magnetic pickups indicate engine speed and crankshaft position. In case of a malfunction of the gas valve, a servo will detect the change, such as in the exhaust gas temperature, prompting it to be shut off. The commands are performed by actuators, which can be electrically controlled solenoid motors and other devices, such as those motorized for air and fuel flow control. Other controls guard against over- and underspeeding and maintaining the required coolant temperature and gas fuel supply pressure. Figure 18.2 shows a schematic layout of the controls of a compressed natural gas (CNG)-fueled dual-fuel engine that employs the regular diesel injection pump to provide for pilot injection, while regulating the fuel gas supply to sufficiently low pressure for it to be fumigated into the engine intake air system.

18.2   SOME FEATURES OF DUAL-FUEL MANAGEMENT

It is less common nowadays to have dual-fuel engines equipped with a throttle control, as commonly fitted to spark ignition automotive engines, including those for fuel gas applications. Instead, the power output of a dual-fuel engine is controlled mainly through changes to the overall quality of the gaseous fuel–air mixture, together with a control of the pilot liquid fuel quantity used. Sometimes, partial throttling of the air component of the intake charge is employed, so as to produce sufficiently richer mixtures for the same amount of gaseous fuel admitted. This would result, at sufficiently light load, in improving the overall fuel utilization, enhancing power output, and reducing the emissions of unburned fuel gas and other partial oxidation products. However, excessive or inappropriate application of throttling can undermine and

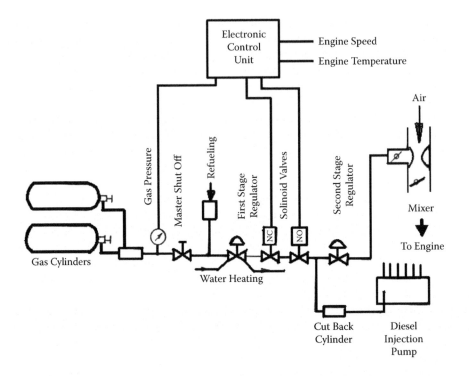

FIGURE 18.2 An arrangement of the controls of a CNG-fueled dual-fuel engine. (From Milken, E., A Simple Retrofit System for the Operation of Diesel Engines on Natural Gas, in *Proceedings of the International Conference on Auto Technology*, Bangkok, Thailand, 1990, pp. 445–449.)

impede pilot ignition, increase cycling variations and pumping losses, and render operation on some fuel mixtures problematic. On this basis, variable application of throttling in dual-fuel operation has been largely confined to low-load operation.

Pilot injection equipment and processes, when employed without modifications from those for diesel operation, will have a limited turn-down ratio and invariably cannot ensure the maintenance of good liquid fuel spray characteristics, such as those of spray atomization, its penetration, and its mixing with the available fuel–air mixture.

Another control measure that has been increasingly followed in multicylinder premixed dual-fuel engine applications, particularly at very light loads, is skip firing. In this approach, all the cylinders are not necessarily fired uninterruptedly, so that those cylinders that are will be run on richer mixtures than had the charge of gaseous fuel and air been uniformly distributed among all the cylinders. The remaining non-fueled cylinders are allowed to be motored without fuel injection. This procedure of skip firing can be made, if necessary, to alternate among the different cylinders. A similar procedure can be applied where all cylinders are fired continuously, except some are run entirely in the diesel mode, while the others are run in the dual-fuel mode using only pilot fuel quantities. This way, there would be only a fraction of the cylinders operating.

In the dual-fuel mode, the overall gas fuel-to-air ratio employed for the dual-fuel-fired cylinders should not drop to a sufficiently low level where significant amounts of unburnt hydrocarbons and carbon monoxide would be discharged into the exhaust. The remaining cylinders continue to operate on diesel fuel only. This approach, although permitting an overall increase in the efficiency at part loads, while not permitting the discharge of unacceptably high levels of emissions, represents a demanding task in engine management and its controls.

Virtually, all diesel engines of recent designs are fitted with turbochargers that help to boost engine power output while often simultaneously improving engine efficiency and emissions. When engines are converted to dual-fuel operation, they usually retain their turbocharged operation. However, the proper matching of the turbocharger performance to that of the engine is critically important. In recent years, variable geometry turbochargers have been increasingly employed to effect better efficiency, higher power output, and improved control of exhaust gas emissions. Moreover, the mass of the exhaust gas passing through the turbine of the charger is controlled over the whole load range through a bypass system of the turbocharger. Some of the air from the compressor may also be recirculated as an added measure of controlling turbocharged engine output and performance at light loads. Such measures add to the complexity of controls and increase costs. But, when implemented properly, especially at the engine manufacturing stage, they can be well worthwhile and cost-effective. It may also render the engine more tolerant of operating on a wider range of gaseous fuels with different compositions.

The operation of the internal combustion engine usually aims at ensuring the filling of the engine cylinder with fresh charge and expelling all combustion products from previous cycles to ensure high power output and efficiency, while avoiding the incidence of knock. However, mainly to reduce emissions from engines, especially those of nitrogen oxides, some controlled variation in the composition of the incoming charge is being increasingly applied. This is done through the planned imperfection in the scavenging process or through employing measures to increase the fraction of trapped residual gases. In dual-fuel engines for transport applications, when there is no independent fuel injection system for the pilot, provision is made to modify the fuel quantity signal from the accelerator, so that only the desired pilot quantity is delivered. In most electromechanical systems, as in mechanical systems, the force required to inject the fuel is provided by the camshaft, either directly by the action of a cam lobe on the injector follower or indirectly by gearing a separate in-line pump to the camshaft drive. Unlike in mechanical control systems, the quantity of the fuel injected and, within limits, the timing of the fuel injection are determined by the action of electronically controlled solenoid valves, which can provide a flexible fuel injection timing.

The gas fuel, which is invariably supplied from high-pressure sources, is passed through a solenoid shutoff valve, before its pressure is reduced and regulated to the lower operational intake pressures. Suitable valves control the flow of the gas into the engine intake system. For example, a stepping motor controls the gas supply valve, where the gas flow rate is a known function of the valve net opening, and an actuator controls the liquid fuel supply through the pump lever. Alternatively, the electronic control calculates the duration the gas valve is to remain open for each working cycle from the output signal of the engine regulator. This allows the engine to respond very quickly to changes in operating conditions and demand. The typical input signals for the engine control system include fuel pump, engine speed, diesel to dual-fuel selection mode, and engine coolant temperature. The output from the processor goes to the gas valve, which controls the amount of gas flowing into the intake of the engine and opens it by the desired amount. A second output from the processor goes to the diesel fuel control valve, which diverts the required amount of the liquid fuel flow.

Various safety control measures are incorporated in the conversion of diesel engines to dual-fuel operation. Examples of these include the shutting of gas injection in case the pilot injection is missed. The injection of the fuel gas is made by a controlling oil pressure activated by the oil pump control system, which is basically the same as that of the liquid fuel injection pump in pure diesel operation. If an electric service is ever interrupted at a location equipped with a dual-fuel standby system, a transfer switch triggers an electric generator that operates on fuel gas. If at any time the gas service is also interrupted, the generator will automatically begin to operate on fuel gas supplied on site from cylinders.

There is a potential for improved control of the combustion process in dual-fuel engines through changes in the composition of the premixed gas fuel–air charge. Residual gas control may be implemented through modification to chamber geometry, valve timing, scavenging, recirculation processes, and exhaust gas processing. This is an approach that, with proper control, can be implemented over a wide range of operating conditions and a variety of engine types and sizes. Also, the use of additives to the gaseous fuel is another possible controlling approach. Such additives may include the use of natural gas admission when operating on low-quality gas fuel mixtures, or the addition of fuel gases that contain some hydrogen. Antiknock additive agents have also been tried.

Changes in the injection characteristics and size of the pilot widely affect almost all aspects of dual-fuel engine operation: properly controlled changes in the value of the pilot size, its injection timing, and other suitable injection characteristics are key measures for controlling the combustion process and engine performance. Since dual-fuel operation tends to use only relatively small quantities of the injected pilot diesel fuel, it may become operationally worthwhile to use a higher-cetane-number fuel than commonly employed. This way, especially when infrequent engine running as a diesel takes place, further improvements in performance, pilot size reduction, or operation on a relatively poor heating value fuel-gas mixture may be achieved. Figure 18.3 shows an example of the extent of enhancement in the ignition point over the operational total equivalence ratio as a higher-cetane-number pilot fuel is employed for both methane and propane dual-fuel operational constant pilot quantity injection. The corresponding variation for diesel operation is also shown.

In general, the engine control system has to be suitably matched to the characteristics of the diesel engine to be converted to dual-fuel operation. This can impose a limitation on the universality of the application of such control equipment, and can be quite restrictive since the corresponding diesel operation is increasingly being subjected to more demanding performance requirements, especially those associated with emissions and fuel quality controls. Some increase in fuel consumption is sometimes tolerated when the cost of the fuel gas is so much lower than that of diesel. Also, the

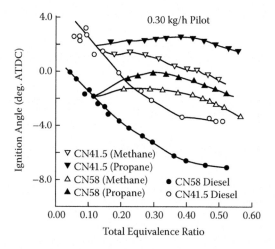

FIGURE 18.3 An example of the variation of ignition point with total equivalence ratio of dual-fuel operation on methane and propane with two different cetane number pilot fuels employed. The corresponding performance as a diesel is also shown. (Adapted from Gunea, C., Examination of the Pilot Quality on the Performance of Gas Fueled Diesel Engine, MSc thesis, University of Calgary, Canada, 1997.)

decision as to how much a pilot is going to be used is often based on economy. Some operators in transport applications may prefer to deal with the simpler control and portability of one fuel and choose to modify the engine so as to operate as an SI, thereby sacrificing many of the positive features and benefits of dual fueling.

REFERENCES AND RECOMMENDED READING

Amoozigar, N., Examination of the Performance of a Dual Fuel Diesel Engine with Particular Reference to the Addition of Some Inerts and Intake Liquid Additives, MSc thesis, University of Calgary, Canada, 1982.

Baur, H., ed., *Diesel Engine Management*, 2nd ed., SAE, Warrendale, PA, 1999.

Bechtold, R.L., *Alternative Fuels Guidebook*, SAE Publishing, Warrendale, PA, 1997.

Beck, J., Karim, G.A., Mirosh, E., and Pronin, E., Bus Fuel Efficiency Local Emissions and Impact on Global Emissions, presented at NGV 96 Conference and Exhibition, Kuala Lumpur, Malaysia, October 1996.

Beck, N., Johnson, W., George, A., Peterson, P., vander Lee, B., and Klopp, G., *Electronic Fuel Injection for Dual Fuel Diesel Methane*, SAE Paper 891652, 1989.

Bergman, H., and Busenthur, B., Facts Concerning the Utilization of Gaseous Fuels in Heavy Duty Vehicles, in *Proceedings of the Conference on Gaseous Fuels for Transportation*, August 1986, pp. 813–849.

Blizzard, D., Schaub, F.S., and Smith, J., Development of the Cooper–Bessemer Clean Burn Gas-Diesel (Dual Fuel) Engine, in *ASME-ICE Division*, 1991, vol. 15, pp. 89–97.

Burn, K.S., The Effect of Cold Intake Temperatures on the Combustion of Gaseous Fuels in a Dual Fuel Engine, MSc thesis, University of Calgary, Canada, 1977.

Carlucci, A.P., Ficarella, A., and Laforgia, D., Control of the Combustion Behavior in a Diesel Engine Using Early Injection and Gas Addition, *Applied Thermal Engineering*, 26, 2279–2286, 2006.

Challen, B., and Barnescu, R., *Diesel Engine Handbook*, 2nd ed., SAE, Warrendale, PA, 1999.

Chrisman, B., Callaham, T., and Chiu, J., Investigation of Macro Pilot Combustion in Stationary Gas Engine, presented at ASME-ICE Division, 1998, paper 98-ICE-106.

Daisho, Y., Takahashi, Y.I., Iwashiro, Y., Nakayama, S., and Saito, T., *Controlling Combustion and Exhaust Emissions in a Direct-Injection Diesel Engine Dual Fuelled with Natural Gas*, SAE Paper 952436, 1995.

Danyluk, P.R., Development of a High Output Dual Fuel Engine, *ASME Journal of Engineering for Gas Turbines and Power*, 115, 728–733, 1993.

Felt, A.E., and Steele, W.A., Combustion Control in Dual Fuel Engines, *SAE Transactions*, 70, 644, 1982.

Gettel, L.E., Perry, G.C., Boisvert, J., and O'Sullivan, P.J., Dual Fuel Engine Control Systems for Transport Applications, presented at ASME-ICE Division, 1987, paper 87-ICE 121.

Gunea, C., Examination of the Pilot Quality on the Performance of Gas Fueled Diesel Engine, MSc thesis, University of Calgary, Canada, 1997.

Gunea, C., Razavi, M.R., and Karim, G.A., *The Effects of Pilot Fuel Quality on Dual Fuel Engine Ignition Delay*, SAE Paper 982453, 1998.

Johnson, W.P., Beck, N.J., Van der Lee, A., Koshkin, V.K., Lovkov, D., and Platov, I.S., *An Electronic Dual Fuel Injection System for the Belarus D144 Diesel Engine*, SAE Paper 901502, 1990.

Klimstra, J., Hatter, C., Nylund, I., and Sillanpää, H., The Technology and Benefits of Skip Firing for Large Reciprocating Engines, presented at ASME-ICE Division, Chicago, 2005, paper ICES2005-1073.

Krepec, T., Giannacopoulos, T., and Miele, D., New Electronically Controlled Hydrogen Gas Injector Development and Testing, *International Journal of Hydrogen Energy*, 12, 855–861, 1987.

Lowe, W., and Williamson, P.B., Combustion and Automatic Mixture Strength Control in Medium Speed Gaseous Fuel Engines, in *Proceedings of the 8th Congress de Machines a Combustion (CIMAC)*, 1996, pp. A14–A56.

MacCarley, C.A., and Vorst, W.D.V., Electronic Fuel Injection Techniques for Hydrogen Powered I.C. Engines, *International Journal of Hydrogen Energy*, 5, 179–203, 1980.

Mayer, R., Meyers, D., Shahed, S.M., and Duggal, V.K., *Development of a Heavy Duty On-Highway Natural Gas Fueled Engine*, SAE Paper 922362, 1992.

Milken, E., A Simple Retrofit System for the Operation of Diesel Engines on Natural Gas, in *Proceedings of the International Conference on Auto Technology*, Bangkok, Thailand, 1990, pp. 445–449.

Needham, J.R., May, M.P., Doyle, D.M., Faulkner, S.A., and Ishiwata, H., *Injection Timing and Rate Control—A Solution for Low Emissions*, SAE Paper 900854, 1990.

Turner, S.H., and Weaver, C.S., *Dual-Fuel Natural/Diesel Engines: Technology, Performance and Emissions*, No. GRI-94/0094, Topical Report Gas Research Institute, November 1994.

19 Dual-Fuel Engine Operation on Alternative Fuels

19.1 DUAL-FUEL OPERATION ON PROPANE, BUTANE, AND LPG

The use of propane, n-butane, and liquefied petroleum gas (LPG) as engine fuels has been widely practiced in spark ignition engines, especially for transport applications. These fuels have a number of attractive features, especially since they can be liquefied much more readily than methane, rendering them more easily portable. However, they are less widely available as a cheap alternative fuel resource than natural or biogases. N-butane has a lower octane number and a higher boiling point than propane, which makes it relatively easier to inject in engines as a liquid. In fumigation applications there is a need to ensure the full vaporization of the fuel and to mix it thoroughly with the engine air.

Much of the information available in the open literature about dual-fuel operation with these fuels tends to relate to relatively small engines with a variety of combustion chamber types of both four- and two-stroke engines and of the direct and indirect injection types. Methane operation, apart from its attractive low cost, is more suitable than LPG fuels for large-bore multicylinder turbocharged engine applications. This is mainly because of its superior resistance to knock. Dual-fuel engine performance with LPG fuels has many similarities to that obtained with methane under similar operating conditions. Significant amounts of unburned fuel are usually observed in the exhaust gas at part load conditions, but at higher loads, the extent of gas utilization improves rapidly, and when knocking conditions are approached, the gas approaches essentially full conversion. Part load brake specific energy consumption (BSEC) at low pilots is significantly improved with an increase in the intake temperature. Light load performance can be improved through measures that would include the employment of larger diesel fuel pilots and resorting to some advance in their point of injection. The effectiveness of such measures at high load becomes limited by the early onset of knock, especially in the case of n-butane and, to a lesser extent, propane. Figure 19.1 shows variations of the mean effective pressure with changes in total equivalence ratio for a variety of increasing size pilots in a normally aspirated propane-fumigated engine. Figure 19.2 shows the corresponding performance for n-butane operation. The increased presence of propane or n-butane with methane results in restricting the knock-free output of the engine considerably, especially with butane, which makes it a less attractive fuel for dual-fuel applications than methane or propane.

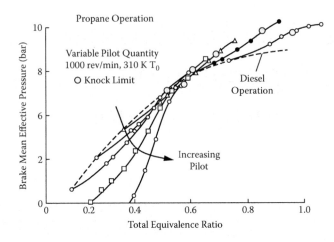

FIGURE 19.1 Variation of the brake mean effective pressure with the total equivalence ratio for a range of pilot injection quantities of an engine fumigated with propane. The corresponding observed knock mixture limits are indicated, together with values obtained with diesel operation under the same operating conditions. (From Karim, G.A., and Rogers, A., *Journal of the Institute of Fuel*, 40, 513–522, 1967.)

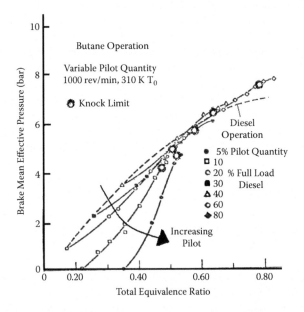

FIGURE 19.2 Variation of the brake mean effective pressure with the total equivalence ratio for a range of pilot injection quantities of an engine fumigated with butane under the same conditions as those for the propane operation of Figure 19.1. The corresponding observed knock mixture limits are indicated, together with values obtained with diesel operation. (From Karim, G.A., and Rogers, A., *Journal of the Institute of Fuel*, 40, 513–522, 1967.)

The operation at light load on propane is associated with a specific energy consumption that is higher than the values of the corresponding operation as a diesel, but better than the corresponding values for methane operation. This is largely due to the reduction in the incomplete combustion of the fuel gas. But with less lean fuel–air mixtures, the specific energy consumption improves considerably and can approach and may even surpass the corresponding values for diesel operation, as shown in Figure 19.3. With the more reactive butane, the BSEC values are lower at low loads and small pilots than the corresponding values for propane operation. With all pilot sizes, higher rates of pressure rise and earlier transition to knocking are encountered with butane, compared to propane or methane operation, producing correspondingly less knock-free power output. However, as the intake mixture temperature is increased, the differences in the operational performance between the fuels are reduced. Also, with relatively low pilot quantities, as knocking conditions are approached, the performance tends to approach that of the corresponding diesel values.

Any size of pilot fuel in excess of around 20% of the full-load diesel value with conventional fuel injection equipment would enter the combustion chamber after ignition and gas combustion have started. As shown in Figure 19.4, its size would not exert much influence on the conditions leading to knock. Also, the performance tends to be dominated by the combustion of the diesel fuel since the gas supplies then contribute only a smaller proportion of the total energy releases.

Figure 19.5 shows the power output variation with changes in the total equivalence ratio when the propane-fueled engine is operating at different intake mixture

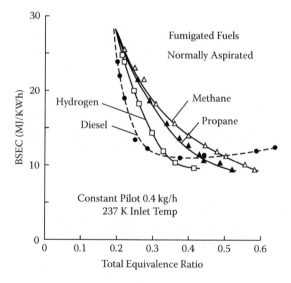

FIGURE 19.3   Variation of the brake specific energy consumption with changes in total equivalence ratio when operating on a number of common fuel gases with constant pilot quantity. The corresponding values with diesel operation are also shown. (Adapted from Burn, K.S., The Effect of Cold Intake Temperatures on the Combustion of Gaseous Fuels in a Dual-Fuel Engine, MSc thesis, University of Calgary, Canada, 1977.)

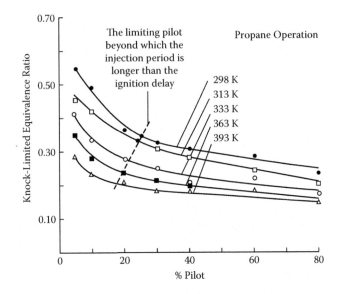

FIGURE 19.4 Variation of the knock-limited power output with changes in the pilot quantity over the whole range of diesel quantity injection for different intake mixture temperatures for an engine operating on fumigated propane. (From Karim, G.A., and Rogers, A., *Journal of the Institute of Fuel*, 40, 513–522, 1967.)

temperatures. It can be seen that the power output is very much lowered with the earlier incidence of knock, as the mixture temperature is increased.

The emissions of CO with propane operation tend to be at a level a little higher than with methane, but at higher loads propane produces less CO. These emissions can become yet higher as the intake mixture temperature is lowered. A bigger pilot quantity has the tendency to reduce the emissions of CO for lean overall equivalence ratios, while smaller pilots behave similarly, but with richer mixtures. On the whole, operation with propane produces larger amounts of exhaust NOx than operation with methane. Expectedly, at a given power output, less smoke emission is produced than with diesel operation.

The knock-free power output is higher with propane than with methane for the same pilot quantity and injection timing. But at light loads, the output with propane tends to be less of a reflection of the relatively longer delays associated with propane operation. The maximum power output for the same engine and operating conditions can be greater with a blend of methane and propane than with either of the two fuels on their own. This is because the introduction of some methane with the propane can delay the onset of knock and improves engine performance up to the point when knock sets in, while with propane on its own, the performance will be inferior in comparison due to the earlier encountering of knock.

The knock-limited power output of an engine when operating on binary fuel gas mixtures such as those of methane and propane or butane and propane or hydrogen

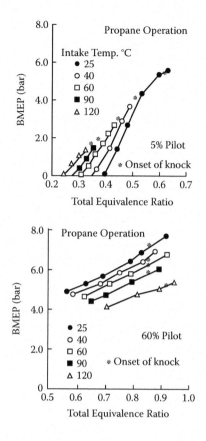

FIGURE 19.5 Variations in the brake mean effective pressure of the propane-fed engine with changes in the total equivalence ratio at a number of inlet mixture temperatures when relatively small pilot quantities of 5% and 10%, as well as a very large pilot of 60%, are employed, showing the corresponding changes in the knock mixture limits. (From Karim, G.A., and Rogers, A., *Journal of the Institute of Fuel*, 40, 513–522, 1967.)

and methane is related to the limits of the two fuels on their own under the same operating conditions according to the following approximate relationship:

$$1/(K.L.Power)_{mix} = \Phi_1/(K.L.Power)_1 + \Phi_2/(K.L.Power)_2 \qquad (19.1)$$

where Φ_1 and Φ_2 are the fractions of the thermal energies of the constituents 1 and 2 in the fuel blend, and *K.L.Power* is the corresponding knock-limited power output of the engine under the same operating condition. Figure 19.6 shows the knock-limited power outputs of an engine operating on methane, propane, and a binary mixture of the fuels, showing good agreement between the experimental values and those obtained according to Equation (19.1). Similarly, the knock-limited power output of a dual-fuel engine operating at a constant pilot on blends of methane and hydrogen, as shown in Figure 19.7, tends to follow equally well the trends represented by the equation.

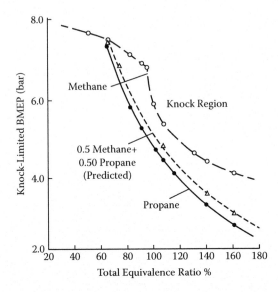

FIGURE 19.6 Variations of the knock-limited brake mean effective pressure with the total equivalence ratio for an engine operating on mixtures of methane and propane at constant pilot injection with the total equivalence ratio. The corresponding values obtained according to Equation (19.1) for a blend of methane and propane are shown. (Adapted from Karim, G.A., *Journal of the Institute of Fuel*, 37, 530–536, 1964.)

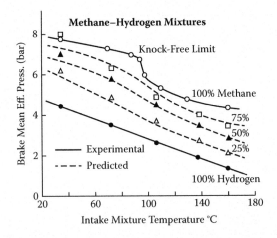

FIGURE 19.7 Variation of the knock-limited power output for different fumigated blends of methane and hydrogen with changes in intake temperature while employing a fixed pilot quantity. The corresponding values predicted according to Equation (19.1) are also shown. (Adapted from Karim, G.A., *Journal of the Institute of Fuel*, 37, 530–536, 1964.)

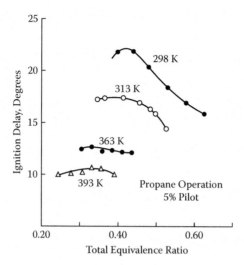

FIGURE 19.8 Variation in the point of ignition with the total equivalence ratio for an engine fumigated with methane at relatively small pilot for a range of elevated intake mixture temperatures. (Adapted from Karim, G.A., and Rogers, A., *Journal of the Institute of Fuel*, 40, 513–522, 1967.)

Intake mixture temperatures higher than ambient, whether encountered as a result of intake air temperature changes, preheating the intake mixture, application of warm exhaust gas recirculation (EGR), or turbocharging, have a considerable influence on the progress of the combustion processes in general, and those contributing to the onset of knock and the nature of exhaust emissions in particular. As shown in Figure 19.8, increasing the intake mixture temperature permitted combustion to be sustained at leaner mixtures and lowered the apparent lean combustion limit and ignition delay. A rise in intake temperature has a detrimental effect on the power output. The knock-free power output with a very large size pilot would be much restricted in comparison to the corresponding values with small pilots. However, in the absence of knock, the full-load power output with butane may exceed that of the corresponding diesel operation. However, as shown in Figure 19.8 for propane operation with a relatively small pilot, the operational mixture range is substantially narrowed with the increase in intake mixture temperature as a result of the increase in delay at low temperatures on one side, and the earlier tendency to encounter knocking conditions at high temperatures on the other side.

19.2 OPERATION ON HYDROGEN WITH PILOT IGNITION

There are a number of challenges, especially in transport applications, to the wide application of hydrogen as an engine fuel in dual-fuel engines. In addition to those of safety, they include the size, weight, and cost of the fuel and its storage options.

The presence of some hydrogen with the methane brings about a very substantial increase in the reaction and flame propagation rates. This would be reflected by the substantial reduction in the ignition delay. The extent of these changes depends markedly on both the temperature and pressure. Figure 19.9 shows an example of the dramatic reduction in the calculated ignition delay of methane–hydrogen mixtures in stoichiometric air under constant volume conditions, with the increased presence of hydrogen at 1000 K.

Hydrogen-fueled dual-fuel engines have a relatively narrow knock-free operational mixture range that narrows rapidly with the increase in intake temperature or the size of the pilot (Figure 19.10). An example of the knock-limited equivalence ratio of operation on methane–hydrogen mixtures in a dual-fuel engine with changes in intake mixture temperature when operating with a constant size pilot is shown in

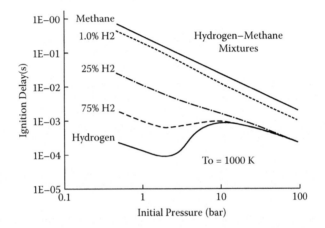

FIGURE 19.9 An example of the calculated logarithmic changes in the ignition delay with pressure at a mixture temperature of 1000 K of various stoichiometric mixtures in air of methane with hydrogen. (From Thiessen, S., Khalil, E., and Karim, G.A., *International Journal of Hydrogen Energy*, 35, 10013–10017, 2010.)

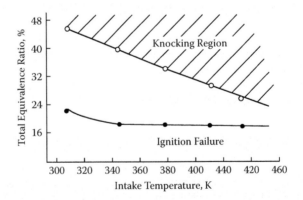

FIGURE 19.10 An example of the changes in the knock-free operational range with mixture temperature for a dual-fuel engine operating on hydrogen with a fixed pilot quantity. (From Karim, G.A., *Progress in Energy and Combustion Science*, 6, 277–285, 1980.)

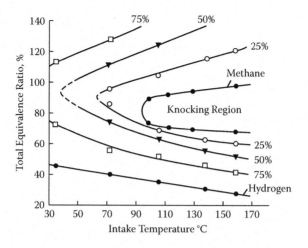

FIGURE 19.11 The knock-limited equivalence ratio variations with intake temperature for hydrogen, methane, and some of their binary mixtures for a dual-fuel engine operating with a constant pilot. (Adapted from Karim, G.A., Klat, S.R., and Moore, N.P.W., Knock in Dual Fuel Engines, in *Proceedings of the Institute of Mechanical Engineers*, vol. 181, 1967, pp. 453–466.)

Figure 19.11. The wider mixture range for engine knocking with hydrogen operation contrasts with that for methane admission. It can be seen that the presence of some hydrogen with methane seriously undermines the excellent knock resistance qualities of methane and substantially brings down its tolerance to increased initial mixture temperature. Nevertheless, the blending of some methane with hydrogen results in increasing the knock-limited power output to become correspondingly higher than with pure hydrogen on its own, but lower than that with only methane.

In general, diesel engines tend to display some improvements in performance with the increased admission of hydrogen to an extent that is better than with the corresponding admission of other common gaseous fuels, such as methane, propane, ethylene, and their mixtures (Figure 19.12). Also, as shown in Figure 19.3, there is much improvement in specific energy consumption with hydrogen admission, which can surpass the corresponding values for diesel engine operation over much of the high load range. The ignition delay is also shortened, with a reduction in the emission of particulates, but with increased NOx emissions (Figure 19.13).

In dual-fuel operation under nonknocking conditions, the blending of small quantities of hydrogen with methane, particularly at low intake temperatures, brings about some improvements to the power output from the corresponding values obtained with only methane. This trend is largely due to the increase in the combustion rates resulting from the addition of hydrogen. However, increasing the admission of hydrogen increasingly further brings about a reduction in power with the increased competition for oxygen between the fuels, resulting in a relative rise in the emissions of carbon monoxide.

The employment of liquid hydrogen in dual-fuel engines represents a number of operational and design challenges, such as those associated with handling the

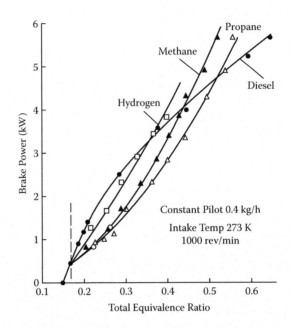

FIGURE 19.12 An example of the variations in the power output of a dual-fuel engine with total equivalence ratio under constant pilot condition for a number of common gaseous fuels. The corresponding diesel operation values are shown. (Adapted from Burn, K.S., The Effect of Cold Intake Temperatures on the Combustion of Gaseous Fuels in a Dual Fuel Engine, MSc thesis, University of Calgary, Canada, 1977.)

exceptionally cold fuel. Following evaporation of the liquid hydrogen, the net effect of operating the engine on LH_2 is similar to that of operating on very chilled intake mixtures of hydrogen and air. The combustion process and its reactions show signs of being seriously chilled, with increased production of quenched products. The use of sufficiently large pilots could ensure better regular ignition. For example, Figure 19.14 shows increased emissions of carbon monoxide arising from the pilot with the lowering of the intake mixture temperature, despite operation on the carbon-free hydrogen.

The direct injection of LH_2 into engines quickly forms a cold gas having a large volume that needs to be injected sufficiently quickly into the high-pressure combustion space and effectively distributed in the relatively short time available. The very high associated difference in the density between the fuel gas and air, the considerably high flame speed, and the different turbulence characteristics all will influence the mixing and combustion processes.

19.3   DUAL-FUEL OPERATION WITH SOME GASOLINE FUMIGATION

The role of the presence in small concentrations of higher hydrocarbon vapors in a fuel gas, as it occurs in some natural and industrial gases, has been the subject of

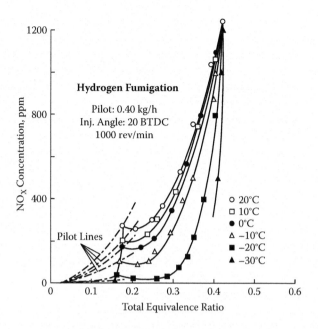

FIGURE 19.13 An example of the variations in the concentrations of NO_x emissions with total equivalence ratio for a hydrogen fumigated dual-fuel engine at different intake mixture temperatures while employing a constant pilot quantity. (Adapted from Burn, K.S., The Effect of Cold Intake Temperatures on the Combustion of Gaseous Fuels in a Dual Fuel Engine, MSc thesis, University of Calgary, Canada, 1977.)

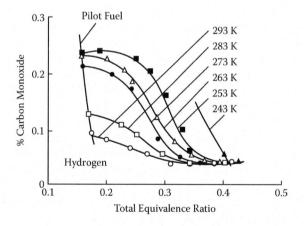

FIGURE 19.14 An example of the variations in the exhaust gas concentration of carbon monoxide with total equivalence ratio of a dual-fuel engine when operating with constant pilot on various chilled hydrogen air mixtures. (Adapted from Burn, K.S., The Effect of Cold Intake Temperatures on the Combustion of Gaseous Fuels in a Dual Fuel Engine, MSc thesis, University of Calgary, Canada, 1977.)

investigations over the years. There are also schemes where the addition of gasoline, alcohols, or other liquid fuels in small proportions to engines has been considered to serve as a potential approach for modifying the ignition and combustion processes and the associated performance of diesel engines. In the case of dual-fuel operation, the small amount of an auxiliary fuel added can be varied with load and other operating conditions, while retaining pilot ignition.

Though the need in dual-fuel engines to provide for two fuel systems may be considered a practical inconvenience, the economic advantages associated with such an operation can be well worthwhile. However, to introduce yet another additional fuel would represent an added complexity that may be difficult to justify economically and operationally. But it may serve as a worthwhile means for modifying dual-fuel engine performance, as well as increasing the diversity of operation on a multitude of fuels that may include some liquid fuel resources, such as those of alcohols or gasolines. There are potential positive features with using a low-octane-number gasoline as the supplementary or even the main auxiliary fuel in dual-fuel type engines since high-compression-ratio engines are used while relying on the pilot for ignition of the very lean fuel–air mixtures. Such approaches have their limitations and appear not to have been commonly employed in recent years.

The fumigation of some gasoline into the intake of a diesel engine while employing its injection system to provide pilot ignition tends to shorten the ignition delay, especially with the employment of larger pilots. The aspirated gasoline vapor into the intake air undergoes slow oxidation reactions during compression, to an extent much greater than with methane or propane. These reactions could aid both thermally and kinetically the injection and ignition processes of the diesel fuel pilot and subsequent combustion. However, when only small amounts of gasoline with small pilots are employed, such activity tends to be of marginal intensity. In a similar manner as with the fumigation of fuel gases, with less lean fuel–air mixtures at high loads, the delay period also decreases with the fumigation of small amounts of liquid fuels, whether of a very-high-octane-number fuel such as benzene or a low-octane-number fuel such as n-hexane. Expectedly, there is an increase in power output, but with an earlier onset of knock, which limits the employment of large pilots in conjunction with increasing the amounts of gasoline, benzene, or n-hexane. Figure 19.15 shows an example of the changes in power output with gasoline fumigation while employing different constant diesel pilot quantities.

The introduction of small amounts of liquid fuels into a dual-fuel engine operating on methane with a fixed diesel pilot quantity shows some improvements in the power output and the extent of methane conversion, which can become practically significant at low loads, where it is generally needed (Figure 19.16). However, at very low equivalence ratios the small additions of gasoline tend to increase the production of CO, which is indicative of the slow oxidation reactions proceeding only partially. In general, the concentration of carbon monoxide in the exhaust gas remains relatively unchanged with such operation, except with richer fuel-to-air ratios.

In the case of the fumigation of a slow-to-react high-octane-number fuel, such as benzene, the delay period is extended, but it can be reduced substantially by employing larger pilots. This is indicative of the relative lack of preignition reaction activity of benzene-air mixtures during much of the compression stroke. However, with the

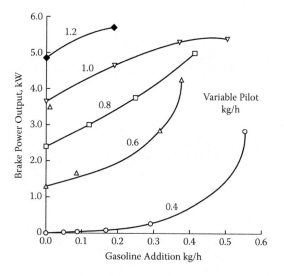

FIGURE 19.15   An example of the variation of the knock-free power output with changes in the amount of gasoline fumigated for a dual-fuel engine operating with different values of the constant pilot employed. (Adapted from Amoozigar, N., Examination of the Performance of a Dual Fuel Diesel Engine with Particular Reference to the Addition of Some Inerts and Intake Liquid Additives, MSc thesis, University of Calgary, Canada, 1982.)

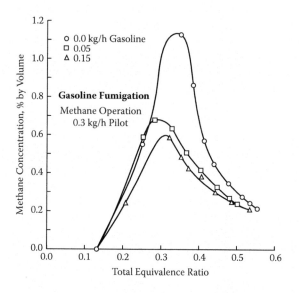

FIGURE 19.16   An example of the variation of the concentration of unconsumed methane with total equivalence ratio of a dual-fuel engine operating on methane with a constant amount of fumigated gasoline vapor while employing a fixed amount of pilot injection. (Adapted from Amoozigar, N., Examination of the Performance of a Dual Fuel Diesel Engine with Particular Reference to the Addition of Some Inerts and Intake Liquid Additives, MSc thesis, University of Calgary, Canada, 1982.)

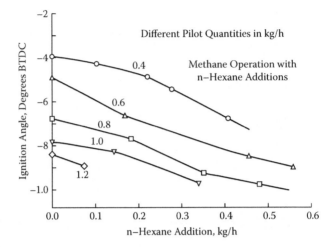

FIGURE 19.17   An example of the changes in the point of ignition of a dual-fuel engine operating on fumigated methane with the extent of n-hexane admission when different pilot quantities of diesel fuel are employed. (Adapted from Amoozigar, N., Examination of the Performance of a Dual Fuel Diesel Engine with Particular Reference to the Addition of Some Inerts and Intake Liquid Additives, MSc thesis, University of Calgary, Canada, 1982.)

fumigation of a more reactive and volatile fuel, such as n-hexane, the delay is shorter, with the oxidation reactions taking place during the compression process, aiding both kinetically and thermally the preignition processes of the pilot (Figure 19.17).

The reduction in the ignition delay and the improvement in the extent of gaseous fuel utilization tend to reduce the specific energy consumption at light loads. However, with the introduction of a relatively large amount of liquid fuels, the specific energy consumption is increased. This results from the corresponding substantial exothermic reactivity of a fuel such as gasoline during compression and its competition with the other two fuels for the available oxygen. Benzene, which is known not to readily undergo preignition exothermic reactions, does not display to the same extent such a trend. Figure 19.18 shows an example of the changes in the specific energy consumption with the total equivalence ratio for a dual-fuel engine operating on methane and constant diesel pilot showing improvements with a small addition of gasoline, but deterioration with a correspondingly larger gasoline addition.

Particulate emissions remain small due to the relative increase in the lean premixed type of combustion. The onset of knock can be delayed up to a point through adjusting pilot injection timing and its quantity.

19.4   DUAL-FUEL OPERATION WITH ALCOHOL FUMIGATION

There are a number of issues, both positive and negative—some have been outlined earlier—associated with the employment of alcohols as fuels in internal combustion engines in general, and in particular their suitability for application in dual-fuel engines, whether on their own or in the presence of methane with diesel fuel pilot ignition. For example, currently produced diesel engines would not run satisfactorily

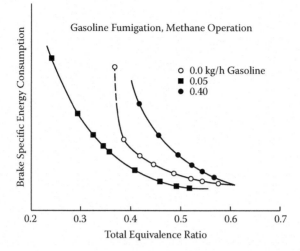

FIGURE 19.18 An example of the variation in the brake specific energy consumption with total equivalence ratio of a dual-fuel operation with methane and the fumigation of a fixed amount of gasoline while using a fixed size of diesel pilot. (Adapted from Amoozigar, N., Examination of the Performance of a Dual Fuel Diesel Engine with Particular Reference to the Addition of Some Inerts and Intake Liquid Additives, MSc thesis, University of Calgary, Canada, 1982.)

on pure methanol due primarily to its extremely low cetane number. Mixtures of methanol or ethanol with diesel fuel in worthwhile proportions display separation occurring when the fuel blends come in contact with traces of water. The biological activity of the raw particulates with alcohols is enhanced. Additional negative issues include the questions of materials and lubricant compatibility and engine reliability and durability.

In a single-cylinder direct injection diesel engine with methane fumigation, the additional induction of methanol or ethanol increases the delay period almost proportionally (Figure 19.19). This is mainly a reflection of the reduction in the compression temperature and pressure due to the high latent heat of vaporization of alcohols and their associated kinetic effects. Increasing the size of the pilot for a certain amount of alcohol fumigation improves the delay and the power output, but results in the onset of knock taking place sooner. Also, the extent of unutilized methane and the exhaust temperature increase as the alcohol concentration in the intake or the quantity of the pilot increases. Generally, the fumigation of alcohols decreases NOx emissions and has only a minor effect on smoke opacity. At higher loads, it can have some beneficial effects on efficiency (Figure 19.20).

The relatively high latent heat of vaporization of alcohols leads to excessive cooling of the charge and working surfaces. The maximum cycle pressure and power decrease as the amount of alcohol is increased for the same equivalence ratio. At low loads, the fumigation of alcohols needs to be reduced, mainly to reduce flame quenching and prevent misfires, while at high loads, to prevent preignition and knock. There are additional concerns in turbocharged engines arising from issues

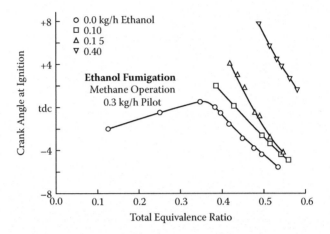

FIGURE 19.19 An example of the variation in the point of ignition with changes in total equivalence ratio of methane fumigated dual-fuel engine with the addition of various amounts of ethanol to the intake charge while employing a fixed amount of pilot fuel. (Adapted from Amoozigar, N., Examination of the Performance of a Dual Fuel Diesel Engine with Particular Reference to the Addition of Some Inerts and Intake Liquid Additives, MSc thesis, University of Calgary, Canada, 1982.)

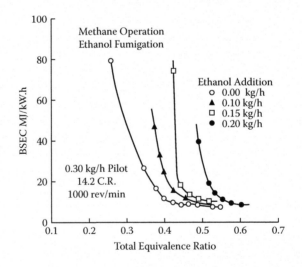

FIGURE 19.20 An example of the variation in the brake specific energy consumption with total equivalence ratio for a dual-fuel engine fumigated with methane with the additional introduction into the intake charge fixed amounts of ethanol while employing constant pilot fuel quantity injection. (Adapted from Amoozigar, N., Examination of the Performance of a Dual Fuel Diesel Engine with Particular Reference to the Addition of Some Inerts and Intake Liquid Additives, MSc thesis, University of Calgary, Canada, 1982.)

such as those of poor vaporization and the maldistribution of mixtures among the different cylinders, which can lead to charger mismatch and damage.

Compared to the corresponding diesel operation, the rates of pressure rise are higher with alcohol fumigation and increase with the increase in ignition delay. The emissions of HC, particulates, and CO increase, but those of NOx are reduced. An increase in the amount of water or ethanol fumigation in a dual-fuel engine increases the extent of unutilized methane. The maximum cycle pressure and power decrease as the amount of water or alcohol is increased for the same equivalence ratio, while the concentrations of CO increase at high loads with the addition of water and alcohol. Using larger pilots tends to moderate the extent of deterioration in the brake power.

19.5 OPERATION ON BIOGASES AND METHANE–DILUENT FUEL MIXTURES

The presence of diluents such as carbon dioxide or nitrogen with methane significantly modifies the thermodynamic, kinetic, and combustion characteristics of the resulting fuel mixtures. In addition to substantially lowering their effective heating value, they bring about adverse changes to the combustion characteristics of the fuel mixtures and their potential usage in dual-fuel engines. The presence of excess nitrogen, whether introduced with the air or the fuel, is usually tolerated relatively better in engine applications than a similar presence of carbon dioxide. Figure 19.21 shows an example of the extension of the ignition delay of a normally aspirated diesel engine when the oxygen partial pressure of the incoming air is reduced through the increased presence of nitrogen and carbon dioxide. The superior tolerance to excess nitrogen relative to carbon dioxide is evident.

The increased presence of carbon dioxide or excess nitrogen within the charge brings about a rapid reduction in the levels of combustion temperatures. The corresponding reduction with the presence of carbon dioxide is significantly greater than that with nitrogen, since the energy released through combustion of the fuel component becomes shared with the diluents present. The relative fraction of the energy

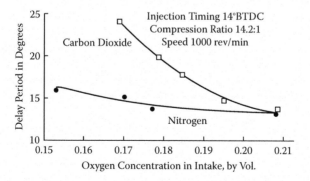

FIGURE 19.21   An example of the variations in the delay period of a diesel engine at ambient intake conditions and constant speed with changes in the relative intake concentration of oxygen resulting from the increased admission of nitrogen and carbon dioxide. (Adapted from Karim, G.A., Gee, D., and Satterford, R., *The Engineer,* 219, 551–556, 1965.)

release taken by these diluents will increase as the temperature is increased since the thermodynamic properties, especially those of enthalpy and internal energy, increase rapidly with temperature. The rates of increase are greater for carbon dioxide than those for nitrogen or oxygen. The relative reduction in temperatures is especially significant for near-stoichiometric mixtures where the high temperatures are and where much of the pilot combustion and diffusional burning take place. At relatively low initial mixture temperatures, carbon dioxide behaves largely as a diluent and appears in the products essentially intact. Figure 19.22 shows the changes in the energy release rate and ignition delay with increasing the presence of carbon dioxide with methane while employing a fixed pilot quantity.

Some general features of the combustion of fuel mixtures of methane in association with either carbon dioxide or nitrogen in relation to dual-fuel engine applications are as follows:

- In general, dilution of methane with CO_2, N_2, or H_2O below the 15% range may be considered as though the diluents are approximately interchangeable with excess diluting air. The heating value of the fuel may be raised when desired through supplementing it with a higher-grade fuel, such as processed natural gas or the use of a larger pilot.
- Operation on methane–diluent mixtures reduces the overall peak combustion temperature and ensures lower emissions of NOx. However, this may lead to some possible increases in greenhouse gas emissions in the form of carbon dioxide, and possibly with higher unconverted methane exhaust concentrations. The emission of particulates tends to remain largely insignificant with a lowering in the noise level.
- Operation on fuel mixtures having the same heat output would require the volumetric flow of the fuel gas to be sufficiently larger than when operating on normal quality fuel gas, such as processed natural gas. Appropriate modification of piping, valving, and controls needs to be made. High-pressure gas injection of the fuel mixtures directly into the engine cylinder remains inefficient and uneconomical.
- The flammability limits or heating values of the fuel mixtures cannot be relied upon for evaluating whether they can be employed effectively in an engine or not. These values can serve merely as rough guidelines for the burning characteristics of the fuel mixtures. Laboratory-scale experiments alone also do not necessarily produce results that are equally applicable to industrial size engines and settings. Full-scale testing needs to be relied on whenever possible.
- Since the engine and its supply infrastructure depend on the volume of the gas that needs to be handled for a certain engine loading, the heating value of the gas-fuel mixture on volume basis can be considered a key performance indicator. The resulting flame temperature is also important, since combustion is more challenging and cylinder temperatures are lower. Increasingly longer retention times are required to complete the combustion.
- A sufficiently high heating value fuel may be required to be used to overcome the operational difficulties that are usually associated with engine

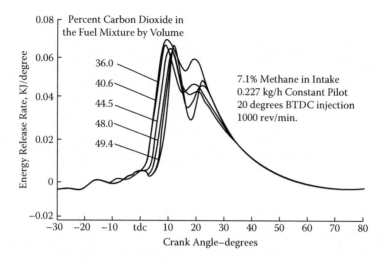

FIGURE 19.22 Variations in the energy release rate with increased concentrations of carbon dioxide in the fuel under constant pilot operation with methane. (Adapted from Karim, G.A., and Khan, M.O., *Journal of Mechanical Engineering Science*, 10, 13–23, 1968.)

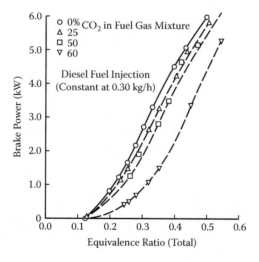

FIGURE 19.23 An example of the variation in the power output of a dual-fuel engine operating on fumigated methane and constant pilot with changes in the total equivalence ratio when the fuel gas mixture contains different concentrations of carbon dioxide. (Adapted from Amoozigar, N., Examination of the Performance of a Dual Fuel Diesel Engine with Particular Reference to the Addition of Some Inerts and Intake Liquid Additives, MSc thesis, University of Calgary, Canada, 1982.)

start-up, transient, and light load changes. Operation on the low-heating-value fuels is then reserved primarily for relatively steady high-load operation. Additional remedial measures may include, whenever possible, preheating the fuel gas, such as through heat circulation from the exhaust gases, consistent with the constraints of materials and equipment used and the increase in the potential pollutants formed.

- Initiating the combustion of low-heating-value fuel mixtures by pilot ignition brings about increased tolerance to high concentrations of diluents in the fuel mixture to be burned, in comparison to spark ignition operation. The judicial recirculation of some hot exhaust gases can positively influence the course of the combustion process.

In summary, the increased presence of diluents in the fuel mixture brings about further deterioration in the inefficient combustion of the lean mixtures at light loads. It increases the ignition delay, results in a drop in power and efficiency, and increases cyclic variations (Figure 19.23). The levels of CO and unconsumed methane in the exhaust gas also increase, but NOx levels decrease. Increasing the pilot injection advance tends to produce some increases in NOx and longer combustion times. In general, there is a limit to engine tolerance of the presence of a diluent in the fuel gas, which depends largely on the equivalence ratio and pilot size employed.

REFERENCES AND RECOMMENDED READING

Al-Garni, M., A Simple and Reliable Approach for the Direct Injection of Hydrogen in Internal Combustion Engines at Low and Medium Pressures, *International Journal of Hydrogen Energy*, 20(9), 723–726, 1995.

Aly, H., and Siemer, G., Experimental Investigation of Gaseous Hydrogen Utilization in a Dual-Fuel Engine for Stationary Power Plants, in ASME-ICE Division, 1993, vol. 20, pp. 67–79.

Amoozigar, N., Examination of the Performance of a Dual Fuel Diesel Engine with Particular Reference to the Addition of Some Inerts and Intake Liquid Additives, MSc thesis, University of Calgary, Canada, 1982.

Amoozigar, N., and Karim, G.A., *Examination of the Performance of a Dual-Fuel Diesel Engine with Particular Reference to the Presence of Some Inert Diluents in the Engine Intake Charge*, SAE Paper 821222, 1982.

Antunes, J.M.G., Mikalsen, R., and Roskilly, A.P., An Experimental Study of a Direct Injection Compression Ignition Hydrogen Engine, *International Journal of Hydrogen Energy*, 14, 6516–6522, 2009.

Baranescu, R., *Fumigation of Alcohol in a Diesel Engine*, SAE Paper 1080, 1980.

Boyce, T.R., Karim, G.A., and Moore, N.P.W., An Experimental Investigation into the Effects of the Addition of Partially Oxidized Reaction Products to the Intake Charge of a Compression Ignition Engine, *Journal of the Institute of Petroleum*, 52, 300–311, 1968.

Burn, K.S., The Effect of Cold Intake Temperatures on the Combustion of Gaseous Fuels in a Dual-Fuel Engine, MSc thesis, University of Calgary, Canada, 1977.

Cruz, I.E., Studies on the Practical Application of Producer Gas from Agricultural Residues as Supplementary Fuel for Diesel Engines, in *Thermal Conversion of Solid Wastes and Biomass*, ed. J.L. Jones and S.B. Radding, ACS Symposium Series 130, American Chemical Society, 1980, pp. 649–669.

Dardalis, D., Matthews, R.D., Lewis, D., and Davis, K., *The Texas Project, Part 5—Economic Analysis: CNG and LPG Conversions of Light-Duty Vehicle Fleets*, SAE Paper 982447, 1998.

Ecklund, E., Bechtold, R., Timbario, T., and McCallum, Alcohol Fuel Use in Diesel Transportation Vehicles, in *IGT 3rd Symposium on Nonpetroleum Vehicular Fuels*, October 1982, pp. 261–313.

Gopal, G., Rao, P.S., Gopalakrishnan, K.V., and Murthy, B.S., Use of Hydrogen in Dual-Fuel Engines, *International Journal of Hydrogen Energy*, 7(3), 267–272, 1982.

Goto, S., Furutani, H., and Delic, R., *Dual Fuel Diesel Engine Using Butane*, SAE Paper 920690, 1992.

Green, R.K., and Glasson, N.D., High-Pressure Hydrogen Injection for Internal Combustion Engines, *International Journal of Hydrogen Energy*, 17(11), 895–901, 1992.

Herdin, G.R., Gruber, F., Plohberger, D., and Wagner, M., Experience with Gas Engines Optimized for H2-Rich Fuels, presented at ASME-ICE Division, 2003, Paper ICES2003-596.

Houser, K.R., Lestz, S.S., Dukovich, M., and Yasbin, R., *Methanol Fumigation of a Light Duty Automotive Diesel Engine*, SAE Paper 801379, 1980.

Ikegami, M., Miwa, K., and Shioji, M., A Study of Hydrogen Fuelled Compression Ignition Engines, *International Journal of Hydrogen Energy*, 7, 341–353, 1982.

Jorach, R., Enderle, C., and Decker, R., Development of a Low-NO_x Truck Hydrogen Engine with High Specific Power Output, *International Journal of Hydrogen Energy*, 22(4), 423–427, 1997.

Karim, G., The Combustion of Low Heating Value Gaseous Fuel Mixtures, in *Handbook of Combustion*, ed. M. Lackner, F. Winter, and A. Agarwal, vol. 3, Wiley Publishers, Weinheim, Germany, 2010, pp.141–163.

Karim, G.A., An Analytical Approach to the Uncontrolled Combustion Phenomena in Dual Fuel Engines, *Journal of the Institute of Fuel,* 37, 530–536, 1964.

Karim, G.A., A Review of Combustion Processes in the Dual Fuel Engine—The Gas Diesel Engine, *Progress in Energy and Combustion Science*, 6, 277–285, 1980.

Karim, G.A., Combustion in Gas Fueled Compression Ignition Engines of the Dual Fuel Type, *ASME Journal of Engineering for Gas Turbines and Power*, 125, 827–836, 2003.

Karim, G.A., *An Investigation of the Combustion in an IDI Diesel Engine with Low Concentrations of Added Hydrogen*, SAE Paper PF11-0875, 2010.

Karim, G.A., The Combustion of Bio-Gases and Low Heating Value Gaseous Fuel Mixtures, *International Journal of Green Energy*, 8, 1–10, 2011.

Karim, G.A., and Amoozigar, N., *Determination of the Performance of a Dual Fuel Diesel Engine with the Addition of Various Liquid Fuels to the Intake Charge*, SAE Paper 830265, 1983.

Karim, G.A., and Burn, K.S., *Combustion of Gaseous Fuels in a Dual Fuel Engine of the Compression Ignition Type with Particular Reference to Cold Intake Temperature Conditions*, SAE Paper 800263, 1980.

Karim, G.A., and Khan, M.O., Examination of Effective Rates of Combustion Heat Release in a Dual Fuel Engine, *Journal of Mechanical Engineering Science*, 10, 13–23, 1968.

Karim, G.A., and Klat, S.R., Knock and Autoignition Characteristics of Some Gaseous Fuels and Their Mixtures, *Journal of the Institute of Fuel*, 39, 109–119, 1966.

Karim, G.A., and Klat, S.R., Hydrogen as a Fuel in Compression Ignition Engines, *Mechanical Engineering*, 98(4), 34–39, 1976.

Karim, G.A., and Rogers, A., Comparative Studies of Propane and Butane as Dual Fuel Engine Fuels, *Journal of the Institute of Fuel*, 40, 513–522, 1967.

Karim, G.A., and Wierzba, I., Comparative Studies of Methane and Propane as Fuels for Spark Ignition and Compression Ignition Engines, *SAE Transactions*, 92, 3677–3688, 1983.

Karim, G.A., and Wierzba, I., *Methane–Carbon Dioxide Mixtures as a Fuel*, SAE Paper 921557, 1992.

Karim, G.A., Gee, D., and Satterford, R., Performance of a C.I. Engine in Unconventional Atmospheres, *The Engineer,* 219, 551–556, 1965.

Karim, G.A., Klat, S.R., and Moore, N.P.W., Knock in Dual Fuel Engines, in *Proceedings of the Institute of Mechanical Engineers*, 1967, vol. 181, pp. 453–466.

Lowi, A., *Supplementary Fueling of Four Stroke Cycle Automotive Diesel Engines by Propane Fumigation*, SAE Paper 41398, 1984.

Mathur, H.B., Das, L.M., and Patro, T.N., Hydrogen-Fuelled Diesel Engine: Performance Improvement through Charge Dilution Techniques, *International Journal of Hydrogen Energy*, 18(5), 421–431, 1993.

Oester, U., and Wallace, J.S., *Liquid Propane Injection for Diesel Engines*, SAE Paper 872095, 1987.

Poonia, M.P., Ramesh, A., and Gaur, R.R., *Effect of Intake Air Temperature on Pilot Fuel Quantity on the Combustion Characteristics of a LPG Dual Fuel Engine*, SAE Paper 982455, 1998.

Sierens, R., and Verhelst, S., Influence of the Injection Parameters on the Efficiency and Power Output of a Hydrogen Fueled Engine, ASME Journal of Engineering for Gas Turbines and Power, 125, 444–449, 2003.

Thiessen, S., Khalil, E., and Karim, G.A., The Autoignition in Air of Some Binary Fuel Mixtures Containing Hydrogen, *International Journal of Hydrogen Energy*, 35, 10013–10017, 2010.

Tomita, E., Fukatani, N., Kawahara, N., and Maruyama, K., Combustion in a Supercharged Biomass Gas Engine with Micro Pilot Ignition Effects of Injection Pressure and Amount of Diesel Fuel, *Journal of Kones Powertrain and Transport*, 14(2), 513–520, 2007.

Tsolakis, A., and Megaritis, A., Partially Premixed Charge Compression Ignition Engine with On Board H2 Production by Exhaust Gas Fuel Reforming of Diesel and Biodiesel, *International Journal of Hydrogen Energy*, 31, 731–745, 2005.

Tsolakis, A., Hernandez, J.J., Megaritis, A., and Crampton, M., Dual Fuel Diesel Engine Operation Using H2 Effect on Particulate Emissions, *Energy and Fuels*, 19, 418–425, 2005.

Turner, S.H., and Weaver, C.S., *Dual-Fuel Natural/Diesel Engines: Technology, Performance and Emissions*, No. GRI-94/0094, Topical Report Gas Research Institute, November 1994.

Varde, K.S., *Propane Fumigation in a Direct Injection Type Diesel Engine*, SAE Paper 831354, 1983.

Varde, K.S., and Frame, G.A., Hydrogen Aspiration in a Direct Injection Type Diesel Engine—Its Effects on Smoke and Other Engine Performance Parameters, *International Journal of Hydrogen Energy*, 8, 549–555, 1983.

Wood, C.D., *Alternative Fuels in Diesel Engines: A Review*, SAE Paper 810248, 1981.

Xiao, F., and Karim, G.A., *Combustion in a Diesel Engine with Low Concentrations of Added Hydrogen*, SAE Paper 2011-01-0676, 2011, pp. 1–17.

Xiao, F., Sohrabi, A., and Karim, G.A., *Reducing the Environmental Impact of Fugitive Gas Emissions through Combustion in Diesel Engines*, SAE Paper 2007-01-2048, 2007.

Yonetani, H., Hara, K., and Fukatani, I., *Hybrid Combustion Engine with Premixed Gasoline Homogeneous Charge and Ignition by Injected Diesel Fuel—Exhaust Emission Characteristics*, SAE Paper 940268, 1994.

Yoshida, K., Shoji, H., and Tanaka, H., *Study on Combustion and Exhaust Emission Characteristics of Lean Gasoline-Air Mixture Ignited by Diesel Fuel Direct Injection*, SAE Paper 982482, 1998.

Zaidi, K., Andrews, G., and Greenhough, J., *Diesel Fumigation Partial Mixing for Reducing Ignition Delay and Amplitude of Pressure Fluctuations*, SAE Paper 980535, 1998.

20 Predictive Modeling of Dual-Fuel Engine Performance

20.1 MODELING COMBUSTION PROCESSES

The proper modeling of cylinder events and associated engine performance involves a number of complex closely coupled physical and chemical processes. These include the transient three-dimensional dynamics of evaporating fuel sprays interacting with flowing multicomponent gases undergoing mixing, ignition chemical reactions, and heat transfer.

Much progress has been made in recent years in developing comprehensive schemes for the oxidation of fuels in air. These schemes vary widely in complexity, detail, and accuracy. When the transient and final concentrations of the reacting species are required, such as when determining the detailed composition of exhaust emissions, the use of a suitably comprehensive scheme is needed, especially when wide ranges of temperature, equivalence ratio, and pressure are to be covered, as normally the case in engine combustion. Throughout, there is a need to be aware of the inherent limitations of such approaches, and the support of experimental validation remains necessary. Figure 20.1 shows, for the case of the adiabatic constant pressure oxidation of methane in air initially at 800 K, how the peak transient values reached of some of the key reactive species vary with equivalence ratio. Some of these species are radicals and have very low concentrations (e.g., H, OH, O, and CH_3), while others do reach substantial concentrations during the course of the reaction. The corresponding peak values of temperature attained during the reaction are also shown.

The vast chemical kinetic information that is becoming increasingly available is being employed in analytical models that aim at presenting engine combustion and associated performance processes. An example of modeling the complex reactions of a higher hydrocarbon fuel such as n-heptane in air in stoichiometric mixtures under isothermal conditions is shown in Figure 20.2. The combustion energy release for a range of temperature values reflects the multistage nature of its reaction. This would not be represented empirically satisfactorily by a single gross reaction.

20.2 PREDICTIVE MODELING OF ENGINE PERFORMANCE

To be able to carry out effective engine design changes, improvements in performance, and operate different engine types on a range of gaseous fuels and liquid

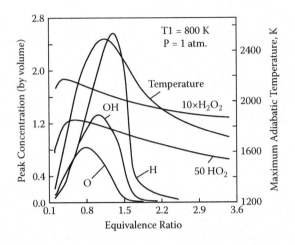

FIGURE 20.1 Variations with equivalence ratio of methane and air at an initial temperature of 800 K. The maximum transient concentrations of a selection of species taking part in the reaction of methane in air and the corresponding maximum temperature reached are also shown. (From Karim, G.A., *Fuels, Energy, and the Environment*, CRC Press, Boca Raton, FL, 2012.)

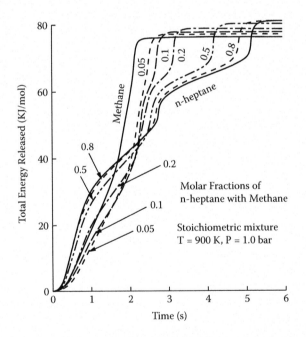

FIGURE 20.2 Calculated isothermal energy release rates for stoichiometric n-heptane-methane binary mixture in air at 900 K and ambient pressure. (Adapted from Samuel, P., Computational and Experimental Investigation of Ignition and Combustion of Liquid Hydrocarbon Fuels in Homogeneous Environment of Fuel and Air, PhD thesis, University of Calgary, Canada, 1994.)

fuel pilots requires a thorough and continuing improvement to our detailed understanding of the complex combustion processes that occur in these engines. If such an effort is to rely almost exclusively on experimental testing approaches, then it would have serious limitations and could be wasteful of resources, costly, and time-consuming. The computational modeling of internal combustion engines in general is being increasingly used as a design support tool and economizing greatly on the need for extensive experimentation. It can also be used for interpolation of design and performance data and serve as an analysis tool for the interpretation of complex experimental results. It can also serve as a developmental tool for the improvement of physical, chemical, or numerical computational submodels, so as to serve as an effective economical approach for improving the design and performance of a wide variety of engines and operating conditions. As an example, models may be employed to simulate the performance in the field of specific engines that have inadequate or insufficient monitoring or metering facilities. Thus, it can serve as a diagnostic tool of operational problems and a source to suggest improvements without incurring excessive expenditure. Accordingly, experimental approaches in engine development are increasingly benefiting from being supplemented by results obtained from the employment of predictive modeling. Also these approaches may be focused specifically on certain operational features of the engine, such as establishing the expected power output or determining the conditions for the potential incidence of knock. Alternatively, they may attempt to provide comprehensive analytical simulations of the whole combustion process and the operational consequences, such as determining the nature and extent of exhaust emissions and the effective changes that can be made to the geometry of the engine cylinder to bring about improvements in performance. Throughout, validation of the prediction needs to be made against corresponding experimental results obtained in suitably monitored engines. In principle, the computational modeling of engines can have the following main objectives:

- As a design tool requiring the least experimental support
- As an interpolative design tool with experimental validation
- As an analysis tool in the interpretation of the results of complex experiments
- As an explorative tool in engine design
- As a developmental tool for improving performance and physical, chemical, or numerical submodels

The complexity of the modeling will be dependent on the accuracy desired from the results. The ability to predict deteriorates as the objectives of the results increase in complexity, such as from merely generating the cylinder pressure temporal development, and hence power output, through rates of fuel consumption. As far as emissions are concerned, the prediction of the extent of formation of unburnt hydrocarbon and particulates tends to be more challenging and yields less accurate results than the corresponding prediction of species such as oxides of nitrogen formation. This is due to the increased need for accurate and sufficiently detailed descriptions of the corresponding processes of mixing and the associated combustion reactions.

Examples of the options available for modeling the performance of engines follow:

- Approaches that invariably assume thermodynamic equilibrium conditions may apply throughout, such that the products, energy releases, and associated charge properties are functions of the prevailing conditions and are independent of time.
- Approaches that assume the conversion of the fuel–air mixture to final products is complete, releasing the associated heating value of the fuel. These are very convenient but are very approximate, limited in applicability, and do not provide sufficiently realistic information about some of the key features of the combustion process, such as the composition of the final exhaust products.
- Models that consider the conversion of the reactant fuel–air mixture proceeds with time at varying rates, producing during the combustion phase changes in composition and varying energy release rates. These modeling approaches have been increasing in sophistication and are accepted as a valuable approach to more realistically represent the progress of the combustion process via its multichain reactions.

20.3   MODELING DUAL-FUEL ENGINE PERFORMANCE

In comparison to the extensive efforts that have been made over the years toward modeling the combustion processes in spark ignition, compression ignition, or homogeneous charge compression ignition (HCCI) engines, much less effort has been devoted to modeling the combustion processes and associated performance of dual-fuel engines. This is a reflection of the relative complexity of these processes with the great number of associated controlling variables and the much fewer numbers of engines produced to operate in the dual-fuel mode in comparison to the other types.

The nature of modeling to be employed and its completeness and complexity can be linked to what is desired to be modeled and the associated accuracy required of the results. It would be uneconomical and wasteful of resources to employ unnecessarily very complex and extensively detailed models if it is desired only to predict the yield in gross performance parameters, such as power output and how it changes, for example, due to modifications in the compression ratio or turbocharging characteristics employed. However, when the consequences of changes in the nature and composition of either of the two systems of fuels need to be established, then a more elaborate modeling procedure is advisable. It should take into account the nature and extent of variations in the influencing variables, such as the nature of the turbulent flow processes and the corresponding combustion reactions. When the effects on performance of other changes are needed, such as those in cylinder geometry or the fuel injection characteristics at very high pressures, detailed CFD modeling is needed. When a simultaneous consideration of the nature and extent of emissions, including those of unburned hydrocarbons and particulate matter, is required, then CFD modeling, combined with a sufficiently detailed accounting of the relevant chemical kinetics of the two fuel systems, is needed. In addition, the spatial nonuniformity in the key properties of the reactive system, such as temperature, including accounting for changes in the trapped exhaust and partially oxidized products, must be considered within the cycle or over numerous consecutive cycles.

It is becoming increasingly possible to formulate guidelines based on the results of predictive modeling to indicate for any dual-fuel engine setup the fuel composition and operating conditions, the corresponding knock-limited performance, concentrations of the exhaust emissions, and the role of associated changes in the many design and operating conditions. Through such approaches, various measures may be identified so as to favorably modify some aspects of the course of the combustion process and associated changes in engine performance. As an example, these can include the employment of multiple pilot and stratified timed injections, suitable modifications to recirculated exhaust gases, varying the residual gas fraction, and of course, various aspects of the performance. This is especially effectively manageable with the relatively much simpler operation of HCCI type devices that do not involve the complexities associated with accounting for the ignition process through pilot injection. Another example may be considering the consequences of measures for the selective modification of the preignition reactions of the charge during compression, such as through the fumigation of an additional fuel or separately prepared partial oxidation products.

The relatively simple single-zone thermodynamic models, when based on zero-dimensional simulation of the combustion process in engines, generally assume that spatially uniform temperature and composition are retained throughout the whole charge at all times. Such a class of modeling has been widely used in the past as an approximate guide in some applications, mainly because of their relative simplicity. A notable example of these is the deriving of effective rates of heat release as representative of the progress of the combustion process. Associated overall features of engine performance can then be derived from experimentally based cylinder pressure development records.

Through the additional consideration of the overall chemical reaction activity of the cylinder charge, an approximate model can be derived to specify the conditions for the onset of knock and the associated knock-limited power output with different fuels (Figure 15.10). While using such a simple approach the predicted knock-limited power output of a dual-fuel engine with a fixed pilot quantity, was shown to be approximately dependent logarithmically on the inverse of the absolute intake mixture temperature. An extension of such an approach, for example, is to assume that the ignition of the small quantity of pilot fuel employed would take place around TDC sufficiently rapidly, for it to be considered taking place under constant volume conditions. The state of the reactive mixture can be modeled then using the corresponding reaction kinetics. Another example is to find the change in the flame spread limits with variations in the size of the pilot or its timing. Such relatively simple approaches could provide an indication of the potential for the incidence of knock in premixed dual-fuel engines and the effects of changes in a limited number of engine operating and design variables on some of the key performance parameters. The prediction of power output, efficiency, and NOx emissions can also be determined approximately, while using variations on such approaches.

Some predictive approaches employ quasi-dimensional combustion models with thermodynamic multizone approaches. These are based on conceiving the whole cylinder charge to be made up of different zones, which change in size and properties during the progress of combustion, as represented schematically in Figure 20.3. These approaches are based on an assumed representation of the sequencing and regions of the combustion processes obtained through appropriate theory, empiricism, and experimentation, while adequately incorporating representative chemical kinetics. For example, such approaches may be used to indicate approximately the conditions that lead to the onset of knock with a certain gaseous and pilot fuels, exhaust gas recirculation (EGR) effects, and associated emissions.

Continuing progress is being made in the simulation of dual-fuel engine operation using 3D-CFD turbulent combustion modeling while accounting in sufficient detail for the flow patterns and the associated kinetics of combustion of the pilot and gaseous fuels. Such approaches are the subject of continuing research and development, while they increase in complexity, sophistication, and reliability. More detailed information can then be obtained about the different features of the combustion process, albeit tending to become extensively complex, but yielding information that serves as an effective guide to what would be observed in direct experimentation.

An example is 3D-CFD-based models, specifically those based on the KIVA III computing program, with reduced detailed chemical kinetics schemes for the complex diesel fuel pilot, and a more detailed chemical kinetics approach for the gaseous fuel component. A turbulence function is incorporated in the simulation of the combustion processes of dual-fuel engines having a variety of types of chambers.

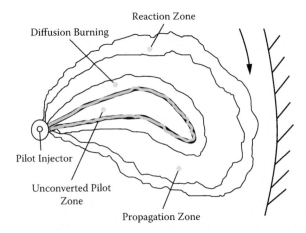

FIGURE 20.3 An example of visualizing the different regimes within the cylinder of a premixed direct injection dual-fuel engine. (Adapted from Liu, Z., and Karim, G.A., *Journal of Power and Energy*, 211, 159–171, 1997.)

20.4 SOME MODELED EXAMPLES

Numerous predictive approaches involving single-zone, multizone, and multidimensional CFD models have been developed and applied to the prediction of the performance of different engine types and fuel systems. For example, the autoignition of methane–air mixtures in a variable compression ratio engine in the absence of an ignition source representing HCCI operation was modeled while using a detailed kinetic scheme for the oxidation reactions of methane in air. The prediction was based on considering the mixture to behave as a single zone and accounting for the effects of heat transfer and residual gases. Figure 20.4 shows good agreement of the predicted values with corresponding experimental results. Additionally, such an approach yields in parallel comprehensive data for the oxidation reactions of the cylinder contents about the concentration changes in the many species taking part in the reaction, whether stable or unstable.

To incorporate comprehensive chemical kinetic modeling for the oxidation reactions of the cylinder contents into 3D-CFD models is a demanding task. A somewhat simpler approach may instead employ empirically reduced kinetic schemes. For example, such approaches can generate comprehensively detailed information of the flow characteristics of the system, but the chemical aspects, in comparison, may remain less well represented. Other approaches have aimed at essentially decoupling the physical and chemical aspects of the modeling such that comprehensively detailed chemical representations are used with a relatively simplified corresponding zero-dimensional modeling of the system. The results are then applied in a fitting

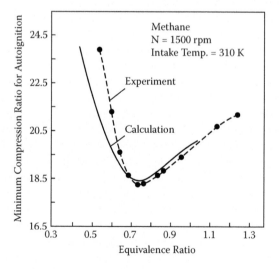

FIGURE 20.4 An example of the calculated minimum compression ration required for autoignition of methane–air mixtures over a range of equivalence ratios. (Adapted from Liu, Z., An Examination of the Combustion Characteristics in Compression Ignition Engines Fueled with Gaseous Fuels, PhD thesis, University of Calgary, Canada, 1995; the experimental results are also shown from Downs, D., Walsh, A.D., and Wheeler, R.W., *Philosophical Transactions of the Royal Society*, 243, 463–524, 1951.)

procedure over a small incremental change to obtain apparent gross reaction rate functions, which are then reapplied into the CFD code to yield further improvements to the results. With such approaches a quasi-combined CFD and detailed kinetics modeling is achieved.

A widely used KIVA code has been developed with internal combustion engine applications in mind. The code has often been incorporated in models that were aimed at calculating engine flows with differently shaped cylinder and piston geometries, including the effects of turbulence and wall heat transfer.

The multidimensional modeling of combustion processes in engines in recent years has become increasingly effective in providing a better understanding of these processes, finding effective measures to overcome some operational problems, evaluating new design concepts, and reducing hardware prototype and development costs. However, there are still not many 3D-CFD simulations of dual-fuel engine operation where much of the modeling relates to the direct injection type. Modeling of the precombustion or swirl chamber type engines tends to lag behind, mainly due to the greater complexity of both the geometry of the combustion chamber and the associated combustion and transport processes.

Computational 3D methods have been used to obtain optimized diesel engine geometry and performance for the reduction of emissions and fuel consumption. Such analyses showed, for example, the need for a precise control of the overall fuel-to-air ratio, the optimum start of injection timing, and EGR levels to ensure sufficiently low NOx emissions. For example, those of the combustion process and its cyclic variation in HCCI engines were predicted using a multizone model with a sufficiently detailed chemical kinetics representation.

An important consideration in dual-fuel engine modeling is that it can be employed to investigate the role of changes in some key influencing parameters where there are no effective economic practical approaches for performing it. An example of the utility of the 3D-CFD modeling approach based on KIVA with full chemistry is investigating the influence of possible changes in the temperatures of the different segments of the cylinder surfaces on dual-fuel engine performance and emissions. An example is an indirect injection (IDI) engine operating on very lean mixtures of hydrogen while employing a fixed mass of diesel fuel injection. Changes in the mean values of the prechamber surface temperature are expected to be more influential than for a corresponding direct injection (DI) engine. The resulting changes, particularly in the values of the ignition delay and rates of energy release, lead in turn to significant changes in emissions and engine performance (Figure 20.5). When different mean surface temperature values were assigned in turn to the cylinder wall, piston, cylinder head, and prechamber surface temperatures, it was demonstrated that in general, the values of the indicated power and maximum cylinder pressure had only relatively small increases when the cylinder wall surface temperature was increased. However, NOx emissions tended to increase slightly as the temperature increased, resulting from the longer exposure of the charge to high surface temperatures, particularly during the expansion. Similarly, increasing piston head surface temperature produced almost no changes. However, changes in cylinder head temperature, including that of the surface of a prechamber, produce significant increases in the maximum cylinder pressure and NOx emissions, which is a reflection of the increased average

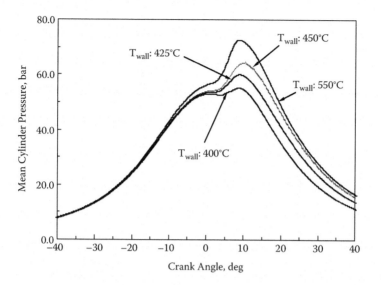

FIGURE 20.5 An example of the effects of changes in cylinder surface temperature on the calculated pressure development in a prechamber dual-fuel engine operating on methane with constant pilot quantity, 1800 rev/min, CR=23, EQdiese=0.19, EQCH4: 0.25. (From Liu, C., An Experimental and Analytical Investigation into the Combustion Characteristics of HCCI and Dual Fuel Engines with Pilot Injection, PhD thesis, University of Calgary, Canada, 2006.)

temperature of the cylinder head surface and the relatively large surface area of the swirl chamber.

REFERENCES AND RECOMMENDED READING

Amsden, A.A., Butler, T.D., and O'Rourke, P.J., *The KIVA-ii Computer Program for Transient Multidimensional Chemically Reactive Flows with Sprays*, SAE Paper 872072, 1987.

Downs, D., Walsh, A.D., and Wheeler, R.W., A Study of the Reactions That Lead to Knock in the Spark Ignition Engine, *Philosophical Transactions of the Royal Society*, 243, 463–524, 1951.

Golovitchev, V.I., Nordin, N., Jarnicki, R., and Chomiak, J., *3-D Diesel Spray Simulation Using a New Detailed Chemistry Turbulent Combustion Model*, SAE Paper 2000-01-1891, 2000.

Hountalas, D.T., and Papagiannakis, R., *Development of a Simulation Model for Direct Injection Dual Fuel Diesel-Natural Gas Engines*, SAE Paper 2000-01-0286, 2000.

Karim, G.A., An Analytical Approach to the Uncontrolled Combustion Phenomena in Dual Fuel Engines, *Journal of the Institute of Fuel*, 37, 530–536, 1964.

Karim, G.A., and Zhoada, Y., An Analytical Model for Knock in Dual Fuel Engines of the Compression Ignition Type, *SAE Transactions*, 3.52–3.62, 1988.

Khalil, E., and Karim, G.A., A Kinetic Investigation of the Role of Changes in the Composition of Natural Gas in Engine Applications, *ASME Journal of Engineering for Gas Turbines and Power*, 124, 404–411, 2002.

Khalil, E., Samuel, P., and Karim, G.A., *An Analytical Examination of the Chemical Kinetics of the Combustion of N-Heptane-Methane Air Mixtures*, SAE Paper 961932, 1996.

Kong, S.C., Han, Z., and Reitz, R.D., *The Development and Application of a Diesel Ignition and Combustion Model for Multidimensional Engine Simulation*, SAE Paper 950278, 1995.

Kusaka, J., Tsazuki, K., and Daisho, Y., *A Numerical Study on Combustion and Exhaust Gas Emissions Characteristics of a Dual-Fuel Natural Gas Engine Using a Multi-Dimensional Model Combined with Detailed Kinetics*, SAE Paper 2002-01-1750, 2002.

Liu, C., An Experimental and Analytical Investigation into the Combustion Characteristics of HCCI and Dual Fuel Engines with Pilot Injection, PhD thesis, University of Calgary, Canada, 2006.

Liu, C., and Karim, G.A., Three Dimensional Computational Fluid Simulation of Diesel and Dual Fuel Engine Combustion, *ASME Journal of Engineering for Gas Turbines and Power*, 131, 128041–128049, 2009.

Liu, C., Karim, G.A., Xiao, F., and Sohrabi, A., *An Experimental and Numerical Investigation of the Combustion Characteristics of a Dual Fuel Engine with a Swirl Chamber*, SAE Paper 07PFL-783, 2007.

Liu, Z., An Examination of the Combustion Characteristics in Compression Ignition Engines Fueled with Gaseous Fuels, PhD thesis, University of Calgary, Canada, 1995.

Liu, Z., and Karim, G.A., *A Predictive Model for the Combustion Process in Dual Fuel Engines*, SAE Paper 952435, 1996.

Liu, Z., and Karim, G.A., Simulation of the Combustion Processes in Gas Fuelled Diesel Engines, *Journal of Power and Energy*, 211, 159–171, 1997.

Ouellette, P., Mtui, P., and Hill, P., *Numerical Simulations of Directly Injected Natural Gas and Pilot Diesel Fuel in a Two Stroke Compression Ignition Engine*, SAE Paper 981400, 1998.

Papagiannakis, R.G., Hountalas, D.T., and Rakopoulos, C.D., Theoretical Study of the Effects of Pilot Fuel Quantity and Its Injection Timing on the Performance and Emissions of a Dual Fuel Diesel Engine, *Energy Conversion and Management*, 48, 2951–2961, 2007.

Samuel, P., Computational and Experimental Investigation of Ignition and Combustion of Liquid Hydrocarbon Fuels in Homogeneous Environment of Fuel and Air, PhD thesis, University of Calgary, Canada, 1994.

Schaub, F.S., and Hubbard, R.L., A Procedure for Calculating Fuel Gas Blend Knock Rating for Large-Bore Gas Engines and Predicting Engine Operation, *ASME Transactions*, 107, 922–930, 1985.

Thyagarajan, V., and Babu, M.K.G., A Combustion Model for a Dual Fuel Direct Injection Diesel Engine, Diagnostics and Modeling of Combustion in Reciprocation Engine, presented at the JSME Proceedings of COMODIA Symposium, Tokyo, 1995.

Wiecman, D., Senecal, P.K., and Reitz, R.D., *Diesel Engine Combustion Chamber Geometry Optimization Using Genetic Algorithms and Multi Dimensional Spray and Combustion Modeling*, SAE Paper 2001-01-0547, 2001.

Zhang, Y., Kong, S.C., and Reitz, R.D., *Modeling and Simulation of a Dual Fuel (Diesel/Natural Gas) Engine with Multi-Dimensional CFD*, SAE Paper 2003-01-0755, 2003.

Zhoada, Y., and Karim, G.A., *An Analytical Model for Knock in Dual Fuel Engines of the Compression Ignition Type*, SAE Paper 880151, 1988.

Zhoada, Y., and Karim, G.A., Modelling of the Combustion Process in a Dual Fuel Direct Injection Engine, *ASME Journal of Energy Resources Technology*, 112, 254–259, 1990.

Zhou, G., and Karim, G.A., An Analytical Examination of Various Criteria for Defining Autoignition within Heated Methane-Air Homogeneous Mixtures, *ASME Journal of Energy Resources Technology*, 116, 175–180, 1994.

Glossary

Adiabatic flame temperature: Final combustion temperature calculated thermodynamically for an adiabatic system, usually related to stoichiometric mixtures, while accounting for the effects of variable properties and dissociation.

Advance: Earlier passage of electric spark or pilot liquid fuel injection into the mixture of an engine.

Alternative fuels: Fuels alternative to conventional gasoline or diesel fuel, such as hydrogen, LPG, or alcohols.

Autoignition: The acceleration of oxidation reaction rates due to the energy release becoming in excess of the prevailing losses to the surrounding without the aid of deliberate external ignition energy sources.

Autoignition temperature: The lowest temperature at which a combustible material ignites in air without the aid of an external spark or a flame.

Bagasse: Gaseous fuel mixture produced from the processing of sugarcane, from which the juice has been extracted.

Barrel: Crude oil volumetric measure that is equivalent to 159 L or 35 imperial gallons or 42 U.S. gallons.

Bi-fuel engine: An engine that alternately utilizes two different fuels while having ignition provided by an external source of energy, such as an electric spark.

Blast furnace gas: A low heating value gas fuel mixture produced as a by-product of blast furnaces in steelmaking.

Bosch smoke number/unit: An empirical measure of smoke density in exhaust gases.

Catalytic converter: A catalytic device fitted to the exhaust system of an engine so as to reduce the concentrations of carbon monoxide, unburned hydrocarbons, particulates, and oxides of nitrogen before being discharged into the outside atmosphere.

Cetane number: A measure of the ignition quality of diesel fuel based on the associated length of the ignition delay under certain specified operating conditions. The higher the cetane number, the shorter is the ignition delay.

Cogeneration: The production of power while simultaneously utilizing the heat rejected by the engine.

Compressed natural gas (CNG): Natural gas compressed to high pressure, so as to be transported in high-strength containers, such as for use in transport applications.

Compression ignition: The ignition of a fuel–air mixture at high temperature and pressure produced by rapid compression, such as in diesel engines.

Compression ratio: The maximum engine cylinder volume (at bottom dead center) divided by the corresponding minimum volume (at top dead center) of a reciprocating internal combustion engine.

Cool flames: Flames associated with the generation of very low thermal energy and light intensity. These are encountered in the oxidation of most hydrocarbon fuels due to low-temperature reactions that take place ahead of the main higher-temperature hot flames.

Cracking: The process of breaking up of large molecules into smaller fragments. Thermal cracking takes place at high temperature in the absence of air. Catalytic cracking takes place at lower temperatures, but in the presence of catalysts.

Cyclic variation: The random none repeatability of the progress of the combustion process in engines under otherwise steady operating conditions.

Diffusion flames: Flames where the fuel and air are initially nonhomogeneously premixed, and the combustion rate becomes governed by the rate of mixing of the fuel and air, such as in a candle or a fuel jet.

Direct injection: The injection of the fuel under pressure directly into the cylinder of a diesel engine and occasionally also applied in spark ignition engines.

Dissociation: The breakup of a single molecule into smaller fragments such as at high temperatures, which may be subsequently capable of recombining with the lowering of temperature to form the original molecule.

Dry natural gas: Processed natural gas where most of its high hydrocarbon components and water have been removed.

Dual-fuel engine: An engine where a mixture of gaseous fuel and air is ignited through the injection of pilot diesel fuel.

Electronic engine control (EEC): The assembly of electronic and electromechanical components that continuously monitor engine operation and vary engine calibration to meet the requirements.

Equivalence ratio: The mass of fuel-to-air ratio relative to the corresponding stoichiometric value.

Exhaust gas recirculation (EGR): The recycling of some exhaust gas back into the inlet manifold so as to lower the overall level of the combustion temperature, and hence reduce the production of oxides of nitrogen.

Fischer–Tropsch diesel: Diesel fuel manufactured from synthesis gases while employing the Fischer–Tropsch method.

Flammability limits: The homogeneous fuel–air mixture that can just support flame propagation from an ignition source. The lean or lower limit relates to the leanest mixture in fuel that is flammable, while the rich or higher limit relates to the richest mixture in fuel that is still combustible.

Free radicals: A group of atoms such as CH_3 or OH that is unstable and has the capacity to react more vigorously despite their low concentrations than stable molecules. Radicals are the main active ingredients in most reactions, especially combustion reactions.

Fumigation: The normally aspirated introduction of the fuel into the intake air via simple carburetion.

Gasification: The process of converting solid or liquid fuels into a gaseous fuel, such as the gasification of coal.

Greenhouse gases: Gases, when present in the atmosphere, that contribute to its warming, such as carbon dioxide and methane.

Homogeneous charge compression ignition (HCCI): The compression ignition of a homogeneous mixture of fuel and air.

Hybrid engines: Engines coupled to an electric motor/generator/storage system that can provide controllable power through these components.

Ignition delay: The period between the start of injection of the fuel in diesel engines and the first detection of ignition.

International Organization for Standardization (ISO): A nongovernmental federation of standard organizations from many countries that drafts rules and regulations for industry.

Knock: The undesirable phenomenon in spark ignition engines where energy is rapidly released due to the premature and uncontrolled autoignition of part of the combustible mixture ahead of the propagating flame originating from the electric spark.

Landfill gas: The gas produced and released relatively slowly from the decaying processes of landfill materials. The main constituents of the gases are methane and carbon dioxide.

Lean mixture: A fuel–air mixture that contains excess air beyond that required to completely oxidize the available fuel.

Load: The amount of power or torque delivered or required.

Misfire: The occasional failure of combustion in one cylinder or one cycle.

Mtoe: Millions of tonnes of oil equivalent.

Octane number: An empirical indicator of the knock resistance quality of fuels, particularly gasolines in spark ignition engines.

Pilot: Usually refers to a small independent flame used to ignite a bulk mixture of fuel and air.

Preignition: The undesirable premature, often uncontrolled ignition of a fuel–air mixture in a combustion chamber before the proper timed passage of the spark.

Primary reference fuels: Blends of n-heptane (ON = 0) and iso-octane (ON = 100) used for the knock rating of liquid fuels in spark ignition engines.

Producer gas: A fuel-gas mixture that is produced through the controlled combustion of a fuel such as coal with air and steam. The product gas contains mainly hydrogen and carbon monoxide with some nitrogen.

Quenching distance: The largest distance under a certain specified condition when a flame cannot propagate due to excessive heat loss, such as quenching through narrow gaps or within a boundary layer.

Refinery gas: The noncondensable gas resulting from the fractional distillation of crude oil. Usually, it is either flared or supplied for mixing with fuel gases for use mainly domestically.

Reforming: The operation in which the chemical structure of a molecule is modified in order to improve its knock-resisting quality, such as via thermal or catalytic processes.

Regenerator: A heat exchanger that is usually reserved to periodic operation where heat is transferred alternately from the gaseous products to the air before combustion.

Retard: Indicates a delayed passage of the spark in spark ignition engines or pilot fuel injection in dual-fuel engines.

Rich mixture: A fuel–air mixture that contains excess fuel beyond the amount that can be burned completely with the available oxygen.

Skip firing: The firing of only some of the cylinders in a running multicylinder engine.

Smoke: Small gas-borne particles of carbon or soot, usually less than 1 μm, such as resulting from the incomplete combustion of carbonaceous materials, and of sufficient number to be observable.

Sour gas: Natural gas that contains acidic compounds, such as hydrogen sulfide and other corrosive compounds.

Spontaneous combustion: Ignition of combustible material following slow oxidation without the application of high temperature from an external energy source.

Squish motion: The radial flow in the cylinder toward the end of compression.

Stoichiometric mixture: The chemically correct mixture that can, in principle, complete combustion without resulting in excess air or excess fuel.

Sweet gas: Fuel gas from which corrosive compounds such as sulfur have been removed.

Swirl: The rotational flow within the cylinder around its axis.

Synthesis gas: Manufactured gas mixture made from the reforming or partial oxidation of fossil fuels, such as coal, natural gas, or oil. The gas, which is made up mainly of hydrogen and carbon monoxide, serves as a raw material for the chemical synthesis of a wide range of fuels.

Tight formations: Sands or carbonates having porosities and permeabilities that are sufficiently low to inhibit the flow of the gas at commercial rates.

Turbocharger: A device fitted to engines to raise the intake charge pressure, and hence mass of the intake charge, through utilizing some of the exhaust gas energy by using a gas turbine that drives an air compressor. Thus, higher power output can be obtained from the same engine.

Water gas: A manufactured gaseous fuel mixture consisting primarily of carbon monoxide and hydrogen, usually made by the action of steam and hot carbon/coal.

Wet natural gas: Natural gas that may still contain small quantities of higher hydrocarbons.

Wobbe number: An indication of the rate of energy release by combustion of a fuel when discharged through an orifice under the action of a specified pressure difference.

Bibliography

Abd Alla, G.H., Soliman, H.A., Badr, O.A., and Abd Rabbo, M.F., *Effect of Pilot Quantity on the Performance of a Dual Fuel Engine*, SAE Paper 1999-01-3597, 1999.

Addy, J.M., Binng, A., Norton, P., Peterson, E., Campbell, K., and Bevillaqua, O., *Demonstration of Caterpillar C10 Dual Fuel Natural Gas Engines in Commuter Buses*, SAE Paper 2000-01-1386, 2000.

Aisho, Y., Yaeo, T., Koseki, T., Saito, T., and Kihara, R., *Combustion and Exhaust Emissions in a Direct Injection Diesel Engine Dual-Fueled with Natural Gas*, SAE Paper 950465, 1995.

Alcock, J.F., and Scott, W.M., Some More Light on Diesel Combustion, in *Proceedings of the Institute of Mechanical Engineers (Auto Division)*, 1962, pp. 179–191.

Al-Garni, M., A Simple and Reliable Approach for the Direct Injection of Hydrogen in Internal Combustion Engines at Low and Medium Pressures, *International Journal of Hydrogen Energy*, 20(9), 723–726, 1995.

Al-Himyary, T.J., A Diagnostic Two Zone Combustion Model for Spark Ignition Engines Based on Pressure Time Data, PhD thesis, University of Calgary, Canada, 1988.

Al-Himyary, T.J., Karim, G.A., and Dale, J.D., *An Examination of the Combustion Processes of Methane Fueled Engine When Employing Plasma Jet Ignition*, SAE Paper 891639, 1989.

Ali, A.I., and Karim, G.A., *The Effects of Low Ambient Temperatures on the Combustion of Natural Gas in a Single Cylinder Spark Ignition Engine*, SAE Paper 730084, 1973.

Ali, A.I., and Karim, G.A., Combustion, Knock and Emission Characteristics of a Natural Gas Fuelled Spark Ignition Engine with Particular Reference to Low Intake Temperature Conditions, in *Proceedings of the Institute of Mechanical Engineers*, London, 1975, vol. 189, pp. 139–147.

Aly, H., and Siemer, G., Experimental Investigation of Gaseous Hydrogen Utilization in a Dual-Fuel Engine for Stationary Power Plants, in *ASME-ICE Division*, 1993, vol. 20, pp. 67–79.

Amano, T., Morimoto, S., and Kawabasta, Y., *Modeling of the Effect of Air/Fuel Ratio and Temperature Distribution on HCCI Engines*, SAE Paper 2001-01-1024, 2001.

Amoozigar, N., Examination of the Performance of a Dual Fuel Diesel Engine with Particular Reference to the Addition of Some Inerts and Intake Liquid Additives, MSc thesis, University of Calgary, Canada, 1982.

Amoozigar, N., and Karim, G.A., *Examination of the Performance of a Dual-Fuel Diesel Engine with Particular Reference to the Presence of Some Inert Diluents in the Engine Intake Charge*, SAE Paper 821222, 1982.

Amsden, A.A., Butler, T.D., and O'Rourk, P.J., *The KIVA-ii Computer Program for Transient Multidimensional Chemically Reactive Flows with Sprays*, SAE Paper 872072, 1987.

Anon., *SAE Handbook: Engine, Fuel, Lubricants, Emissions, and Noise*, Society of Automotive Engineers, Warrendale, PA, 1993, vol. 3.

Antunes, J.M.G., Mikalsen, R., and Roskilly, A.P., An Experimental Study of a Direct Injection Compression Ignition Hydrogen Engine, *International Journal of Hydrogen Energy*, 14, 6516–6522, 2009.

Arbon, I.M., Worldwide Use of Biomass in Power Generation and Combined Heat and Power Schemes, *Journal of Energy and Power*, 216, 41–57, 2002.

Attar, A., and Karim, G.A., Knock Rating of Gaseous Fuels, *ASME Journal of Gas Turbine and Power*, 125, 500–504, 2003.

Austen, A.E., and Lyn, W.T., Relation between Fuel Injection and Heat Release in a Direct Injection Engine and the Nature of the Combustion Processes, in *Proceedings of the Institute of Mechanical Engineers (AD)*, London, 1961, no. 1, pp. 47–62.

Azzouz, D., Some Studies of Combustion Processes in Dual Fuel Engines—The Role of Pilot Liquid Injection Characteristics, MSc thesis, Department of Mechanical Engineering, Imperial College of Science and Technology, University of London, 1966.

Bade Shrestha, O.M., and Karim, G.A., The Operational Mixture Limits in Engines Fuelled with Alternative Gaseous Fuels, *ASME Journal of Energy Resources Technology*, 128, 223–228, 2006.

Bade Shrestha, S.O., Wierzba, I., and Karim, G.A., An Approach for Predicting the Flammability Limits of Fuel-Diluent Mixtures in Air, *Journal of Institute of Energy*, 122–130, 1998.

Badr, O., El-Sayed, N., and Karim, G.A., An Investigation of the Lean Operational Limits of Gas Fuelled Spark Ignition Engines, *ASME Journal of Energy Resources Technology*, 118(2), 159–163, 1996.

Badr, O., Karim, G.A., and Liu, B., An Examination of the Flame Spread Limits in a Dual Fuel Engine, *Applied Thermal Engineering*, 19, 1071–1080, 1999.

Baranescu, R., *Fumigation of Alcohol in a Diesel Engine*, SAE Paper 1080, 1980.

Barbour, T.R., Crouse, M.E., and Lestz, S.S., *Gaseous Fuel Utilization in a Light Duty Diesel Engine*, SAE Paper 860070, 1986.

Bartok, W., and Sarofim, A.F., *Fossil Fuel Combustion*, John Wiley & Sons, New York, 1991.

Baur, H., ed., *Diesel Engine Management*, 2nd ed., SAE, Warrendale, PA, 1999.

Bechtold, R.L., *Alternative Fuels Guidebook*, SAE Publishing, Warrendale, PA, 1997.

Beck, J., Karim, G.A., Mirosh, E., and Pronin, E., Bus Fuel Efficiency Local Emissions and Impact on Global Emissions, presented at NGV 96 Conference and Exhibition, Kuala Lumpur, Malaysia, October 1996.

Beck, J., Karim, G.A., Pronin, E., and Mirosh, E., The Diesel Dual-Fuel Engine—Practical Experience and Future Trends Update, paper no. 96EL070, *Proceedings of the International Symposium on Automotive Technology and Automation,* ISATA, Florence, Italy, pp. 1–6, 1996.

Beck, N., Johnson, W., George, A., Peterson, P., vander Lee, B., and Klopp, G., *Electronic Fuel Injection for Dual Fuel Diesel Methane*, SAE Paper 891652, 1989.

Bell, S., *Natural Gas as a Transportation Fuel*, SAE Paper 931829, 1993.

Beppu, O., Fukuda, T., Komoda, T., Miyake, S., and Tanaka, T., Service Experience of Mitsui Gas Injection Diesel Engines, in *Proceedings of CIMAC Congress*, Copenhagen, 1998, pp. 187–202.

Bergman, H., and Busenthur, B., Facts Concerning the Utilization of Gaseous Fuels in Heavy Duty Vehicles, in *Proceedings of the Conference on Gaseous Fuels for Transportation*, August 1986, pp. 813–849.

Bittner, R.W., and Aboujaoude, F., Catalytic Control of NOx, CO, and NMHC Emissions from Stationary Diesel and Dual Fuel Engines, *ASME Journal of Engineering for Gas Turbines and Power*, 114, 597–601, 1992.

Blizzard, D., Schaub, F.S., and Smith, J., Development of the Cooper–Bessemer Clean Burn Gas-Diesel (Dual Fuel) Engine, in *ASME-ICE Division*, 1991, vol. 15, pp. 89–97.

Blyth, N., Development of the Fairbanks Morse Enviro-Design Opposed Piston Dual Fuel Engine, presented at ASME-ICE Division, 1994, Paper 100375.

Boisvert, J., Gettel, L.E., and Perry, G.C., Particulate Emissions of a Dual Fuel Caterpillar 3208 Engine, in *ASME-ICE Division*, 1988, pp. 1–7, Paper 88-ICE-18.

Bols, R.E., and Tuve, G.L., eds., *Handbook of Tables for Applied Engineering Science*, CRC Press, Cleveland, OH, 1970.

Borman, G.I., and Ragland, K., *Combustion Engineering*, int. ed., McGraw Hill, New York, 1998.

Boyce, T.R., Karim, G.A., and Moore, N.P.W., An Experimental Investigation into the Effects of the Addition of Partially Oxidized Reaction Products to the Intake Charge of a Compression Ignition Engine, *Journal of the Institute of Petroleum*, 52, 300–311, 1968.

Boyce, T.R., Karim, G.A., and Moore, N.P.W., The Effects of Some Chemical Factors on Combustion Processes in Diesel Engines, in *Proceedings of the Institute of Mechanical Engineers*, 1970, pp. 123–132, Paper 14.

Boyer, R.L., *Status of Dual Fuel Engine Development*, SAE Paper 4900, 1949.

Brekken, M., and Durbin, E., *An Analysis of the True Efficiency of Alternative Vehicle Power Plants and Alternative Fuels*, SAE Paper 981399, 1998.

Brogan, T.R., Graboski, M.S., Macomber, J.R., Helmich, M.J., and Schaub, F.S., Operation of a Large Bore Medium Speed Turbocharged Dual Fuel Engine on Low BTU Wood Gas, in *ASME-ICE Division*, 1993, vol. 20, pp. 51–66.

Burn, K.S., The Effect of Cold Intake Temperatures on the Combustion of Gaseous Fuels in a Dual Fuel Engine, MSc thesis, University of Calgary, Canada, 1977.

Callahan, T., Survey of Gas Engine Performance and Future Trends, presented at Proceedings of ASME Conference, ASME-ICE Division, Salzburg, Austria, 2003, Paper ICES2003-628.

Carlucci, A.P., Ficarella, A., and Laforgia, D., Control of the Combustion Behavior in a Diesel Engine Using Early Injection and Gas Addition, *Applied Thermal Engineering*, 26, 2279–2286, 2006.

Challen, B., and Barnescu, R., *Diesel Engine Handbook*, 2nd ed., SAE, Warrendale, PA, 1999.

Checkel, M., Newman, P., Van der Lee, B., and Pollak, I., *Performance and Emissions of a Converted RABA 2356 Bus Engine in Diesel and Dual Fuel Diesel/Natural Gas Operation*, SAE Paper 931823, 1993.

Chen, K., and Karim, G.A., *An Examination of the Effects of Charge Inhomogeneity on the Compression Ignition of Fuel–Air Mixtures*, SAE Paper 982614, 1998.

Chen, Z., Konno, M., Oguma, M., and Yanai, T., *Experimental Study of CI Natural Gas/DME Homogeneous Charge Engine*, SAE Paper 2000-01-0329, 2000.

Choi, S., Yoon, Y.K., Kim, S., Yeo, G., and Han, H., *Development of Urea-SCR System for Light Duty Diesel Passenger Car*, SAE Paper 2001-01-0519, 2001.

Chrisman, B., Callaham, T., and Chiu, J., Investigation of Macro Pilot Combustion in Stationary Gas Engine, presented at ASME-ICE Division, 1998, Paper 98-ICE-106.

Christensen, M., Johansson, B., and Einewall, P., *Homogeneous Charge Compression Ignition Using Isooctane, Ethanol and Natural Gas*, SAE Paper 9728774, 1997.

Coward, H.F., and Jones, G.W., *Limits of Flammability of Gases and Vapours*, Bulletin 503, U.S. Bureau of Mines, 1952.

Cruz, I.E., Studies on the Practical Application of Producer Gas from Agricultural Residues as Supplementary Fuel for Diesel Engines, in *Thermal Conversion of Solid Wastes and Biomass*, ed. J.L. Jones and S.B. Radding, ACS Symposium Series 130, ACS, 1980, pp. 649–669.

Cummins, L., *Internal Fire*, SAE Publishing, Warrendale, PA, 1989.

Czerwinski, J., and Comte, P., *Influences of Gas Quality on Natural Gas Engine*, SAE Paper 2001-01-1194, 2001.

Daisho, Y., Takahashi, Y.I., Iwashiro, Y., Nakayama, S., and Saito, T., *Controlling Combustion and Exhaust Emissions in a Direct-Injection Diesel Engine Dual Fuelled with Natural Gas*, SAE Paper 952436, 1995.

Danyluk, P.R., Development of a High Output Dual Fuel Engine, *ASME Journal of Engineering for Gas Turbines and Power*, 115, 728–733, 1993.

Dardalis, D., Matthews, R.D., Lewis, D., and Davis, K., *The Texas Project, Part 5— Economic Analysis: CNG and LPG Conversions of Light-Duty Vehicle Fleets*, SAE Paper 982447, 1998.

Diesel, R., Method of Igniting and Regulating Combustion for Internal Combustion Engines, U.S. Patent 673,160, April 1901.

Diggins, D., *CNG Fuel Cylinder Storage, Efficiency and Economy in Fast Fill Operations*, SAE Paper 981398, 1998.

Ding, X., and Hill, P.G., *Emissions and Fuel Economy of a Prechamber Diesel Engine with Natural Gas Dual Fuelling*, SAE Paper 860069, 1988.

Douville, B., Ouellette, P., Touchette, A., and Ursu, B., *Performance and Emissions of a Two-Stroke Engine Fueled Using High Pressure Direct Injection of Natural Gas*, SAE Paper 981160, 1998.

Downs, D., Walsh, A.D., and Wheeler, R.W., A Study of the Reactions That Lead to Knock in the Spark Ignition Engine, *Philosophical Transactions of the Royal Society*, 243, 463–524, 1951.

D'Souza, M.V., and Karim, G.A., The Combustion of Methane with Reference to Its Utilization in Power Systems, *Journal of the Institute of Fuel*, 335–339, 1972.

Dumitrescu, S., Hill, P.G., Li, G.G., and Ouellette, P., *Effects of Injection Changes on Efficiency and Emissions of a Diesel Engine Fuelled by Direct Injection of Natural Gas*, SAE Paper 2000-01-1805, 2000.

Ebert, K., Beck, N.J., Barkhimer, R.L., and Wong, H., *Strategies to Improve Combustion and Emission Characteristics of Dual Fuel Pilot Ignited Natural Gas Engines*, SAE Paper 971712, 1997.

Ecklund, E., Bechtold, R., Timbario, T., and McCallum, P., Alcohol Fuel Use in Diesel Transportation Vehicles, in *IGT 3rd Symposium on Nonpetroleum Vehicular Fuels*, October 1982, pp. 261–313.

Einang, P.H., Engja, H.E., and Vestergren, R., Medium Speed 4 Stroke Diesel Engine Using High Pressure Gas Injection Technology, in *Proceedings of the 8th International Congress on Combustion Engines, CIMAC*, Tienjing, PRC, 1989, pp. 916–932.

Eke, P., and Walker, J.H., Gas as an Engine Fuel, in *Proceedings of the Institute of Gas Engineers*, 1970, pp. 121–138.

Elliot, O., and Davis, R.E., Dual Fuel Combustion in Diesel Engines, *Industrial and Engineering Chemistry*, 43, 2854–2863, 1951.

Energy Mines and Resources, *Propane Carburetion*, Government of Canada, Ministry of Supplies and Services, 1984, p. 41.

Ericson, R., Campbell, K., and Morgan, D., Application of Dual Fuel Engine Technology for On-Highway Vehicles, presented at *Proceedings of ASME-ICE Division,* Salzburg, Austria, 2003, Paper ICES2003-586.

Felt, A.E., and Steele, W.A., Combustion Control in Dual Fuel Engines, *SAE Transactions*, 70, 644, 1982.

Foxwell, G.E., *The Efficient Use of Fuel*, British Ministry of Technology, HMSO, London, 1958.

Gas Research Institute, *Technology Today,* Chicago, IL, 1991.

Gee, D., and Karim, G.A., Heat Release in a Compression-Ignition Engine, *The Engineer*, 222, 473–479, 1966.

Gee, D., Satterford, R., and Karim, G.A., Performance of a Compression Ignition Engine in Unconventional Atmospheres, *The Engineer*, 219, 551–556, 1965.

Geiss, R., Burkmyre, M., and Langan, J., *Technical Highlights of the Dodge Compressed Natural Gas Ram Van/Wagon*, SAE Paper 921551, 1992.

Gettel, L., and Perry, G.C., *Natural Gas Conversion Systems for Heavy Duty Truck Engines*, SAE Paper 911663, 1991.

Gettel, L.E., Perry, G.C., Boisvert, J., and O'Sullivan, P.J., Dual Fuel Engine Control Systems for Transport Applications, presented at ASME-ICE Division, 1987, Paper 87-ICE 121.

Golovitchev, V.I., Nordin, N., Jarnicki, R., and Chomiak, J., *3-D Diesel Spray Simulation Using a New Detailed Chemistry Turbulent Combustion Model*, SAE Paper 2000-01-1891, 2000.

Golvoy, A., and Blais, E.J., *Natural Gas Storage on Activated Carbon*, SAE Paper 831678, 1983.

Gopal, G., Rao, P.S., Gopalakrishnan, K.V., and Murthy, B.S., Use of Hydrogen in Dual-Fuel Engines, *International Journal of Hydrogen Energy*, 7(3), 267–272, 1982.

Goto, S., Furutani, H., and Delic, R., *Dual Fuel Diesel Engine Using Butane*, SAE Paper 920690, 1992.

Goto, S., Nishi, Y., and Nakayama, S., High Density Gas Engine with Micro Pilot Compression Ignition Method, presented at Proceedings of ASME-ICE Division, Salzburg, Austria, 2003, Paper ICES2003-679.

Green, R.K., and Glasson, N.D., High-Pressure Hydrogen Injection for Internal Combustion Engines, *International Journal of Hydrogen Energy*, 17(11), 895–901, 1992.

Grosshans, G., Development of a 1200 kW/Cyl. Low Pressure Dual Fuel Engine for LNG Carriers, in *Proceedings of CIMAC 1998*, 1998, pp. 1417–1428.

Gunea, C., Examination of the Pilot Quality on the Performance of Gas Fueled Diesel Engine, MSc thesis, University of Calgary, Canada, 1997.

Gunea, C., Razavi, M.R., and Karim, G.A., *The Effects of Pilot Fuel Quality on Dual Fuel Engine Ignition Delay*, SAE Paper 982453, 1998.

Hanna, M.A., The Combustion of Diffusion Jet Flames Involving Gaseous Fuels in Atmospheres Containing Some Auxiliary Gaseous Fuels, PhD thesis, University of Calgary, Canada, 1983.

Harrington, J., Munashi, S., Nedelcu, C., Ouellette, P., Thompson, J., and Whitfield, S., *Direct Injection of Natural Gas in a Heavy-Duty Engine*, SAE Paper 2002-01-1630, 2002.

Heenan, J., and Gettel, L., *Dual Fueling Diesel/NGV Technology*, SAE Paper 881655, 1988.

Herdin, G.R., Gruber, F., Plohberger, D., and Wagner, M., Experience with Gas Engines Optimized for H2-Rich Fuels, presented at ASME-ICE Division, 2003, Paper ICES2003-596.

Heywood, J., *Internal Combustion Engines*, McGraw Hill, New York, 1988.

Higgins, B.S., Mueller, C.J., and Siebers, D., *Measurements of Fuel Effects on Liquid-Phase Penetration in DI Sprays*, SAE Paper 1999-01-0519, 1999.

Hlousek, J., Common Rail Fuel Injection System for High Speed Large Diesel Engines, in *ASME-ICE Division*, 1998, vol. 31-1, Paper 98-ICE-122.

Hodgins, K.B., Gunawan, H., and Hill, P.G., *Intensifier-Injector for Natural Gas Fuelling Diesel Engines*, SAE Paper 921553, 1992.

Hosseinzadeh, A., and Saray, R.K., An Availability Analysis of Dual-Fuel Engines at Part Loads: The Effects of Pilot Fuel Quantity on Availability Terms, *Journal of Power and Energy*, 223(8), 903–912, 2009.

Hountalas, D.T., and Papagiannakis, R., *Development of a Simulation Model for Direct Injection Dual Fuel Diesel-Natural Gas Engines*, SAE Paper 2000-01-0286, 2000.

Houser, K.R., Lestz, S.S., Dukovich, M., and Yasbin, R., *Methanol Fumigation of a Light Duty Automotive Diesel Engine*, SAE Paper 801379, 1980.

Hsu, D., *Practical Diesel Engine Combustion Analysis*, SAE Publishing, Warrendale, PA, 2002.

Ikegami, M., Miwa, K., and Shioji, M., A Study of Hydrogen Fuelled Compression Ignition Engines, *International Journal of Hydrogen Energy*, 7, 341–353, 1982.

Imitrescu, S., and Hill, P.G., *Effects of Injection Changes on Efficiency and Emissions of a Diesel Engine Fuelled by Direct Injection of Natural Gas*, SAE Paper 2000-01-1805, 2000.

In, C.B., Kim, S.H., Kim, C.D., and Cho, W.S., *Catalyst Technology Satisfying Low Emission of Natural Gas*, SAE Paper 970744, 1997.

International Panel on Climate Change (IPCC), *Climate Change 2001: The Scientific Basis*, Cambridge University Press, Cambridge, 2001.

Ishida, M., Chen, Z.L., Luo, G.F., and Ueki, H., *The Effect of Pilot Injection on Combustion in a Turbocharged D.I. Diesel Engine*, SAE Paper 841692, 1994.

Ishyama, T., Shioji, M., Mitani, S., Shibata, H., and Ikegami, M., *Improvement of Performance and Exhaust Emissions in a Converted Dual Fuel Natural Gas Engine*, SAE Paper 2000-01-1866, 2000.

Ito, K., Abraham, M., Jensen, L., and Karim, G.A., *An Examination of the Role of Formaldehyde in the Ignition Processes of a Dual Fuel Engine*, SAE Paper 912367, 1981.

Jacobs, T., Assanis, D., and Filipi, Z., *The Impact of Exhaust Gas Recirculation on Performance and Emissions of a Heavy-Duty Diesel Engine*, SAE Paper 2003-01-1068, 2003.

Jeong, D.S., Suh, W.S., Oh, S., and Choi, K.N., *Development of a Mechanical CNG-Diesel Dual-Fuel Supply System*, SAE Paper 931947, 1993.

Johnson, W.P, Beck, N.J., Van der Lee, A., Koshkin, V.K., Lovkov, D., and Platov, I.S., *An Electronic Dual Fuel Injection System for the Belarus D144 Diesel Engine*, SAE Paper 901502, 1990.

Jones, W., and Burn, K.S., The Effect of Cold Intake Temperature on the Combustion of Gaseous Fuels in a Dual-Fuel Engine, MSc thesis, University of Calgary, Canada, 1977.

Jones, W., Liu, Z., and Karim, G.A., *Exhaust Emissions from Dual Fuel Engines at Light Load*, SAE Paper 932822, 1993.

Jones, W., Raine, R.R., and Karim, G.A., *An Examination of the Ignition Delay Period in Dual Fuel Engines*, SAE Paper 892140, 1989.

Jorach, R., Enderle, C., and Decker, R., Development of a Low-NO$_x$ Truck Hydrogen Engine with High Specific Power Output, *International Journal of Hydrogen Energy*, 22(4), 423–427, 1997.

Karim, G.A., A Review of Combustion Processes in the Dual Fuel Engine—The Gas Diesel Engine, *Progress in Energy and Combustion Science*, 6, 277–285, 1980.

Karim, G.A., An Analytical Approach to the Uncontrolled Combustion Phenomena in Dual Fuel Engines, *Journal of the Institute of Fuel*, 37, 530–536, 1964.

Karim, G.A., *An Examination of Some Measures for Improving the Performance of Gas Fuelled Diesel Engines at Light Load*, SAE Paper 912366, 1991.

Karim, G.A., *An Investigation of the Combustion in an IDI Diesel Engine with Low Concentrations of Added Hydrogen*, SAE Paper PF11-0875, 2010.

Karim, G.A., Combustion in Dual Fuel Engines—A Status Report, in *Proceedings of the 8th International Congress on Combustion Engines (CIMAC)*, Brussels, 1968, pp. 59–85.

Karim, G.A., Combustion in Gas Fueled Compression Ignition Engines of the Dual Fuel Type, *ASME Journal of Engineering for Gas Turbines and Power*, 125, 827–836, 2003.

Karim, G., *Combustion in Gas-Fueled Compression Ignition Engines of the Dual Fuel Type*, in *Handbook of Combustion*, ed. M. Lackner, F. Winter, and A. Agarwal, vol. 3, Wiley Publishers, Weinheim, Germany, 2010, pp. 213–235.

Karim, G.A., *Dual Fuel Engines of the Compression Ignition Type—Prospects, Problems and Solutions—A Review*, SAE Paper 830173, 1983.

Karim, G.A., *Fuels, Energy and the Environment*, CRC Press, Boca Raton, FL, 2012.

Karim, G.A., Production of Synthesis Gas, *British Chemical Engineering*, 8, 392–396, 1963.

Karim, G.A., Some Aspects of the Utilization of LNG for the Production of Power in Internal Combustion Engines, presented at Proceedings of the International Conference on Liquefied Natural Gas, Institute of Mechanical Engineers, 1969.

Karim, G.A., *Some Considerations of the Safety of Methane (CNG) as an Automotive Fuel—Comparison with Gasoline, Propane and Hydrogen Operation*, SAE Paper 830267, 1983.

Karim, G.A., The Combustion of Bio-Gases and Low Heating Value Gaseous Fuel Mixtures, *International Journal of Green Energy*, 8, 1–10, 2011.

Karim, G.A., The Combustion of Low Heating Value Gaseous Fuel Mixtures, in *Handbook of Combustion*, ed. M. Lackner, F. Winter, and A. Agarwal, vol. 3, Wiley Publishers, Weinheim, Germany, 2010, pp. 141–163.

Karim, G.A., *The Dual Fuel Engine*, ed. Robert L. Evans, Automotive Engine Alternatives, Plenum Press, NY, 1987.

Karim, G.A., The Ignition of a Premixed Fuel and Air Charge by Pilot Fuel Spray Injection with Reference to Dual-Fuel Combustion, *SAE Transactions*, 77, 3017–3024, 1968.

Karim, G.A., The Simultaneous Production of Synthesis Gas and Power in a Reciprocating Internal Combustion Engine, PhD thesis, London University, UK, 1960.

Karim, G.A., and Amoozigar, N., *Determination of the Performance of a Dual Fuel Diesel Engine with the Addition of Various Liquid Fuels to the Intake Charge*, SAE Paper 8302, 1983.

Karim, G.A., and Burn K.S., *Combustion of Gaseous Fuels in a Dual Fuel Engine of the Compression Ignition Type with Particular Reference to Cold Intake Temperature Conditions*, SAE Paper 800263, 1980.

Karim, G.A., and Khan M.O., Examination of Effective Rates of Combustion Heat Release in a Dual Fuel Engine, *Journal of Mechanical Engineering Science*, 10, 13–23, 1968.

Karim, G.A., and Khan, M.O., Examination of Some of the Errors Normally Associated with the Calculation of Apparent *Rates of Combustion Heat Release in Engines*, SAE Paper 710135, 1971.

Karim, G.A., and Klat, S.R., Knock and Autoignition Characteristics of Some Gaseous Fuels and Their Mixtures, *Journal of the Institute of Fuel*, 39, 109–119, 1966.

Karim, G.A., and Klat, S.R., Hydrogen as a Fuel in Compression Ignition Engines, *Mechanical Engineering*, 98(4), 34–39, 1976.

Karim, G.A., and Lui, Z., *A Prediction Model of Knock in Dual Fuel Engines*, SAE Paper 921550, 1992.

Karim, G.A., and Moore, N.P.W., The Production of Synthesis Gas and Power in a Compression Ignition Engine, *Journal of the Institute of Fuel*, 36, 98–105, 1963.

Karim, G.A., and Moore, N.P.W., *Examination of Rich Mixture Operation of a Dual Fuel Engine*, SAE Paper 901500, 1990.

Karim, G.A., and Moore, N.P.W., *The Production of Hydrogen by the Partial Oxidation of Methane in a Dual Fuel Engine*, SAE Paper 901501, 1990.

Karim, G.A., and Rogers, A., Comparative Studies of Propane and Butane as Dual Fuel Engine Fuels, *Journal of the Institute of Fuel*, 40, 513–522, 1967.

Karim, G.A., and Ward, S., The Examination of the Combustion Processes in a Compression-Ignition Engine by Changing the Partial Pressure of Oxygen in the Intake Charge, *SAE Transactions*, 77, 3008–3016, 1968.

Karim, G.A., and Watson, H.C., Experimental and Computational Consideration of the Compression Ignition of Homogeneous Fuel-Oxidant Mixtures, *SAE Transactions*, 80, 450–459, 1971.

Karim, G.A., and Wierzba, I., Comparative Studies of Methane and Propane as Fuels for Spark Ignition and Compression Ignition Engines, *SAE Transactions*, 92, 3677–3688, 1983.

Karim, G.A., and Wierzba, I., *Comparative Studies of Methane and Propane as Fuels for Spark Ignition and Compression Ignition Engines*, SAE Paper 831196, 1983.

Karim, G.A., and Wierzba, I., *Experimental and Analytical Studies of the Lean Operational Limits in Methane Fuelled Spark Ignition and Compression Ignition Engines*, SAE Paper 891637, 1989.

Karim, G.A., and Wierzba, I., *Methane—Carbon Dioxide Mixtures as a Fuel*, SAE Paper 921557, 1992.

Karim, G.A., and Wierzba, I., Safety Measures Associated with the Operation of Engines on Various Alternative Fuels, *Reliability Engineering and Systems Safety Journal*, 37, 93–98, 1993.

Karim, G.A., Klat, S.R., and Moore, N.P.W., Knock in Dual Fuel Engines, *Proceedings of the Institute of Mechanical Engineers*, 181, 453–466, 1967.

Karim, G.A., and Zhoada, Y., An Analytical Model for Knock in Dual Fuel Engines of the Compression Ignition Type, *SAE Transactions*, 3.52–3.62, 1988.

Karim, G.A., and Zhoada, Y., Modeling of the Combustion Process in a Dual Fuel Direct Injection Engine, *ASME Journal of Energy Resources Technology*, 112, 34–42, 1990.

Karim, G.A., Gee, D., and Satterford, R., Performance of a C.I. Engine in Unconventional Atmospheres, *The Engineer*, 219, 551–556, 1965.

Karim, G.A., Wierzba, I., and Soriano, B., The Limits of Flame Propagation within Homogeneous Streams of Fuel and Air, *ASME Journal of Energy Resources Technology*, 108, 183–187, 1983.

Karim, G.A., Wierzba, I., Metwally, M., and Mohan, K., The Combustion of a Fuel Jet in a Stream of Lean Gaseous Fuel–Air Mixtures, in *Proceedings of the Combustion Institute International*, 1981, vol. 18, pp. 977–991.

Keller, E., *International Experience with Clean Fuels*, SAE Paper 931831, 1993.

Khalil, E., Modeling the Chemical Kinetics of Combustion of Higher Hydrocarbon Fuels in Air, PhD dissertation, University of Calgary, 1998.

Khalil, E., and Karim, G.A., A Kinetic Investigation of the Role of Changes in the Composition of Natural Gas in Engine Applications, *ASME Journal of Engineering for Gas Turbines and Power*, 124, 404–411, 2002.

Khalil, E., Samuel, P., and Karim, G.A., *An Analytical Examination of the Chemical Kinetics of the Combustion of N-Heptane-Methane Air Mixtures*, SAE Paper 961932, 1996.

Khan, M.O., Dual Fuel Combustion Phenomena, PhD thesis, London University, Imperial College of Science and Technology, 1969.

Khanna, S.L., and Karim, G.A., *The Effect of Very Low Air Intake Temperature on the Performance and Exhaust Emission Characteristic of a Diesel Engine*, SAE Paper 7407, 1974.

Kibrya, M.G., and Karim, G.A., Turbulent Jet Diffusion Flames in Environments Containing Some Premixed Fuel, *ASME Journal of Energy Resources Technology*, 110(3), 146–150, 1989.

Kibrya, M.G., Karim, G.A., Lapucha, R., and Wierzba, I., *Examination of the Combustion of a Fuel Jet in Homogeneously Pre-Mixed Lean Fuel–Air Stream*, SAE Paper 881662, 1988.

Klimstra, J., *Catalytic Converters for Natural Gas Fueled Engines—A Measurement and Control Problem*, SAE Paper 872165, 1987.

Klimstra, J., Hattar, C., Nylund, I., and Sillanpaa, H., The Technology and Benefits of Skip Firing for Large Reciprocating Engines, presented at ASME-ICE Division, Chicago, April 2005, Paper ICES2005-1073.

Klimstra, J., Heranaez, A.B., Gerard, A., Karti, B., Quinto, V., Roberts, G., and Schollmeyer, H., Classification Methods for Knock Resistance of Gaseous Fuels, in *ASME-ICE Division*, 1999, vol. 33-1, Paper 99-ICE-214.

Kong, S.C., Ayoub, N., and Reitz, R.D., *Modeling Combustion in Compression Ignition Homogeneous Charge Engines*, SAE Paper 920512, 1992.

Kong, S.C., Han, Z., and Reitz, R.D., *The Development and Application of a Diesel Ignition and Combustion Model for Multidimensional Engine Simulation*, SAE Paper 950278, 1995.

Kong, S.C., Marriott, C.D., and Reitz, R.D., *Modeling and Experiments of HCCI Engine Combustion Using Detailed Chemical Kinetics with Multidimensional CFD*, SAE Paper 2001-01-1026, 2001.

Krepec, T., Giannacopoulos, T., and Miele, D., New Electronically Controlled Hydrogen Gas Injector Development and Testing, *International Journal of Hydrogen Energy*, 12, 855–861, 1987.

Kubish, J., and Brehob, D.D., *Analysis of Knock in a Dual Fuel Engine*, SAE Paper 922367, 1992.

Kubish, J., King, S.R., and Liss, W.E., *Effect of Gas Composition on the Octane Number of Gaseous Fuels*, SAE Paper 922359, 1992.

Kukkonen, C.A., and Shelef, M., *Hydrogen as an Alternative Automobile Fuel*, SAE Paper 940766, 1994.

Kumar, M., Senthil, M., Ramesh, A., and Nagalingam, B., Use of Hydrogen to Enhance the Performance of a Vegetable Oil Fuelled Compression Ignition Engine, *International Journal of Hydrogen Energy*, 28, 1143–1154, 2003.

Kurtz, E.M., Mather, D.K., and Foster, D.E., *Parameters That Affect the Impact of Auxiliary Gas Injection in a DI Diesel Engine*, SAE Paper 2000-010233, 2000.

Kusaka, J., Daisho, Y., Shimonagata, T., Kihara, R., and Saito, T., Combustion and Exhaust Characteristics of a Diesel Engine Dual-Fuelled with Natural Gas, in *Proceedings of the 7th International Conference and Exhibition on Natural Gas Vehicles*, Yokohama, Japan, October 17–19, 2000, pp. 23–31.

Kusaka, J., Tsazuki, K., and Daisho, Y., *A Numerical Study on Combustion and Exhaust Gas Emissions Characteristics of a Dual-Fuel Natural Gas Engine Using a Multi-Dimensional Model Combined with Detailed Kinetics*, SAE Paper 2002-01-1750, 2002.

Lambe, S.M., and Watson, H.C., Low Pollution, Energy Efficient C.I. Hydrogen Engine, *International Journal of Hydrogen Energy*, 17(7), 513–525, 1992.

Lapucha, R., and Karim, G.A., *Flame Flashback for Low Reynolds Number Flows*, eds. A.L. Kuhl et al., AIAA Progress in Astronautics and Aeronautics Progress Series, vol. 113, AIAA, Washington, DC, 1988, pp. 367–383.

Larson, C.R., Bushe, W.K., Hill, P.G., and Munshi, S.R., Relative Injection Timing Effects in a Diesel Engine Fuelled with Pilot-Ignited, Directly Injected Natural Gas, presented at Canadian Section of the Combustion Institute (CICS) Spring Technical Meeting, 2003.

Lee, J.T., Kim, Y.Y., and Caton, J.A., The Development of a Dual Injection Hydrogen Fuelled Engine with High Power and High Efficiency, presented at Proceedings of the ASME Fall Technical Conference, ASME-ICE Division, September 8–11, 2002.

Lee, J.T., Kim, Y.Y., Lee, C.W., and Caton, J.A., An Investigation of a Cause of Backfire and Its Control Due to Crevice Volumes in a Hydrogen Fueled Engine, *ASME Journal of Engineering for Gas Turbines and Power*, 123, 204–210, 2001.

Leiker, M., Christoph, K., Rankl, M., Cartellion, W., and Pfiefer, U., Evaluation of Anti-Knock Property of Gaseous Fuels by Means of the Methane Number and Its Practical Application to Gas Engines, *ASME Transactions*, 72-DGP-4, 1–15, 1973.

Li, G., Ouellette, P., Dumitrescu, S., and Hill, P., *Optimization Study of Pilot Ignited Natural Gas Direct Injection in Diesel Engines*, SAE Paper 1999-01-3556, 1999.

Li, H., and Karim, G.A., Experimental Investigation of the Knock and Combustion Characteristics of CH4, H2, CO and Some of Their Mixtures, *Journal of Energy and Power*, 220, 459–473, 2005.

Li, H., and Karim, G.A., Modeling the Performance of a Turbo-Charged Spark Ignition Natural Gas Engine with Cooled Exhaust Gas Recirculation, *ASME Journal of Engineering for Gas Turbines and Power*, 130, 328041-10, 2008.

Li, S.C., and Williams, F.A., *A Reduced Reaction Mechanism for Predicting Knock in Dual Fuel Engines*, SAE Paper 2000-01-0957, 2000.

Lin, Z., and Su, W., *A Study on the Determination of the Amount of Pilot Injection and Lean and Rich Boundaries of the Premixed CNG-Air Mixtures for a CNG/Diesel Dual Fuel Engine*, SAE Paper 2003-01-0765, 2003.

Liss, W.E., and Thrasher, W.H., *Natural Gas as a Stationary Engine and Vehicular Fuel*, SAE Paper 912364, 1991.

Liu, B., and Checkel, D., *Experimental and Modeling Study of Variable Cycle Time for a Reversing Flow Catalytic Convertor of Natural Gas/Diesel Dual Fuel Engines*, SAE Paper 2000-01-0213, 2000.

Liu, C., An Experimental and Analytical Investigation into the Combustion Characteristics of HCCI and Dual Fuel Engines with Pilot Injection, PhD thesis, University of Calgary, Canada, 2006.

Liu, C., and Karim, G.A., The Effects of Intake Flow Swirl on the Combustion Characteristics of Hydrogen Fuelled HCCI Engines, presented at Proceedings of ASME-ICEF04 Fall Technical Conference, 2004, Paper ICEF2004-932.

Liu, C., and Karim, G.A., A Simulation of the Combustion of n-Heptane Fueled HCCI Engines Using a 3-D Model with Detailed Chemical Kinetics, presented at Proceedings of ASME 2005 Fall Technical Conference, 2005, Paper ICEF05-1220.

Liu, C., and Karim, G.A., *A 3D Simulation with Detailed Chemical Kinetics of Combustion and Quenching in an HCCI Engine*, SAE Paper 08SFL-0027, 2008.

Liu, C., and Karim, G.A., *A Simulation of the Combustion of Hydrogen in HCCI Engines Using a 3D Model with Detailed Chemical Kinetics*, International Journal of Hydrogen Energy, 33, 3863–3875, 2008.

Liu, C., and Karim, G.A., Three Dimensional Computational Fluid Simulation of Diesel and Dual Fuel Engine Combustion, *ASME Journal of Engineering for Gas Turbines and Power*, 131, 128041–128049, 2009.

Liu, C., Karim, G.A., Xiao, F., and Sohrabi, A., *An Experimental and Numerical Investigation of the Combustion Characteristics of a Dual Fuel Engine with a Swirl Chamber*, SAE Paper 07PFL-783, 2007.

Liu, Z., An Examination of the Combustion Characteristics in Compression Ignition Engines Fuelled with Gaseous Fuels, PhD thesis, University of Calgary, Canada, 1995.

Liu, Z., and Karim, G.A., *An Analytical Examination of the Preignition Processes within Homogeneous Mixtures of a Gaseous Fuel and Air in a Motored Engine*, SAE Paper 942039, 1994, pp. 127–133.

Liu, Z., and Karim, G.A., *The Ignition Delay Period in Dual Fuel Engines*, SAE Paper 950466, 1995.

Liu, Z., and Karim, G.A., Knock Characteristics of Dual-Fuel Engines Fuelled with Hydrogen Fuel, *International Journal of Hydrogen Energy*, 20(11), 919–924, 1995.

Liu, Z., and Karim, G.A., *A Predictive Model for the Combustion Process in Dual Fuel Engines*, SAE Paper 952435, 1996.

Liu, Z., and Karim, G.A., *An Examination of the Role of Residual Gases in the Combustion Processes of Motored Engines Fuelled with Gaseous Fuels*, SAE Paper 961081, 1996.

Liu, Z., and Karim, G.A., Examination of the Exhaust Emissions of Gas Fuelled Diesel Engines, in *Proceedings of the 18th ASME Fall Conference, ASME-ICE Division*, 1996, Part 3, pp. 9–22.

Liu, Z., and Karim, G.A., Simulation of the Combustion Processes in Gas Fuelled Diesel Engines, *Journal of Power and Energy*, 211, 159–171, 1997.

Liu, Z., and Karim, G.A., An Examination of the Ignition Delay Period in Gas Fuelled Diesel Engines, *ASME Journal of Gas Turbines and Power*, 120, 225–231, 1998.

Lom, E.J., and Ly, K.H., *High Injection of Natural Gas in a Two Stroke Diesel Engine*, SAE Paper 902230, 1990.

Lom, W.L., *Liquefied Natural Gas*, Applied Science Publishers, London, 1974.

Lowe, W., and Williamson, P.B., Combustion and Automatic Mixture Strength Control in Medium Speed Gaseous Fuel Engines, in *Proceedings of the 8th Congress de Machines a Combustion—CIMAC*, 1996, pp. A14–A56.

Lowi, A., *Supplementary Fueling of Four Stroke Cycle Automotive Diesel Engines by Propane Fumigation*, SAE Paper 41398, 1984.

Lyn, W.T., Calculation of the Effect of the Rate of Heat Release on the Shape of Cylinder Pressure Diagram and Cycle Efficiency, in *Proceedings of the Institute of Mechanical Engineers*, London, 1961, pp. 34–37.

Lynch, F.E., Backfire Control Techniques for Hydrogen Fueled Internal Combustion Engines, *International Journal of Hydrogen Energy*, 686–696, 1974.

MacCarley, C.A., and Vorst, W.D.V., Electronic Fuel Injection Techniques for Hydrogen Powered I.C. Engines, *International Journal of Hydrogen Energy*, 5, 179–203, 1980.

MacLean, H.L., and Lave, L.B., Evaluating Automobile Fuel/Propulsion System Technologies, *Progress in Energy Combustion Science*, 29, 1–69, 2003.

Maiboom, A., Tauzia, X., and Hétet, J.F., Experimental Study of Various Effects of Exhaust Gas Recirculation (EGR) on Combustion and Emissions of an Automotive Direct Injection Diesel Engine, *Energy*, 33, 22–34, 2008.

Masood, M., and Ishrat, M.M., Computer Simulation of Hydrogen–Diesel Dual Fuel Exhaust Gas Emissions with Experimental Verification, *Fuel*, 87, 1372–1378, 2008.

Masood, M., Ishrat, M.M., and Reddy, A.S., Computational Combustion and Emission Analysis of Hydrogen–Diesel Blends with Experimental Verification, *International Journal of Hydrogen Energy*, 32(13), 2539–2547, 2007.

Masood, M., Mehdi, S.N., and Reddy, P.R., An Experimental Investigation on a Hydrogen-Diesel Dual Fuel Engine at Different Compression Ratios, *ASME Journal of Engineering for Gas Turbines and Power*, 129, 572–578, 2007.

Matheson, *Guide to Safe Handling of Compressed Gases*, 2nd ed., Matheson Gas Products, New York, 1983, p. 94.

Mathur, H.B., Das, L.M., and Patro, T.N., Hydrogen-Fuelled Diesel Engine: Performance Improvement through Charge Dilution Techniques, *International Journal of Hydrogen Energy*, 18(5), 421–431, 1993.

Maxwell, T., and Jones, J., *Alternative Fuels: Emissions, Economics and Performance*, SAE, Warrendale, PA, 1995.

Mayer, R., Meyers, D., Shahed, S.M., and Duggal, V.K., *Development of a Heavy Duty On-Highway Natural Gas Fueled Engine*, SAE Paper 922362, 1992.

Meyer, D., High Tech Fuel Management and Fuel Control Systems, in *IGT Conference Proceedings on Gaseous Fuel for Transportation*, Vancouver, BC, August 1986.

Milken, E., A Simple Retrofit System for the Operation of Diesel Engines on Natural Gas, in *Proceedings of the International Conference on Auto Technology*, Bangkok, Thailand, 1990, pp. 445–449.

Miller, W., Klein, J., Mueller, R., Doelling, W., and Zuerbig, J., *The Development of Urea—SCR Technology for US Heavy Duty Trailers*, SAE Paper 2000-01-0190, 2000.

Moore, N.P.W., and Mitchell, R.W.S., Combustion in Dual Fuel Engines, in *Proceedings of the Joint Conference on Combustion*, ASME and Institute of Mechanical Engineers, 1955, pp. 300–309.

Moore, N.P.W., and Roy, B.N., Comparative Studies of Methane and Propane as Engine Fuels, in *Proceedings of the Institute of Mechanical Engineers*, 1956, vol. 170, p. 1137.

Mtui, P., Pilot Ignited Natural Gas Combustion Diesel Engines, PhD thesis, University of British Columbia, Canada, 1996.

Mustafi, N.N., Particulates Emissions of a Dual Fuel Engine, PhD thesis, University of Auckland, New Zealand, January 2008.

Naber, J.D., and Siebers, D.L., Hydrogen Combustion under Diesel Engine Conditions, *International Journal of Hydrogen Energy*, 23, 363–371, 1998.

Needham, J.R., May, M.P., Doyle, D.M., Faulkner, S.A., and Ishiwata, H., *Injection Timing and Rate Control—A Solution for Low Emissions*, SAE Paper 900854, 1990.

Nielson, O.B., Qvale, B., and Sorenson, S., *Ignition Delay in the Dual Fuel Engine*, SAE Paper 870589, 1987.

Nylund, I., Gas Engine Development at Wartila NSD, in *ASME-ICE Division*, 2000, vol. 35-2, pp. 131–137, Paper 2000-ICE 330.

Obert, E.E., *Internal Combustion Engines and Air Pollution*, Harper and Row, New York, 1973.

Oester, U., and Wallace, J.S., Liquid Propane Injection for Diesel Engines, SAE Paper 872095, 1987.

Ogden, J.M., Williams, R.H., and Larson, E.D., Societal Lifecycle Costs of Cars with Alternative Fuels/Engines, *Energy Policy*, 32, 7–27, 2004.

Ouellette, P., High Pressure Direct Injection (HPDI) of Natural Gas in Diesel Engines, in *Proceedings of the 7th International Conference and Exhibition on Natural Gas Vehicles*, Yokohama, Japan, 2000, pp. 235–242.

Ouellette, P., Mtui, P., and Hill, P., *Numerical Simulations of Directly Injected Natural Gas and Pilot Diesel Fuel in a Two Stroke Compression Ignition Engine*, SAE Paper 981400, 1998.

Packer, J.P., Advanced Packaged Co-Generation: Gas Engines and Co-Generation, in *Proceedings of the Institute of Mechanical Engineers*, 1980, pp. 25–32.

Papagiannakis, R.G., Hountalas, D.T., and Rakopoulos, C.D., Theoretical Study of the Effects of Pilot Fuel Quantity and Its Injection Timing on the Performance and Emissions of a Dual Fuel Diesel Engine, *Energy Conversion and Management*, 48, 2951–2961, 2007.

Parik, P.P., Bhave, A.G., and Shash, I.K., Performance Evaluation of a Diesel Engine Dual Fuelled on Process Gas and Diesel, in *Proceedings of National Conference on ICE and Combustion*, National Small Industries Corp. Ltd., New Delhi, 1987, pp. Af179–Af186.

Park, T., Traver, M.L., Atkinson, R., Clark, N., and Atkinson, C.M., *Operation of a Compression Ignition Engine with a HEUI Injection System on Natural Gas with Diesel Pilot Injection*, SAE Paper 1999-01-3522, 1999.

Patterson, D.J., and Henein, N.A., *Emissions from Combustion Engines and Their Control*, Ann Arbor Science Publishers, Ann Arbor, MI, 1972.

Pollock, E., ed., *NGV Resource Guide*, RP Publishing, Denver, CO, 1985.

Poonia, M.P., Ramesh, A., and Gaur, R.R., *Effect of Intake Air Temperature on Pilot Fuel Quantity on the Combustion Characteristics of a LPG Dual Fuel Engine*, SAE Paper 982455, 1998.

Poonia, M.P., Ramesh, A., and Gaur, R., *Experimental Investigation of the Factors Affecting the Performance of LPG Dual Fuel Engine*, SAE Paper 1999-01-1123, 1998.

Pounder, C.C., ed., *Diesel Engine Principles and Practice*, George Newens Ltd., London, 1955.

Prabhukuma, G.P., Swaminathan, R.S., Nagalingam, B., and Gopalakrishnan, K.V., Water Induction Studies in a Hydrogen-Diesel Dual-Fuel Engine, *International Journal of Hydrogen Energy*, 12, 177–186, 1987.

Quadfleig, H., From Research to Market Application, Experience with the German Hydrogen Fuel Project, *International Journal of Hydrogen Energy*, 13, 363–374, 1988.

Quigg, D., Pellegrin, V., and Rey, R., *Operational Experience of Compressed Natural Gas in Heavy Duty Transit Buses*, SAE Paper 931786, 1993.

Rain, R.R., and McFeatures, J.S., *New Zealand Experience with Natural Gas Fuelling of Heavy Transport Engines*, SAE Paper 892136, 1989.

Rao, B.H., Shrivastava, K.N., and Bhakta, H.N., Hydrogen for Dual Fuel Engine Operation, *International Journal of Hydrogen Energy*, 8(5), 381–384, 1983.

Rente, T., Golovichev, V.I., and Denbratt, I., *Effect of Injection Parameters on Auto-Ignition and Soot Formation in Diesel Sprays*, SAE Paper 2001-01-3687, 2001.

Ricardo, H., *The High Speed Internal Combustion Engine*, 4th ed., Blackie and Son Ltd., London, 1953.

Rose, J.W., and Cooper, J.R., eds., *Technical Data on Fuels*, 7th ed., British National Committee of World Energy Conference, London, 1977.

Ryan, T.W., Callahan, T.J., and King, S.R., Engine Knock Rating of Natural Gases—Methane Number, *ASME Journal of Engineering for Gas Turbines and Power*, 115, 922–930, 1985.

Ryan, T.W., Lestz, S.S., and Meyer, E., *Extension of the Lean Misfire Limit and Reduction of Exhaust Emission of a S.I. Engine by Modification of Ignition and Intake Systems*, SAE Paper 740105, 1974.

Saito, H., Sakurai, T., Sakaoji, T. Hirashima, T., and Karnno, K., *Study on Lean Burn Gas Engine Using Pilot Oil as the Ignition Source*, SAE Paper 2001-01-0143, 2001.

Sakai, T., Choi, B.C., Ko, Y., and Kim, E., *Unburned Fuel and Formaldehyde Purification Characteristics of Catalytic Convertors for Natural Gas Fueled Automotive Engine*, SAE Paper 920596, 1992.

Samuel, P., Computational and Experimental Investigation of Ignition and Combustion of Liquid Hydrocarbon Fuels in Homogeneous Environment of Fuel and Air, PhD thesis, University of Calgary, Canada, 1994.

Samuel, P., and Karim, G.A., *An Analysis of Fuel Droplets Ignition and Combustion within Homogeneous Mixtures of Fuel and Air*, SAE Paper 940901, 1994.

Samuel, P., and Karim, G.A., A Numerical Study of the Unsteady Effects of Droplet Evaporation and Ignition in Homogeneous Environments of Fuel and Air, *ASME Journal of Energy Resources Technology*, 116, 194–200, 1994.

Schaub, F.S., and Hubbard, R.L., A Procedure for Calculating Fuel Gas Blend Knock Rating for Large-Bore Gas Engines and Predicting Engine Operation, *ASME Transactions*, 107, 922–930, 1985.

Schiffgens, H.J., Brandt, D., Dier, L., and Glauber, R., Development of the New Man B&W 32/40 Dual Fuel Engine, in *ASME-ICE Division*, 1996, vol. 27-3, pp. 33–45.

Schiffgens, H.J., Brandt, D., Rieck, K., and Heider, G., Low NOx-Gas Engines from MAN B&W, in *CIMAC Congress*, Copenhagen, 1998, pp. 1399–1414.

Selim, M., Thermal Loading and Temperature Distribution of a Precombustion Chamber Diesel Engine Running on Gasoil/Natural Gas, in *ASME-ICE Division*, 1998, vol. 31-3, pp. 113–130, Paper 98-ICE-159.

Sheppard, C.G.W., Tolegano, S., and Woolley, R., *On the Nature of Autoignition Leading to Knock in HCCI Engines*, SAE Paper 2002-01-2831, 2002.

Shioji, M., Ishiyama, T., and Ikegami, M., Approaches to High Thermal-Efficiency in High Compression-Ratio Natural-Gas Engines, in *Proceedings of 7th International Conference on Natural Gas Vehicles*, Yokohama, Japan, 2000, pp. 13–21.

Shudo, T., Improving Thermal Efficiency by Reducing Cooling Losses in Hydrogen Combustion Engines, *International Journal of Hydrogen Energy*, 32, 4285–4293, 2007.

Siebers, D.L., *Scaling Liquid-Phase Fuel Penetration in Diesel Sprays Based on Mixing Limited Vaporization*, SAE Paper 1999-01-0528, 1999.

Sierens, R., and Verhelst, S., Influence of the Injection Parameters on the Efficiency and Power Output of a Hydrogen Fueled Engine, *ASME Journal of Engineering for Gas Turbines and Power*, 125, 444–449, 2003.

Sinclair, M.S., and Haddon, J.J., *Operation of a Class 8 Truck on Natural Gas/Diesel*, SAE Paper 911666, 1991.

Singal, S.K., Pundit, B.P., and Mehta, P.S., *Fuel Spray–Air Motion Interaction in DI Diesel Engine: A Review*, SAE Paper 930604, 1993.

Singh, R., and Karim, G.A., A Thermodynamic Investigation into the Combustion of Methane, *Journal of the Institute of Fuel*, 40, 447–455, 1967.

Singh, S., Krishnan, S.R., Srinivasan, K.K., Midkiff, K.C., and Bell, S.R., Effect of Pilot Injection Timing, Pilot Quantity and Intake Charge Conditions on Performance and Emissions for an Advanced Low-Pilot-Ignited Natural Gas Engine, *International Journal of Engine Research*, 5(4), 329–348, 2004.

Stanglmaier, R., Ryan, T.W., and Sounder, J.S., *HCCI Operation of a Dual-Fuel Natural Gas Engine for Improved Fuel Economy and Ultra-Low NOx Emissions at Low to Moderate Engine Loads*, SAE 2001-01-1897, 2001.

Steiger, A., Large Bore Sulzer Dual Fuel Engines, Their Development, Construction and Fields of Application, *Sulzer Technical Review*, 3, 1–8, 1970.

Stokes, B., *Compressed Natural Gas Fuel Composition Issues*, SAE Paper 91828, 1993.

Stone, R., *Introduction to Internal Combustion Engines*, 2nd ed., SAE, Warrendale, PA, 1995.

Stull, D.R., and Prophet, H., *JANAF Thermochemical Tables*, 2nd ed., NSRDS-NBS 37, U.S. Department of Commerce/National Bureau of Standards, 1971.

Stumpp, G., and Ricco, M., *Common Rail—An Attractive Fuel Injection System for Passenger Car DI Diesel Engines*, SAE Paper 960870, 1996.

Taylor, C.F., and Taylor, E.S., *The Internal Combustion Engine*, International Textbook Co., Scranton, PA, 1962.

Thiessen, S., Karim, G.A., and Azad, S., Constant Volume Autoignition of Premixed Methane-Carbon Dioxide Mixtures, *International Journal of Green Energy*, 4, 535–547, 2007.

Thiessen, S., Khalil, E., and Karim, G.A., The Autoignition in Air of Some Binary Fuel Mixtures Containing Hydrogen, *International Journal Hydrogen Energy*, 35, 10013–10017, 2010.

Thyagarajan, V., and Babu, M.K.G., A Combustion Model for a Dual Fuel Direct Injection Diesel Engine, Diagnostics and Modeling of Combustion in Reciprocation Engine, presented at JSME Proceedings of COMODIA Symposium, Tokyo, 1995.

Tomita, E., Fukatani, N., Kawahara, N., and Maruyama, K., Combustion in a Supercharged Biomass Gas Engine with Micro Pilot Ignition Effects of Injection Pressure and Amount of Diesel Fuel, *Journal of Kones Powertrain and Transport*, 14(2), 513–520, 2007.

Tomita, E., Kawahara, N., Piao, Z., Fujita, S., and Hamamoto, Y., *Hydrogen Combustion and Exhaust Emissions Ignited with Diesel Oil in a Dual Fuel Engine*, SAE Paper 2001-01-3503, 2001.

Tsang, P., and Karim, G.A., Flame Propagation through Atmospheres Involving Concentration Gradients Formed by Mass Transfer Phenomena, *ASME Journal of Fluid Engineering*, 97, 615–617, 1975.

Tsang, P., Sarpal, S., Badr, O., and Karim, G.A., A Fundamental Study into Flame Propagation through Stratified Mixtures, in *Proceedings of the International Conference on Stratified Charge Engines*, Institute of Mechanical Engineers, London, 1979, pp. 121–126, Paper C255/76.

Tsolakis, A., and Megaritis, A., Partially Premixed Charge Compression Ignition Engine with On Board H2 Production by Exhaust Gas Fuel Reforming of Diesel and Biodiesel, *International Journal of Hydrogen Energy*, 31, 731–745, 2005.

Tsolakis, A., Megaritis, A., and Wyszynski, M.L., Application of Exhaust Gas Fuel Reforming in Compression Ignition Engines Fueled by Diesel and Biodiesel Fuel Mixtures, *Energy and Fuels*, 17, 1464–1473, 2003.

Tsolakis, A., Hernandez, J.J., Megaritis, A., and Crampton, M., Dual Fuel Diesel Engine Operation Using H2 Effect on Particulate Emissions, *Energy and Fuels*, 19, 418–425, 2005.

Turner, S.H., and Weaver, C.S., *Dual-Fuel Natural/Diesel Engines: Technology, Performance and Emissions*, No. GRI-94/0094, Topical Report Gas Research Institute, November 1994.

Umierski, M., and Stommel, P., *Fuel Efficient Natural Gas Engine with Common-Rail Micro-Pilot Injection*, SAE Paper 2000-01-3080, 2000.

Ursu, B., and Perry, C., *Natural Gas Powered Heavy Duty Truck Demonstration*, SAE Paper 961669, 1996.

U.S. Energy Information Administration, U.S. Department of Energy, Washington, DC, 2014, http://www.eia.doe.gov/emeu/international/reserves.html.

USEPA, *Inventory of U.S. Greenhouse Gas Emissions and Sinks (1990–2002)*, USEPA/30/R/04/003, 2004.

Varde, K.S., *Propane Fumigation in a Direct Injection Type Diesel Engine*, SAE Paper 831354, 1983.

Varde, K.S., and Frame, G.A., Hydrogen Aspiration in a Direct Injection Type Diesel Engine—Its Effects on Smoke and Other Engine Performance Parameters, *International Journal Hydrogen Energy*, 8, 549–555, 1983.

Varde, K.S., and Frame, G.A., Development of a High-Pressure Hydrogen Injection for SI Engine and Results of Engine Behavior, *International Journal Hydrogen Energy*, 10, 743–748, 1985.

Von Ruden, M.S., Cheng, W.K., and Heywood, J., Diesel Engine Operating in a Partially Premixed Mode, presented at ASME-ICE Division, 1993, Paper 93-ICE-23.

Weaver, C., and Turner, S., *Dual Fuel Natural Gas/Diesel Engines: Technology, Performance and Emissions*, SAE Paper 940548, 1994.

Wiecman, D., Senecal, P.K., and Reitz, R.D., *Diesel Engine Combustion Chamber Geometry Optimization Using Genetic Algorithms and Multi Dimensional Spray and Combustion Modeling*, SAE Paper 2001-01-0547, 2001.

Wierzba, I., Metwally, M., and Karim, G.A., The Ignition of a Stream of Premixed Fuel and Air Near the Flammability Limit by a Diffusion Pilot Flame, *Journal of Fire and Flammability*, 11, 221–230, 1990.

Wierzba, P., Karim, G.A., and Wierzba, I., An Analytical Examination of the Combustion of a Turbulent Jet in an Environment of Air Containing a Premixed Fuel or a Diluent, *ASME Journal of Energy Resources Technology*, 117(3), 234–238, 1995.

Winsor, R.E., and Patterson, D.E., *Mixture Turbulence—A Key Cyclic Combustion Variation*, SAE Paper 730086, 1973.

Wong, J.K.S., Compression Ignition of Hydrogen in a Direct Injection Diesel Engine Modified to Operate as a Low-Heat-Rejection Engine, *International Journal of Hydrogen Energy*, 15, 507–514, 1990.

Wong, W.Y., Midkiff, K.C., and Bell, S.R., *Performance and Emissions of a Natural Gas Fuelled Indirect Injected Diesel Engine*, SAE Paper 911766, 1991.

Wong, Y., and Karim, G.A., *An Analytical Examination of the Effects of Exhaust Gas Recirculation on the Compression Ignition Process of Engines Fuelled with Gaseous Fuels*, SAE Paper 961396, 1998, pp. 45–53.

Wong, Y., and Karim, G.A., An Analytical Examination of the Effects of Hydrogen Addition on Cyclic Variations in Homogeneously Charged Compression-Ignition Engines, *International Journal of Hydrogen Energy*, 25, 1217–1224, 2000.

Wong, Y., and Karim, G.A., *A Kinetic Examination of the Effects of Recycled Exhaust Gases on the Autoignition of Homogeneous N-Heptane-Air Mixtures in Engines*, SAE Paper 2000-01-2037, 2000, pp. 1–11.

Wong, Y.K., and Karim, G.A., A Kinetic Examination of the Effects of the Presence of Some Gaseous Fuels and Preignition Reaction Products with Hydrogen in Engines, *International Journal of Hydrogen Energy*, 24, 473–478, 1999.

Wood, C.D., *Alternative Fuels in Diesel Engines: A Review*, SAE Paper 810248, 1981.

Woschni, G., A Universally Applicable Equation for the Instantaneous Heat Transfer Coefficient in the Internal Combustion Engine, *SAE Transactions*, 76, 3065–3083, 1987.

Xiao, F., Experimental and Numerical Investigation of Diesel Combustion Processes in Homogeneously Premixed Lean Methane or Hydrogen–Air Mixtures, PhD thesis, University of Calgary, Canada, 2011.

Xiao, F., and Karim, G.A., *Combustion in a Diesel Engine with Low Concentrations of Added Hydrogen*, SAE Paper 2011-01-0676, 2011, pp. 1–17.

Xiao, F., Lu, C., Hu, Y., and Yang, B., *Experimental Study on Diesel Nitrogen Oxide Reduction by Exhaust Gas Recirculation*, SAE Paper 2000-05-0335, 2000.

Xiao, F., Sohrabi, A., and Karim, G.A., *Reducing the Environmental Impact of Fugitive Gas Emissions through Combustion in Diesel Engines*, SAE Paper 2007-01-2048, 2007.

Xiao, F., Sohrabi, A., and Karim, G.A., Effect of Small Amounts of Fugitive Methane in the Air on Diesel Engine Performance and Its Combustion Characteristics, *International Journal of Green Energy*, 5, 334–345, 2008.

Yang, X., Takamoto, Y., Okajima, A., Obokata, T., and Long, W., *Comparison of Computed and Measured High-Pressure Conical Diesel Sprays*, SAE Paper 2000-01-0951, 2000.

Yonetani, H., Hara, K., and Fukatani, I., *Hybrid Combustion Engine with Premixed Gasoline Homogeneous Charge and Ignition by Injected Diesel Fuel—Exhaust Emission Characteristics*, SAE Paper 940268, 1994.

Yoshida, K., Shoji, H., and Tanaka, H., *Study on Combustion and Exhaust Emission Characteristics of Lean Gasoline-Air Mixture Ignited by Diesel Fuel Direct Injection*, SAE Paper 982482, 1998.

Zabetakis, M., *Flammability Characteristics of Combustible Gases and Vapors*, Bureau of Mines Bulletin 627, U.S. Department of the Interior, Washington, DC, 1965.

Zaidi, K., Andrews, G., and Greenhough, J., *Diesel Fumigation Partial Mixing for Reducing Ignition Delay and Amplitude of Pressure Fluctuations*, SAE Paper 980535, 1998.

Zhang, Y., Kong, S.C., and Reitz, R.D., *Modeling and Simulation of a Dual Fuel (Diesel/Natural Gas) Engine with Multi-Dimensional CFD*, SAE Paper 2003-01-0755, 2003.

Zhaoda, Y., and Karim, G.A., *An Analytical Model for Knock in Dual Fuel Engines of the Compression Ignition Type*, SAE Paper 880151, 1988.

Zhaoda, Y., and Karim, G.A., Modelling of the Combustion Process in a Dual Fuel Direct Injection Engine, *ASME Journal of Energy Resources Technology*, 112, 254–259, 1990.

Zhaoda, Y., and Karim, G.A., Modelling of Autoignition and Knock in a Compression Ignition Engine of the Dual Fuel Type, in *Proceedings of the Institute of Mechanical Engineering*, 1991, pp. 141–148.

Zheng, M., Mirosh, E.A., Ulan, D.A., Klopp, W.E., Pardell, M.E., Newman, P.E., and Nishimura, A., A Novel Reverse Flow Catalytic Converter Operated on an Isuzu-6HH1 Diesel Dual Fuel Engine, presented at ASME-ICE Fall Technical Conference, Ann Arbor, MI, October 1999.

Zhou, G., and Karim, G.A., The Uncatalyzed Partial Oxidation of Methane for the Production of Hydrogen with Recirculation, *ASME Journal of Energy Resources Technology*, 115, 307–313, 1993.

Zhou, G., and Karim, G.A., An Analytical Examination of Various Criteria for Defining Autoignition within Heated Methane-Air Homogeneous Mixtures, *ASME Journal of Energy Resources Technology*, 116, 175–180, 1994.

Zhou, G., Hanafi, A., and Karim, G.A., A Kinetic Investigation of the Oxidation of Low Heating Value Fuel Mixtures of Methane and Diluents, *ASME Journal of Energy Resources Technology*, 115, 301–306, 1993.

Zuo, C., and Yang, M., *Opening Characteristics and Description of a Dual Fuel Engine for Diesel-Natural Gas Heavy-Duty Operation*, SAE Paper 1999-01-3523, 1999.

Index